Primeval Kinship

Primeval Kinship

How Pair-Bonding Gave Birth to Human Society

Bernard Chapais

HARVARD UNIVERSITY PRESS
Cambridge, Massachusetts
London, England
2008

Printed in the United States of America

Library of Congress Cataloging-in-Publication Data

Chapais, Bernard.
Primeval kinship : How pair-bonding gave birth to human society
Bernard Chapais.
p. cm.
Includes bibliographical references and index.
ISBN-13: 978-0-674-02782-4 (alk. paper)
1. Kinship. 2. Human evolution. I. Title.
GN487.C43 2008
306.83—dc22 2007034108

To my father,
for imprinting me with his fascination
for human origins

Contents

Preface

Scientific disciplines have a strong leaning toward territoriality; they thrive and prosper within boundaries, however historically contingent and relatively arbitrary these may be. Like populations isolated on separate islands over centuries, scientific specialties have acquired different habits, beliefs, conventions, and dialects and have become, as a result, barely able to communicate with one another. In some branches, disciplinary boundaries have operated as important checks on the advancement of knowledge. This is particularly true in the behavioral and social sciences, where intellectual territoriality has been extremely pronounced, owing to fundamental divergences such as the old nature–culture opposition. The idea that schools of thought and scientific disciplines cannot feed one another because of impervious epistemological barriers is still very much alive, but it is bound to become vestigial in the long run, for it is consistently invalidated empirically. More and more interdisciplinary bridges are being established, and the ensuing benefits are abundant. But the bridging of two scientific specialties or two long-standing schools of thought carries with it a formidable requirement: one must first grasp the exact nature of the discrepancies causing dissension. No interdisciplinary articulation or reconciliation is possible without understanding why distinct disciplines may conceive of the very same problem in such discrepant and seemingly incongruous manners and without identifying precisely the numerous points where the shoe pinches. To achieve this understanding, one really has no choice but to enter the other's realm. One must temporarily abandon one's own viewpoint and adopt the other's. Interdisciplinary research is basically detailed comparative research about different ways of looking at the same problem, about different sets of data, distinct methods, and disparate concepts and theories. It is also, fundamentally, integrative research. The conceptions and explanations emanating from the integration of two disciplines are often distinct from those of each specialty taken separately. Interdisciplinary research, therefore, cannot be a cannibalistic endeavor, brushing aside the views of one discipline and replacing them with the views of the other. It cannot be so for one simple reason: researchers in a given branch of knowledge or from a particular persuasion can hardly have it all wrong collectively.

This work is an essay in interdisciplinary integration. Specifically, it is an attempt to compare, articulate, and unite behavioral primatology and social anthropology in relation to the study of human society's deep structure and evolutionary origins. I am a primatologist by formation. I came to anthropology as a student with one objective in mind: to study primates as a means to understand human evolution. But in retrospect, I took a twenty-five year detour through primate studies before tackling my initial objective. I spent most of my professional career conducting empirical research on nonhuman primates, much of it on kinship and its social correlates. Over all these years, I could not find the time to write about human evolution or, more likely, did not feel ready to do so, because I found the task formidable and full of pitfalls. For I wanted to understand the evolutionary origins of human society, of such phenomena as kinship, descent, marriage, residence, and the family. I was interested in aspects of human evolution that have left barely any traces in the fossilized bones of our ancestors or in their artifacts. Moreover, because these phenomena are deeply embedded in symbolic meanings and are extremely variable cross-culturally, discussion of them from an evolutionary perspective has consistently been somewhat controversial. Working and teaching in a department of anthropology along with a majority of sociocultural anthropologists, I have been confronted daily with the difficulties of interdisciplinary communication. This led me to realize that there was but a single solution to that problem: to immerse myself in the other discipline and try to understand its viewpoint. This being said, I have no training as an ethnographer or as an ethnologist, and I do not pretend to have become one. My sole claim is that my level of understanding of anthropological concepts is sufficient to allow me to articulate the two realms somewhat congruently. My success will ultimately be up to the sociocultural anthropologists to judge.

I wrote this book with two major audiences in mind: on the one hand, biological anthropologists, primatologists, paleoanthropologists, evolutionary psychologists, behavioral ecologists, and other biologically oriented readers; on the other, sociocultural anthropologists, sociologists, archaeologists, and social scientists in general. My biggest challenge was to give biologically oriented readers enough background information about the concepts of sociocultural anthropology without being unduly tedious to those readers already familiar with them and, reciprocally, to provide social scientists with enough information about primate behavior and evolutionary biology without annoying the informed audiences. My solution was to focus on the points of connection between the two disciplines, to stick to general principles, and to concentrate on the essential concepts.

The present book owes a great deal to the work of Claude Lévi-Strauss and Robin Fox, to whom it would have been dedicated but for the human mind's inclination to favor one's close kin. Robin Fox's pioneering work on the evolutionary origins of human kinship is a model of comparative interdisciplinary research. Long before I knew about this work, I found his classic book *Kinship and Marriage* an illuminating synthesis, with its prominently deductive organization and consistent focus on general principles in a complex domain filled with cultural peculiarities, that of human kinship. Lévi-Strauss's work on the "elementary structures of kinship," in particular his theory of marriage alliance and the concept of reciprocal exogamy, stands as a masterpiece of theoretical abstraction in the extraordinarily intricate realm of ethnographical data. Without his work on the common structural principles underlying human societies, it is doubtful whether my own interspecific comparative analysis would have been feasible.

A number of people have read the entire manuscript and made numerous invaluable comments. I am greatly indebted to Shona Teijeiro, my primary affine, for her staunch support, her endurance and heartening attitude during periods of doubt and perplexity on my part, her built-in intolerance for the slightest lack of clarity in the numerous versions of the text that she read, and her shared fascination with human origins. My deep gratitude goes to Robert Crépeau for the stimulating comments of a sociocultural anthropologist from the very early stages of the project and thereafter and for our numerous and fruitful discussions, which confirmed my belief that a real dialogue with sociocultural anthropology could be highly productive. I also wish to thank another colleague, Bernard Bernier, whose open-mindedness and enthusiasm for this comparative analysis were especially meaningful to me. I owe much to Carol Berman, my long-term primatologist friend, for her willingness to read various versions of the manuscript and for her clarity of thought and intellectual rigor. I very much enjoyed the perceptive and colorful comments and metaphors of Michel Lecomte, in particular, about my intellectual approach in this book. I got insightful comments from Annie Bissonnette, who challenged me with some penetrating questions that helped me clarify my reasoning. Finally the detailed reports of the two anonymous reviewers for Harvard University Press contained numerous helpful comments and suggestions that served to improve the final product substantially.

A number of people read preliminary drafts of various chapters and I am very grateful to them: Michelle Drapeau, a paleoanthropologist colleague who also helped with some references; Jean-Claude Muller, whose extensive knowledge of the work of Lévi-Strauss reassured me about my own understanding of his work; and Anne Pusey, whose expertise on chimpanzees was most welcome. This book grew out of an article that was never submitted for publication because it evolved into the present form. I wish to thank Richard Wrangham, Lars Rodseth, Robert Crépeau, and Norman Clermont for their detailed and helpful comments on that crammed manuscript, and Richard Wrangham again, for suggesting that I write such a book in the first place. My thanks also to John Mitani and Kevin Langergraber for letting me see their unpublished data on sibling relations in chimpanzees. Louis-Bruno Théberge, Julie Cascio, and Constance Dubuc provided technical assistance with the references and figures, and they did so with much professionalism. Aileen Fenton, with her charming Scottish accent, assiduously answered my queries relating to the English language. I am also very much obliged to all my graduate students for their stimulating input over the last twenty years, and to Jean Prud'homme and Carole Gauthier, my long-term collaborators in the laboratory of behavioral primatology at the Université de Montréal, with whom I had innumerable fruitful discussions about primate behavior.

Above all I am deeply indebted to my children, Catherine and Louis-Charles, for being around and for salutarily distracting me, often without mercy, from my obsession with this book.

In closing I would like to thank Peg Anderson for a very careful and accurate editing job, Liz Duvall and Anne Zarella for their diligence in handling matters relating to the preparation of the manuscript, and Michael Fisher, my editor at Harvard University Press, who was enthusiastic about the project right from the outset and whose convictions about the importance of interdisciplinary research were particularly refreshing. Finally, I acknowledge the contribution of the Natural Sciences and Engineering Research Council of Canada for its constant financial support over the last twenty-five years.

Primeval Kinship

1

The Question of the Origin of Human Society

> Exogamy lies far back in the history of man, and perhaps no observer has ever seen it come into existence, nor have the precise conditions of its origin yet been inferred.
>
> *Edward Burnett Tylor (1889a, 267)*

Adolph Schultz (1969) produced a number of particularly telling illustrations of the comparative anatomy of primates. One of these depicts the human foot alongside the feet of twenty-two other primates, all reduced to the same length (Figure 1.1). Each foot represents the "average" for that species, a sort of paradigmatic foot. Ever since comparative anatomists began looking for evolutionary connections between species, they have tracked and ceaselessly documented the remarkable continuity of structures across their cumulative transformation over time. The human foot is specifically adapted to upright locomotion and possesses unique traits in this regard, yet its resemblance to other primate feet is patent. Now let us imagine a comparison featuring not primate feet but diagrams of primate social structures, an illustration depicting the "paradigmatic human society" alongside the paradigmatic societies of other primate species. On the one hand, the distinctiveness and uniqueness of human society would stand out clearly against the primate background; on the other, the connection of human society with other primate social structures would be no less plain. This book is about the very nature of human society—our unique way of integrating kinship, reproduction, residence patterns, and the like—the relationship of this particular social configuration to other primate societies and its evolutionary history.

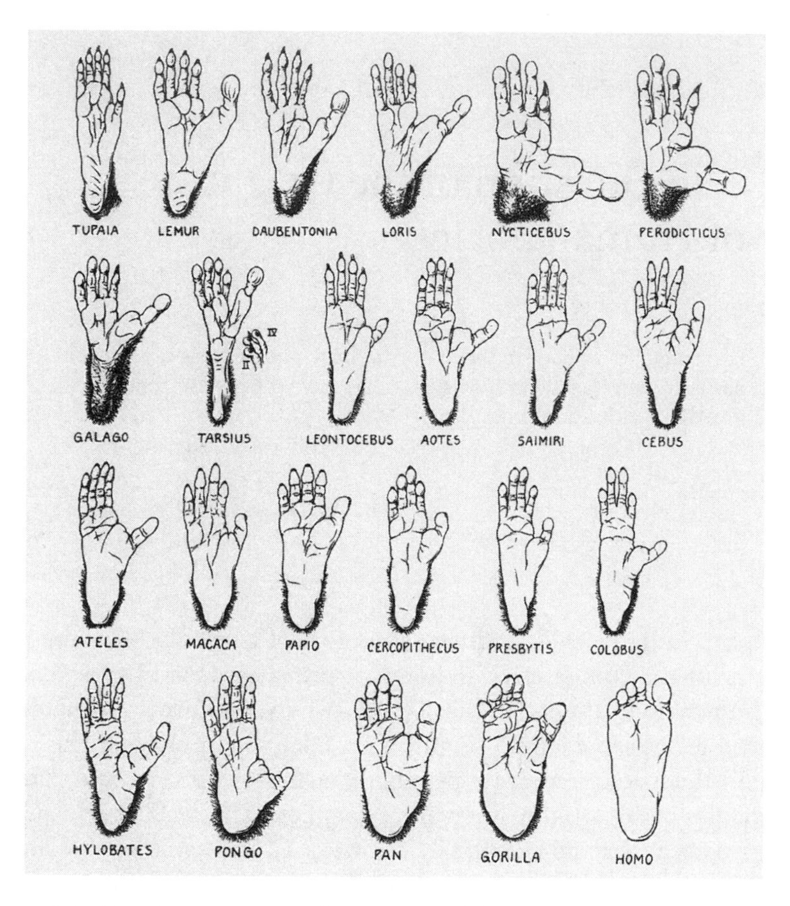

Figure 1.1. Paradigmatic primate feet. Reprinted from Schultz (1969) with permission from Weidenfeld and Nicolson, the Orion Publishing Group.

Considering the great diversity of human societies, the very concept of a paradigmatic one may sound nonsensical, and indeed it is to a large extent. There is no real society that represents all others in the way that a standard foot can represent all human feet. The paradigmatic human society that I have in mind corresponds to no particular human group, even less to some average, standard, or normalized social structure. It stands rather as a universal set of organizational principles, a *deep structure* that runs through all human societies and pervades social organization as a whole. The deep structure of human society is also, as I will argue, a remnant of our evolutionary past: all human societies are built around the same basic structure because they all originated from the same archetypal framework.

One might think that uncovering the unique set of principles out of which all human societies diversified would count among the most elementary objectives of sociocultural anthropology. Yet this problem is not an integral part of the discipline's research program; the deep structure of human society is not really an issue in anthropology. Several factors account for this. First, anthropologists are experts in diversity. And cultural diversity is certainly one of the most complex subjects tackled by scholars since the rise of modern science. Even setting aside huge portions of cultural diversity to concentrate on the narrower topic of social relations and social systems, the complexity is overwhelming. To this plain fact one must add the powerful conceptual and historical influences that have diverted anthropology away from the search for universal structures. For example, Franz Boas, one of the most influential figures in the history of anthropology, explicitly denied that anthropology's objectives included the formulation of general laws governing culture and society. And one of the most dominant theoretical movements in the history of anthropological thought, the American culture and personality school, with figures like Ruth Benedict and Margaret Mead, has consistently emphasized the originality and uniqueness of every culture and has focused accordingly on differences rather than similarities between societies. In that vein, American cultural anthropology has always stressed the importance of understanding cultures from the inside and has kept away from large-scale cross-cultural comparisons. From such a perspective, the search for humankind's deep social structure was inherently a nonissue.

Other major schools of anthropological thought did conceive of anthropology as a comparative science whose principal objective was to arrive at general laws about culture or society, namely, the British structural functionalists, led by Alfred Radcliffe-Brown, and the French structuralists, led by Claude Lévi-Strauss. At first sight, these two schools of thought would seem ideally positioned to search for the deep structure of human society and, having characterized it, to pose the problem of its evolutionary origins. Interestingly, however, they could not do so for one simple reason: both schools purposefully excluded history and, a fortiori, evolutionary history from their theoretical frameworks. They did so not so much because they denied the importance of history but because they strategically posited that social structures were systems whose workings could be understood without reference to the past. For example, the structuralists sought to uncover the universal structures of the human mind, notably in the social sphere. To them, such structures were at work, and hence ought to be manifest, in the workings of any society at any time. Therefore, mental structures could be, and ought to be, abstracted from strictly structuralist and synchronic (unchronological) analyses of human societies.

Another reason why the functionalists and structuralists removed any evolutionary consideration from their analyses is that, like the American cultural anthropologists, for that matter, they sought at all costs to dissociate themselves from the nineteenth-century school of historical evolutionism and its attempts to explain the evolution of human societies. So speculative and erroneous had been those prior attempts (discussed below) that early in the twentieth century, mixing evolution with the study of culture and society had extremely negative connotations. If only for this reason, the study of the evolutionary origins of human society was simply out of the question.

In any case, regardless of their conception of the role of evolution in understanding human society, the comparativists did not arrive at a consensus about the laws of human society. As a result, despite several decades of comparative analysis, contemporary anthropologists still barely agree on the meaning, and in some cases even the reality, of some of their most basic concepts, such as marriage, residence, kinship, descent, and family. Not surprisingly, human societies still defy any transculturally valid characterization or unitary definition of their fundamental nature. If the absence of transculturally valid concepts is not necessarily a problem in itself, difficulties arise when the magnitude of cultural diversity is taken to mean that universally valid principles might be mere illusions and creations of the researchers themselves. Humankind's behavioral unity exists, but it lies deeply buried under several thousand years of cumulative cultural evolution and is barely visible from within the human realm. The best way to uncover this unity is to step out of that realm and look at human groups from the perspective of nonhuman forms of social organization. From such an outlook, differences between human groups give way to their similarities. Compared to the social organization of other species, human societies collectively share numerous original attributes. But more importantly, they display a unique combination of traits, not all of which are original, and it is this particular configuration that sets them apart from other animal societies. From this angle, therefore, humankind's deep social structure readily stands out. It differs clearly from that of chimpanzee or gorilla societies, for instance. In this book I set out to characterize that specific set of features and to describe how it progressively came about in the course of evolutionary time.

There is another reason that the deep structure of human society could not be recognized as a research topic in anthropology. Somewhat paradoxically at first sight, the evolutionary history of human society is out of reach of the two anthropological disciplines whose concerns appear most closely connected to it: paleoanthropology and archaeology. The commencement of human society lies beyond the methodological capabilities of both disciplines. Archaeology does not go back in time far enough. The evolutionary history of human society took place over several million years, and the archaeological record becomes extremely tenuous beyond a few hundred thousand years. Paleoanthropology does go back in time far enough, but it is basically silent about the evolution of social organization—marriage, kinship, and the like—even though it reveals some aspects of human behavior. For that matter, the topic does not belong to any particular scientific discipline. In fact, it can be apprehended only through interdisciplinary research, specifically through the junction of social anthropology and comparative social primatology. And here lies a further reason that the deep structure of human society is still in want of a definition: very few researchers have adopted the topic and embarked upon the necessary comparative work on key concepts such as kinship, descent, and exogamy. Unidisciplinary research is time-consuming on its own, and interdisciplinary research is even more demanding.

To all intents and purposes, then, the evolutionary history of human society is an orphan research topic. The upshot is that sociocultural anthropology is a science whose study objects (societies) are evolved entities—that is, whose unitary core structure has a phylogenetic history—but whose research program largely neglects this basic fact. It is in the position of a planetary science that would concentrate on planets and their interrelations and overlook the question of the origin and evolution of the solar system. Yet the origin of human society is no less sound a research theme than the origin of language or bipedal locomotion, for example. It is a subject for which methods, hard data, concepts, hypotheses, and partial answers do exist. It is also a fundamental question in the true sense of the word; it is foundational.

A Forsaken Quest

The birth of human society has not always lacked students. There was a time when social anthropology itself was very much interested in the issue. In the latter half of the nineteenth century, anthropologists of the historical evolutionism school were working hard to reconstruct the universal history of humankind. Many evolutionary schemes were proposed and their relative values hotly debated over decades. Included in some of these schemes were descriptions of the hypothesized earliest stage in the evolution of human social structure, that of the "primeval ape-man." To reconstruct vanished periods in the history of human society, historical evolutionists relied on the idea that these periods had left traces still discernible in the most primitive of existing societies. Thus, by correctly interpreting the right telltale signs one could infer the past. Essentially, the evolutionists counted on "that great class of facts" called *survivals*, which Edward Tylor described as these "processes, customs, opinions, and so forth, which have been carried on by force of habit into a new state of society different from that in which they had their original home, and . . . thus remain as proofs and examples of an older condition of culture" (1889a, 16).

For example, Lewis Henry Morgan hypothesized that the most ancient period in the formation of human society featured group marriage and sexual promiscuity among all members of ego's generation. This included incestuous unions between brothers and sisters, what he called the consanguine family. Morgan's proof of the existence of that marriage system lay among the Hawaiians, whose "system of consanguinity and affinity . . . has outlived for unnumbered centuries the marriage custom in which it originated, and which remains to attest the fact that such a family existed" ([1877] 1974, 410). The Hawaiians, Morgan noted, used the same term, "father," to refer to ego's father, the father's brothers, and the father's cousins. Such a merging term meant, Morgan reasoned, that all the men it referred to were the *potential* fathers of ego, hence that they were all, in effect, married to ego's mother. Proceeding along those lines with every Hawaiian kin term, Morgan found that they all made sense when construed as survivals of a primeval stage of consanguine group marriage.

Clearly the evolutionists' schemes rested on a scanty empirical basis, and as a result they could differ drastically from one another. For example, one of Morgan's contemporaries, John McLennan, proposed a scenario featuring wife capture, the acquisition of wives by force or theft, as a prominent aspect of "ape-man society." Wife capture, McLennan thought, had survived until today in the form of ritualized ceremonies "so widely spread" that it must be "connected with some universal tendency of mankind" (McLennan [1865] 1970, 20) and so significant that "in the whole range of legal symbolism there is no symbol more remarkable than that of capture in marriage ceremonies" (11). Wife capture between tribes, he reasoned, implied that marriages were commonly contracted between, not within, tribes, and hence that the tribes were *exogamous*, a term coined by McLennan himself (22). The question then became: What is the origin of exogamous wife capture? McLennan speculated that the harsh living conditions endured by primeval human hordes had caused males to be more valued than females, which in turn had brought about female infanticide (another survival). The ensuing shortage of women compelled men to share wives, resulting in polyandry, which McLennan saw as the primeval form of marriage. Eventually men had to compensate for the lack of women in their own group by stealing women in other hordes; they were thereby practicing exogamy.

Significantly, descriptions of bygone stages in the evolution of human society took up extremely small segments in the evolutionists' chronological sequences. Above all, historical evolutionists were concerned with existing societies, which, they proposed, could all be ordered within a unilinear and universal continuum of growing complexity. All over the world, they contended, each society had gone through, and would go through, the same evolutionary sequence. Thus human beings, wherever they thrived, were supposedly obeying similar universal forces, a conception that implicitly bespoke the immense potency of humankind's psychic unity in organizing social life. As ethnographic data accumulated early in the twentieth century, the evolutionists' schemes could not withstand the outpouring of criticism, both empirical and theoretical, that fell upon them—in particular, the arguments of Boas and his students, and of Branislaw Malinowski, Radcliffe-Brown, and the functionalist school. Evolutionary sequences as a whole fell into disfavor, and with them efforts to characterize vanished stages in the genesis of human society. Such efforts were by then deemed conjectural at best, most often quixotic and vain. For example, in the 1931 edition of the *Encyclopaedia of the Social Sciences,* Margaret Mead stated: "All of these attempts to reconstruct the earlier forms of organization of the family remain at best only elaborate hypotheses. Contemporary refutations rest upon . . . a methodological refusal to admit the discussion of a question upon which *there is not and cannot be any valid evidence*" (cited by White 1959, 71; my emphasis).

In rejecting one particular school of anthropological thought that had erred to a considerable extent, early twentieth-century anthropologists were rejecting a whole research topic: the evolutionary history of humankind. By rightly dissociating themselves from historical evolutionism, they were simultaneously turning away from evolutionary history as a whole. In retrospect, anthropologists were rejecting the baby with the bath water. They would stick to this approach for decades, eschewing any evolutionary consideration relating to human society.

That the evidence needed to reconstruct the origin of human society was dramatically lacking in the 1930s cannot be denied. But it was premature to state that such evidence should never be gathered. True, even though physical anthropology, paleontology, and archaeology were already providing hard data on human evolution, the question of the evolutionary origins of human social structure still eluded anthropologists. Aspects of the social realm such as postmarital residence patterns, kinship networks, unisexual descent, incest prohibitions, and a host of others do not leave significant physical traces, whether in fossils or in artifacts. Thus, by the middle of the twentieth century, speculations about human evolution were perforce giving way to interpretations based on fossils and artifacts, but it was still not possible to address satisfactorily the question of the evolution of human social structure. The necessary information, namely data on the behavior of nonhuman primates, was just beginning to accumulate. As researchers began to collect data on monkeys and apes, it became clear to many that these kinds of data were uniquely useful to enlighten behavioral aspects, especially the social dimension, of human evolution and that they provided a powerful means to avoid far-fetched speculations about the origins of human society. Retrospectively, had the nineteenth-century evolutionists been aware that primates commonly form polygynous breeding units but very rarely polyandrous ones, that they have sex-biased residence patterns, constitute kin groups, discriminate categories of relatives, breed outside their natal group, and avoid incestuous matings, they might not have posited the existence of a primeval stage of polyandry in early humans or consanguine group marriage.

But early in the twentieth century such data were simply nonexistent. In their place, anecdotes on the behavior of monkeys and apes flourished. As noted by the anatomist Solly Zuckerman, anecdotal reports "were regarded in the most generous light, and it was believed that many animals were highly rational creatures, possessed of exalted ethical codes of social behaviour" (1932, 9). Clearly, going beyond anthropomorphic interpretations of animal behavior would require systematic studies of primates living in undisturbed groups in their natural habitats. The first field studies were carried out as early as the 1920s and 1930s by a few pioneers such as Henry Nissen (1931) and especially Clarence Carpenter (1934, 1935, 1940). Still, before 1950, studies on the behavior of wild primates were extremely fragmentary. Certainly the existing ones had provided some general principles about the social life of primates (Zuckerman 1932; Carpenter 1942), but these principles were abstracted from a very small number of populations and species, and they suffered innumerable exceptions and erroneous statements.

Primate studies began to grow at a steady pace only during the second half of the twentieth century, and only in the 1970s did they reach quantitative and qualitative levels that would make it possible to carry out detailed cross-specific comparisons. In this context, it is noteworthy that among the first researchers to seize upon the primate data to ponder the origin of human social structure were social anthropologists, whose contribution to this topic is more important than that of contemporary primatologists and physical anthropologists. In the 1940s and 1950s, sociocultural anthropology was going through a revival of interest in questions relating to the long-term social and cultural evolution of humankind with the work of Leslie White, Gordon Childe, Julian Steward, and others. These second-wave cultural evolutionists duly acknowledged the great vulnerability and major weaknesses of their nineteenth-century predecessors, but they also recognized the intrinsic interest of some of the questions they had asked, notably the "search for cultural origins" and the "original social condition of man." White, in particular, soon realized the potential import of the still scanty body of knowledge on primate behavior for understanding the origin of human society. In the late 1950s he explicitly stated the principles underlying what amounts to a whole research program on the evolutionary history of human society, based on the use of primate data (White 1959). He was followed by others, including Elman Service (1962) and, most notably, Robin Fox (1972, 1975, 1979, 1980, 1993).

In retrospect, it makes perfect sense that the reconstruction of the origins of human society was initially the province of social anthropology. Only social anthropologists were familiar enough with the complexity of human societies to carry out interspecific comparisons that encompassed many important dimensions of social structure. Physical anthropologists, primatologists, and paleoanthropologists were certainly hard at work exploiting the same data on primate behavior. But they were focusing on other, more circumscribed, though equally important, topics. Beginning with the work of Sherwood Washburn and Irven DeVore (1961), they were looking into the evolution of tool use, cultural transmission, territoriality and aggression, pair-bonding, the sexual division of labor, hunting, food sharing, and, along with comparative psychologists, language and cognitive abilities. They were not working on the origin of human society as a whole.

The Deep Structure of Human Societies

The first step in any attempt to reconstruct the origins of human society is to define its deep structure, the features that are common to all human societies and at the same time uniquely human—to find the lowest common denominator, so to speak, of all human societies. That deep structure must necessarily be abstracted from cross-cultural comparative analyses; it may be defined only by comparative social anthropology. But very few anthropologists have tackled this problem explicitly, let alone arrived at a comprehensive conclusion. One did, however, though he did not necessarily conceive of his work in this way. In his magnum opus *Les Structures Élémentaires de la Parenté (The Elementary Structures of Kinship)* first published in 1949, Lévi-Strauss proposed that *reciprocal exogamy* was a universal structural feature of human society, one that embodied primitive social organization, was uniquely human, and, moreover, marked the "transition from nature to culture." Marshaling a host of ethnographic facts and reasoning from a strictly structuralist viewpoint, Lévi-Strauss argued compellingly that the exchange of women between kin groups as a means to forge alliances was a primary principle governing primitive human social organization. He also asserted that all aspects of reciprocal exogamy were inseparable elements of a normative whole that had been culturally elaborated at some point in our evolutionary past. That is to say, he believed that kin groups, incest prohibitions, marriage, exogamy rules, and such were cultural and institutionalized constructs, devoid of any evolutionary history.

To my knowledge, no anthropologist other than Lévi-Strauss has attempted to abstract mankind's unity with regard to social organization or proposed a comprehensive and empirically founded theory about the deep structure of human society. This is a remarkable fact in itself, which certainly reflects to a large extent the formidability of the task. Unquestionably, Lévi-Strauss was building upon the work of his predecessors and contemporaries, and as we shall see, some of his most important ideas were already in the air. But he was alone in the scope of his vision and the breadth of his theoretical objectives. Although Lévi-Strauss himself never referred to reciprocal exogamy as the deep structure of human society, his characterization of reciprocal exogamy fits particularly well with my definition of such a structure. In stating that reciprocal exogamy is both structurally foundational and the key pattern that defines the human primate in the social sphere, Lévi-Strauss is providing us with a rare working hypothesis about the deep structure of human society, one that can be tested from an evolutionary perspective. One may indeed undertake to assess (1) whether reciprocal exogamy is truly uniquely human when compared to other animal societies; (2) if it is, whether it does, in effect, embody the deep structure of human society; and (3) whether it is a cultural construct devoid of phylogenetic roots, as asserted by Lévi-Strauss. These are three different statements. In theory, all three might be true or all three might be false. Or the first statement might be validated while the two others are disproved. Or the first two might be confirmed while the third is not.

In this book I argue that the comparative sociology of human and nonhuman primates vindicates the first two claims but not the third. I attempt to show that the deep structure of human society hinges on the combination of a limited set of traits that I refer to as the *exogamy configuration*. The most important features of that configuration are the following: stable kin groups, enduring breeding bonds between particular males and females, a dual-phase system of residence patterns (premarital and postmarital), incest avoidance among coresident close kin, recognition of kin on both the mother's and the father's sides, kinship networks that extend beyond the local group, lifetime bonds between brothers and sisters (and a number of related phenomena such as avuncular relationships and marriage between cross-cousins), recognition of in-laws (affinal kinship), and matrimonial exchange (exchange of spouses between groups). Significantly, the evolutionary originality of the exogamy configuration—of human society—lies not so much in each of the traits taken separately as in the fact that all traits co-occur in the same species. In short, human societies would be unique in integrating a number of local groups through bonds of kinship and affinity.

Lévi-Strauss's concept of reciprocal exogamy and the present "exogamy configuration" refer basically to the same thing, except that the latter concept describes reciprocal exogamy after it has been decomposed into its evolutionarily meaningful components, as will be amply discussed. It should be clear that the term "exogamy" here does not mean mere outbreeding (breeding outside one's group), as the word is sometimes used by nonanthropologists, in particular by many primatologists. Nor does it mean solely outmarriage, as in George Murdock's classic definition: "the rule of marriage which forbids an individual to take a spouse from within the local, kin, or status group to which he himself belongs" (1949, 18). Rather, exogamy refers to the *binding* aspect of marriage and, more specifically, to between-group binding, regardless of the nature of the entities that are bound, whether families, lineages, clans, tribes, nations, or others. Acknowledgment of the role of exogamy in binding human groups has a long history, dating back to the nineteenth century, in particular to Tylor and his famous "marrying-out or being killed-out" aphorism. But the significance of exogamy as a major divide between human and animal societies is more recent, and no researcher has argued more comprehensively in this sense than Lévi-Strauss. The convergence between the primate data and Lévi-Strauss's so-called alliance theory—or "marriage alliance theory," as Louis Dumont (1997) aptly called it—is far from self-evident, and much effort has been devoted to articulating as clearly as possible the numerous points of junction between the two conceptual/empirical realms. But it should be noted at this point that the convergence does not bear on all aspects of alliance theory. It rests on some of the theory's most basic principles, such as the role of marital bonds in binding distinct social groups and that of kinship in structuring marital alliances.

Remarkably, the idea that exogamy embodies the essence of human society has received very little attention, and its implications have been little explored. At the same time, the fact that exogamy was not born in an evolutionary vacuum, that it has deep phylogenetic roots observable in other species, is no less clear and needs to be better appraised. Indeed, though no other animals display the complete exogamy configuration, many primate species exhibit one or several of its building blocks, from outbreeding and the formation of kin groups to incest avoidance and pair-bonding. Several major ingredients of exogamy thus lie scattered across the primate order, and it is now clear that their conjunction and fusion in the human species did not take place out of the blue. The views of Lévi-Strauss and those of comparative primatology about the *origins* of exogamy are thus fundamentally divergent. The exogamy configuration appears to be the outcome of a specific concatenation of events that took place over evolutionary time. It is the product of a particular conjunction of otherwise typically primate phenomena. It is also likely that a primitive version of the exogamy configuration thrived as a set of behavioral regularities well before culture took hold of it, ultimately generating a considerably more sophisticated version, one that was normative and institutionalized. The very origins of human society would remain forever out of reach if it were not for the behavioral archives of the exogamy configuration displayed by our closest living relatives.

I

Primatologists as Evolutionary Historians

2

Primatology and the Evolution of Human Behavior

If no organic being excepting man had possessed any mental power, or if his powers had been of a wholly different nature from those of the lower animals, then we should never have been able to convince ourselves that our high faculties had been gradually developed.

Charles Darwin ([1871] 1981, 34)

This book is an exercise in comparative sociology, but one carried out at the interspecific level. In what follows I discuss some epistemological and methodological considerations relating to this particular type of knowledge about our species. The significance of primate studies for social anthropology is clouded by a misconception that is common enough to merit some clarification. The principal contribution of primate data to social anthropology lies not so much in what they reveal about the processes and mechanisms underlying human behavior as in what they tell us about its *origins*. This distinction is fundamental. The import of primatology to social anthropology is akin to that of history to the social sciences in general. Primatology, like history, sheds light on how the human patterns came to be rather than on how the patterns actually work. In this sense, comparative primatologists are evolutionary (or phylogenetic) historians. This is not to say that nonhuman primates cannot help understand the proximate and developmental causes of human behavior. Primate models provide important insights about the physiological, neurobiological, and psychological bases of human behavior (Hinde 1987; Nadler and Phoenix 1991; Nicolson 1991; Parker, Mitchell, and Boccia 1994; Dixson 1998; Maestripieri and Wallen 2003; Kalin and Shelton 2003; Maestripieri and Roney 2006). But as far as social relationships and social organization are concerned, primate studies are bound to provide utterly incomplete, and thus reductionist, models of human phenomena—assuming that one would want to use such models. Nevertheless, in shedding light on the origins of human behavior, primate studies point to some of the factors—phylogenetically primitive ones—that may still affect the behavior of contemporary humans along with a host of more recent factors. I return to this issue later.

How, then, do comparative primatologists reconstruct the past? Interestingly enough, they can hardly do otherwise. Primatologists who carry out systematic comparisons of nonhuman primate species with one another and with humans inevitably end up thinking in terms of phylogenetic history. This is so whether they compare morphological traits or behavior patterns. The same applies to ethologists in general inasmuch as they compare closely related species. Comparative primatologists and ethologists are invariably drawn toward a diachronic conception of their material, one that indicates some sort of chronological sequence running through the behavior patterns they compare. I refer to this aspect of the comparative exercise as the *phylogenetic decomposition principle*. It is an integral part of any phylogenetic reconstruction.

The Phylogenetic Decomposition Principle

Figure 2.1 portrays the phylogenetic relationships of humans and other anthropoid primates (apes, Old World monkeys, and New World monkeys). This phylogeny is the likeliest one, based on several sets of molecular data (Goodman et al. 1998, 2005; Enard and Pääbo 2004). It indicates that the two chimpanzee species (common chimpanzees and bonobos, or *Pan*) are more closely related to humans *(Homo)* than to gorillas. That is to say, humans and chimpanzees form a monophyletic group (or clade) relative to gorillas, rather than chimpanzees and gorillas forming a clade relative to humans (Sibley and Ahlquist 1987; Goodman et al. 1990, 1998; Ruvolo 1997; Chen and Li 2001). Humans and chimpanzees thus share an exclusive common ancestor. The speciation event out of which the lineage leading to chimpanzees diverged from that leading to humans is commonly referred to as the *Pan–Homo* divergence. Traditionally, all members of the human lineage—species of the genus *Australopithecus* and *Homo*, among others—were classified in the same family (Hominidae) and called hominids. But with accumulating

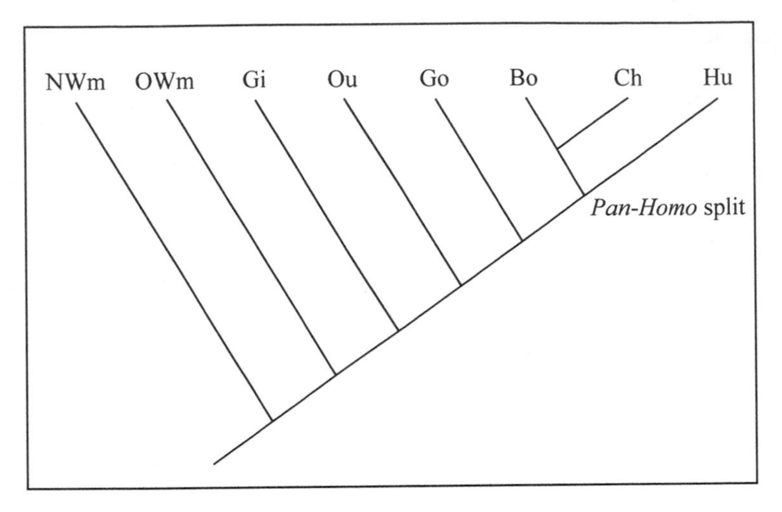

Figure 2.1. Phylogenetic relationships of humans and other anthropoid primates as assessed by several sets of molecular data. NWm: New World monkeys; OWm: Old World monkeys; Gi: gibbons and siamangs; Ou: orangutans; Go: gorillas; Bo: bonobos; Ch: chimpanzees; Hu: humans.

molecular data showing that humans are more closely related to the great apes than previously believed, new classifications were proposed. One solution was to place chimpanzees along with humans in Hominidae. Accordingly, the members of the human lineage belong to the same subfamily (Homininae) and are called hominins. Other authors have proposed to reduce even further the taxonomic rank of the human lineage and to limit it to a single subgenus called *Homo (Homo)*, in which case the genus *Homo* includes not only humans but australopithecines and even chimpanzees (Goodman et al. 1998; Wildman et al. 2003). Regardless of whether the latter classification is adopted or not, it has the merit of emphasizing the extremely close degree of genetic similarity between humans and chimpanzees, which is presently estimated at about 99 percent (Chen and Li 2001; Wildman et al. 2003; see also Varki and Altheide 2005). In any case, the important point is that most authors agree on the basic phylogeny portrayed here even though they may disagree on how to classify and name the species. The issue is far from settled, and for the sake of clarity I use the traditional "hominid" denomination to refer to the members of the human lineage after the *Pan–Homo* divergence.

The absolute date of the *Pan–Homo* split is still debated. The earliest known fossil member of the hominid lineage (*Sahelanthropus*) is dated at 6–7 million years. It shows a mosaic of primitive and derived traits that place it close to the *Pan–Homo* divergence (Brunet et al. 2002). Other estimates based on molecular data give somewhat earlier branching dates, at about 5–6 million years (Goodman et al. 1998, Chen and Li 2001). Whatever the exact date, some molecular data sets indicate that the time span between the gorilla divergence and the *Pan–Homo* divergence would be about one-third the time span since the *Pan–Homo* split; that is, the gorilla speciation would have taken place some 1.5–2 million years before the *Pan* and *Homo* lineages began their independent evolution (Chen and Li 2001).

Ever since well-informed comparisons between humans and other animals have been performed, they have disclosed some elements of the evolutionary history of the phenomenon under consideration. Remarkably, the same basic principle underlies all of the comparative analyses, whether in the area of language (King 1999; Snowdon 2001; Hauser and Fitch 2003; Hurford 2003; Cheney and Seyfarth 2005), cultural transmission (McGrew 1992, 1998, 2001; Boesch and Tomasello 1998; Whiten et al. 1999; de Waal 2001; van Schaik et al. 2003; van Schaik 2004), cognitive abilities and intelligence (Byrne and Whiten 1988; Byrne 1995, 2001; Tomasello and Call 1997), self-consciousness (Parker et al. 1994; Keenan, Gallup, and Falk 2003), sexuality (Nadler and Phoenix 1991; Wolfe 1991; Dixson 1998), aggression and power (Chapais 1991; Wrangham and Peterson 1996; Wrangham 1999), sexual coercion and patriarchy (Smuts 1992, 1995; Hrdy 1997), maternal behavior (Nicolson 1991; Pryce et al. 1995; Hrdy 1999), kinship (Fox 1972, 1975, 1979, 1980, 1993; Rodseth et al. 1991, Reynolds 1994; Rodseth and Wrangham 2004), or morality (de Waal 1996; Killen and de Waal 2000). Every comparison has, in effect, forced a breaking down, or decomposition, of the human phenomenon into a number of basic components or properties that had not necessarily been identified as distinct before the comparison was carried out. Decomposition is an integral aspect of the comparative exercise itself. This is because some building blocks of the phenomenon are observed in other species, sometimes in a large number of them, but sometimes only in a few species, whereas other components of the phenomenon are uniquely human.

It should be noted right from the outset that the decomposition principle is not limited to the evolution of behavior, and even less so to human evolution. In fact it is one of the most basic principles of evolutionary theory and one whose range of applications over the last century has been extremely wide. Phylogenetic decomposition has underlain the construction of evolutionary trees ever since Charles Darwin introduced these in 1859; it has supported the fields of comparative anatomy and systematics to this day; and it is still the main tool for reconstructing the evolutionary history of a species' characters. Initially limited to morphological traits, interspecific comparisons were extended to behavioral traits by the early ethologists in the middle of the last century, and more recently they came to encompass the molecular, genetic, and chromosomal levels of analysis. Whatever the trait under consideration, interspecific comparisons consistently break it down into variants whose taxonomic distributions betray their phylogenetic connections. What is specific to the present discussion is the application of the comparative method to complex behavioral systems such as language, morality, and kinship. When DNA and proteins are compared across species, similarities and differences may be quantified precisely in terms of nucleotide or amino acid sequences. As a result, phylogenetic connections between molecular structures belonging to related species (homologous similarities) may be assessed with a relatively high degree of accuracy. When anatomical traits are compared, the assessment of homologies reaches a higher level of difficulty because anatomical traits are less easily quantifiable compared to nucleotides and also because anatomy may change in the course of development. One must therefore abstract from the observed intraspecific variation some sort of common denominator that represents the species' norm. The situation is similar in the case of relatively "simple" behavior patterns such as courtship displays in insects or birds. Although behavioral phenotypes are substantially more subject to environmental influences, and hence more variable, homologies may still be identified (for example, see Prum 1990).

But what about complex behavioral systems such as language or kinship? The degree of difficulty is certainly higher in these situations, but the principles are basically the same. One starts with the most complex form among all those being compared, the human form, and reconstructs its evolutionary trajectory by "cumulatively" integrating several building blocks discerned in related species. Let us consider, for example, the evolution of language, perhaps the first complex behavioral system to have undergone a process of phylogenetic decomposition. Interspecific comparisons have shown that some dimensions of language are widespread among primates. For example, the capacity to perceive a continuum of speech sounds as a series of distinct categories, an ability originally thought to be uniquely human, is in fact present in many species of primates and nonprimates. A widespread distribution across several related taxa strongly suggests a common evolutionary origin, and one from an ancient ancestor. The so-called categorical perception of speech sounds is thus a primitive aspect of the whole set of language features. In contrast, other aspects of language display a much narrower distribution. For example, some aspects of both the naming and syntactic dimensions of semanticity characterize a small number of species, notably our closest relatives, the great apes. As in the previous example, such a distribution suggests a common evolutionary origin, but one that is more recent, dating back at least to our most recent common ancestor with these species. Finally, other properties of language have no nonhuman counterparts. Examples include the use of complex syntax and the capacity to generate new words and new meanings almost infinitely. Uniquely human characters presumably evolved after the *Pan–Homo* split.

Comparative studies thus reveal that language is a multicomposite system whose elementary building blocks and properties have various taxonomic distributions. And it is these distributions taken as a whole that translate into a diachronic sequence, hence into a phylogenetic history. The same conclusion imposes itself whatever the phenomenon under consideration, be it cultural transmission, morality, or sexual coercion. When compared across species, each system breaks down into a number of constituent blocks and properties whose phylogenetic connection lends itself to analysis. The inferred diachronic sequences, of course, are extremely gross and incomplete. But this is no major problem when one's aim is not to reconstruct the exact phylogenetic history of a human phenomenon but, more modestly, to sort the phenomenon's components into two categories: those that characterized the first specimens of the human lineage when it began its independent evolution immediately after the *Pan–Homo* split and those that evolved after that point in time. The first category includes traits that we have in common, minimally, with chimpanzees and bonobos, and possibly with many species belonging to more inclusive taxonomic units. These similarities, which we owe to descent from a common ancestral species, are called *homologies*. Taken as a whole, homologous traits compose the biological heritage bequeathed to us by the last common ancestor that we shared with any other living species. As we shall see, a number of components of the exogamy configuration, such as the fundamentals of uterine kinship and incest avoidance, belong to that particular portion of our primate legacy.

But humankind's phylogenetic heritage is not confined to the features already present immediately after the *Pan–Homo* split; it is not limited to homologies. Other components of the exogamy configuration evolved more recently, and some of these nonetheless have their own primate counterparts, but ones that are absent in chimpanzees and present in more distantly related species. Cases in point are pair-bonding and simple forms of unisexual descent. In such a situation, a common evolutionary origin from our last common ancestor with the species exhibiting the similarity can hardly be invoked because the trait would then be present in a majority of species descended from that ancestor. But the comparative exercise is no less informative because similarities between species may be the product of similar selective pressures that acted separately on each of them, producing *homoplasies* through convergent evolution. Homoplasious similarities provide hypotheses about the evolutionary origins of human phenomena. The word "origin" is important here. To argue, for example, that pair-bonds in humans and some other primate species are homoplasies would not be to say that primate pair-bonds and human marital unions are homoplasious. It would not imply that marriage and primate pair-bonding have the same Darwinian function. It would mean that the selective pressures responsible for the *initial evolution* of pair-bonding in these species were similar. This is a hypothesis about the evolutionary origins, primitive function, and initial form of pair-bonding in hominids, not about the present form taken by marital unions in humans. Marital unions as we know them are the product of several hundred thousand, if not million, years of further evolutionary change, both biological and cultural. In sum, whether comparative primatology points to homologous or homoplasious similarities between human and nonhuman primates, it is mainly concerned with the evolutionary history of human behavior, not with its proximate causes and contemporary adaptive aspects.

Finally, homologies and homoplasies are not necessarily mutually exclusive, a point that is not sufficiently stressed in discussions about human behavioral evolution. Similarities between species may arise from the two processes operating concurrently. Indeed, assuming that two closely related species, chimpanzees and humans, for example, have undergone similar selective pressures on a given trait, the resulting homoplasious similarities would have developed out of homologous structures present in their last common ancestor. The actual similarities would thus reflect the combined contributions of common descent and convergent evolution; they would exemplify so-called *parallel* evolution. In theory, similar selective pressures acting on two closely related species should produce higher levels of overall similarity compared to the same selective pressures acting on two more distantly related species. In closely related species, convergent evolution has the shared-ancestry lead, so to speak. Possible cases of parallel evolution between human and chimpanzees will be discussed in relation to dispersal patterns and violence patterns.

In Part II of this book I use the phylogenetic decomposition principle to analyze the deep structure of human society. I proceed historically, beginning with Lévi-Strauss's description of reciprocal exogamy, which was set in an evolutionary vacuum. I then show how human society was progressively replaced within the evolutionary realm by a few authors who undertook the necessary comparative work. This historical sketch and my own analysis lead me to argue that reciprocal exogamy resulted from the amalgamation of twelve major features in the course of hominid evolution. These features are a multimale-multifemale group composition, kin-group outbreeding, a mating system based on enduring breeding bonds, kin recognition along the maternal line (uterine kinship), kin recognition along the paternal line (agnatic kinship), incest avoidance among coresident close kin, the brother–sister kinship complex, the recognition of in-laws (affinal kinship), a dual-phase pattern of residence, the structural prerequisites of unilineal descent, matrimonial alliances, and a supragroup (or tribal) level of social structure. Several of these features are present in various forms in other primates, while some are uniquely human. But even in the latter situation, some of the traits' phylogenetic roots may be found in the behavior of primates.

Each feature is a complex, multidimensional behavioral system. For example, uterine kinship takes on various behavioral forms across groups and species, it affects both the content of dyadic relationships and the group's social structure, and it involves a whole range of emotional, motivational, cognitive, and developmental processes. Because several of these processes have their own determinants and evolutionary history, one important point follows. For any of the twelve features, the overall resemblance between human and nonhuman forms most often involves a number of *separate* homologous or homoplasious similarities. It is a composite of homologous and/or homoplasious traits. It would thus be simplistic to conceive of any of the twelve building blocks of the exogamy configuration in terms of a single unitary homology or a single unitary homoplasy in relation to its primate counterparts. I will have ample opportunities to illustrate this point in the rest of the book.

Reconstructing the Exogamy Configuration

Breaking down reciprocal exogamy into its constituent building blocks is only the first step in reconstructing the evolutionary history of human society. One must also recompose the whole system and address two questions in particular: How did the original features of the exogamy configuration, those that have no counterparts in the primate order, evolve, and how did all the building blocks come to be assembled in the same hominid species? These questions are the object of Part III. I shall argue that the answers lie in the merging over evolutionary time of the ancestral multimale-multifemale group composition—an otherwise common pattern in nonhuman primates—with a novel mating system featuring stable breeding bonds, a combination that produced *multifamily groups*. I use the expression "stable breeding bonds" to characterize the mating system of the human species. I could have used instead the term "pair-bonding," but to most people a pair-bond means an enduring breeding relationship between a single male and a single female; it excludes an important fraction of human mating arrangements, namely, polygynous and polyandrous unions. "Stable breeding bonds" is thus a less ambiguous term, but it should be noted that the two expressions are synonymous, a topic I return to in detail later on. As to the word "family," I use it here in a broad sense to refer to several types of marital unions, including monogamy, polygyny, and polyandry, and to several types of composition, from nuclear to extended families.

As emphasized by Rodseth and colleagues (1991), the multifamily community is the modal type of human group. Humans may form other types of groups, but these correspond to a "few ethnographic oddities," such as autonomous families and religious groups of celibates. In all such cases, "it may be argued that the modal pattern has been abandoned only temporarily or by some proportion of the population" (227; see also Foley and Lee 1989). The multifamily group is extremely uncommon in the animal world. In the vast majority of primate species, when several males do coreside with several females, as humans do, mating is promiscuous: males copulate with several females, females copulate with several males, and there are no stable breeding bonds. And when primate species display families, the family most often constitutes the whole social group, as in gibbons, which form territorial monogamous units, or in gorillas, which form autonomous polygynous units. Several families do not associate to form a cohesive multifamily group, except in the rare type of group exemplified by hamadryas and gelada baboons, who form so-called multiharem groups. As I will argue at length in Part III, the humanlike multifamily community probably originated in the transition from sexual promiscuity to stable breeding bonds, which took place in the ancestral multimale-multifemale group. After the new mating system had evolved, several changes cascaded, resulting ultimately in an embryonic version of the exogamy configuration, one that was presymbolic and nonnormative and manifest entirely in the form of behavioral regularities.

One basic principle that will be amply illustrated throughout the book is that several components of the configuration emerged merely from the combination of other components and as byproducts of the merging itself. They were not the outcome of specific selective pressures, and therefore one need not invoke specific selective pressures to account for them. The idea that evolution proceeds from the constraints imposed by prior adaptations and that it produces novel features out of preexisting parts and functions is as old as the idea of evolution itself. The concept of *preadaptation* is central here. Preadaptations have been documented repeatedly ever since Darwin introduced the concept of adaptation by natural selection. To take a simple, nonsocial example relating to human evolution, the fact that the human hand is commonly used to carry objects while walking originates in the hand's preadaptation for grasping. Well before the hands could be used to carry things around—that is, well before the evolution of bipedal locomotion—they were preadapted to do so through their use in grasping and manipulating objects. Bipedal locomotion may have evolved as a response to a number of selective pressures, for example as a more efficient way to move around, as a posture reducing heat stress, as a feeding posture, or as a mode of locomotion enabling the transportation of objects (for a review, see Stanford 2003). But whatever the exact selective pressures at work, as soon as the hands were freed from the constraints imposed by quadrupedal locomotion, they could be used in a wide range of contexts unrelated to their prior functions. Accordingly, they could undergo new selective pressures relating to the gathering, extraction, processing, and sharing of food, to the fabrication, transportation, manipulation, or exchange of various types of tools, or to nonverbal communication. In short, the hand's primitive grasping function was coopted for several new uses after it was put to use in a new context, that of bipedalism.

This principle seems to have played a major role in the evolutionary construction of human society. Several cognitive abilities, developmental processes, and social patterns, which otherwise are basic attributes of nonhuman primates, acted as crucial preadaptations that came to be expressed in new contexts upon merging with just a few novel elements. As alluded to earlier, the most important such novel element through which old preadaptations could find a new life and produce original features was a new mating system based on enduring breeding bonds between males and females. The effect of that change on hominid society was like the effect of bipedal locomotion on the use of the hand. For example, I shall argue that agnatic kinship emerged as a byproduct of the combination of stable breeding bonds with preexisting adaptations relating to uterine kinship. The stable breeding bond between a father and mother provided a reliable means for the father to recognize his offspring and for an offspring to recognize his father. In this context, the cognitive abilities and developmental processes involved in the recognition of uterine kin enabled hominids to recognize their kin on the father's side. In short, old abilities were coopted for use in a new context. In retrospect, therefore, the abilities involved in the recognition of uterine kin were preadapted for agnatic kinship recognition. Likewise, several other major features of the exogamy configuration originated as ancillary consequences of other elements merging, thus as nonadaptive traits. From the time they appeared, however, the new features provided novel social environments upon which natural selection could operate.

It should be clear by now that although my purpose is to reconstruct the evolutionary history of human society, I am not concerned with imposing a time frame onto that history nor with fleshing out, behaviorally speaking, specific populations of hominids living at certain times and locations, for example some species of australopithecines. My approach has little to do with paleoanthropological data, absolute dating, and the reconstruction of the behavioral ecology of hominids (see, for example, Washburn and Lancaster 1968; Tanner 1987; Tooby and de Vore 1987; Foley 1989, 1992, 1996, 1999; Dunbar 1993, 2001; Moore 1996; Zihlman 1996; Stanford 1999, 2001; Wrangham et al. 1999; Wrangham 2001; Wrangham and Pilbeam 2001). I am concerned instead with the *sequential order* governing the evolution of the exogamy configuration, regardless of its absolute chronology, and I infer that sequence essentially from the constraints dictated by a structural analysis of human and nonhuman primate societies. To take just one example among many, I shall argue on strictly structural grounds that the evolution of pair-bonding preceded the evolution of alliances between local groups (the tribal level of organization) and that the reverse sequence, the tribe preceding the evolution of pair-bonds, is structurally and logically unlikely.

3

The Uterine Kinship Legacy

> The earliest human groups can have had no idea of kinship. We do not mean to say that there ever was a time when men were not bound together by a feeling of kindred. The filial and fraternal affections may be instinctive. They are obviously independent of any theory of kinship . . . and they may have existed long before kinship became an object of thought.
>
> *John McLennan ([1865] 1970, 63)*

A major assumption underlies the phylogenetic decomposition principle and the very possibility of reconstructing the evolutionary history of the exogamy configuration. This assumption is that there are evolutionary connections between each of the building blocks of the exogamy configuration and their nonhuman counterparts, that there are connections between incest avoidance in nonhuman primates and the incest taboo in humans, between primate kinship and human kinship, between primate dispersal patterns and human postmarital residence patterns, and between primate pair-bonds and human marital unions, among other phenomena. The word "connection" is used in a broad sense here. It refers to homologous similarities, homoplasious similarities, and various mixtures of the two processes. Obviously, if a human phenomenon and its nonhuman counterpart were evolutionarily disconnected—if they were independent phenomena—there would be no point in comparing them with the aim of better understanding the human form. Now the fact is that evolutionary connections between distinct forms of complex behavioral systems cannot just be taken for granted. In some cases, they are far from obvious. Moreover, even granting an evolutionary connection between, say, primate kinship and human kinship, its exact meaning and its consequences for understanding human behavior are complex, debatable, and sometimes controversial.

Primatological Theories and Primate Legacies

Stating, for instance, that primate incest avoidance and human incest prohibitions are phylogenetically connected is tantamount to saying that the processes underlying incest avoidance were an integral part of humankind's primate heritage. The reasoning underlying that claim needs to be spelled out. The comparative study of any phenomenon across the whole primate order is likely to generate a body of principles governing the observed variation. For example, a comparison of incest avoidance in primates in general may be expected to produce a set of principles about the social contexts in which mating avoidance takes place; its distribution and magnitude in relation to various factors such as age, generation, and degree of kinship; the proximate mechanisms and developmental processes involved; and its adaptive aspects—in other words, some sort of "primatological theory" about incest avoidance. Similarly, primate data could be used to generate primatological theories about uterine kinship, pair-bonding, unisexual filiation, residence patterns, and so forth. Importantly, because any such theory would have been abstracted from data on a large number of species belonging to the same taxonomic order but differing in their ecological, demographic, life-history, and social traits, its explanatory value should apply equally to all members of the order. This includes our primate forebears, the early hominids who lived shortly after the *Pan–Homo* divergence. Indeed, cognitively and socially speaking, early hominids were typical primates who were no more sophisticated than the great apes. The implications are straightforward: primatological theories about uterine kinship or incest avoidance help characterize humankind's primate legacies. They allow us to delineate our biological heritage.

This is certainly a fundamental contribution of primate studies to anthropology, but one that brings us to the next question, the crucial one as far as the significance of our biological heritage is concerned. This question concerns the fate of primate legacies in the course of 6–7 million years of hominid evolution. With regard to incest, for example, how do we know whether primatelike incest avoidance had any bearing on human incest prohibitions? The two phenomena might be merely superficially similar. Between animal behavior and human behavior stands the symbolic capacity and the realm of normative rules, and here lies most of the controversy about the significance of our biological heritage. A common view among social scientists is that the gap between the behavioral regularities observed in nonhuman primates, on the one hand, and the institutionalized rules of conduct that govern human affairs, on the other, is so great that at least one of the two following claims must be right. According to the first claim, primate regularities cannot be at the origin of human rules because the latter are cultural constructs built on a foundation of their own. Hence the similarities between incest avoidances and incest prohibitions would be shallow and essentially metaphorical. The second, less drastic, claim is that even granting that interactional regularities in primates might be at the origin of some human rules, they cannot explain some of their fundamental dimensions, notably their moral content. Human rules reflect collective values about what is good or bad, about what *ought* to be done or not done. The moral dimension of rules would illustrate the phylogenetic discontinuity and fundamental divide between normative rules and primatelike behavioral recurrences.

Both claims concern, ultimately, the meaning of the primate legacy for understanding human behavior. The first claim hardly conceives of primate behavior as a significant legacy. The second acknowledges our primate heritage but conceives of it as some interesting vestigial organ, one that reveals the phylogenetic origins of our behavior but has no real bearing on some of the factors that affect our social life. It is with these issues that I am concerned in this and the next three chapters. I address both claims by focusing on two themes that theoretically and historically best illustrate the issues at stake: kinship and incest. I begin with kinship.

Primate studies best attest to the validity of John McLennan's insight: Kinship affected social relations long before it became an "object of thought." The most telling evidence here comes from data on kinship relations through females—uterine kinship. In multimale-multifemale primate groups, the reckoning of kinship through the father, or paternal kinship, is either absent or limited both in its extent and in the intensity of its manifestations (for reviews see Strier 2004; Rendall 2004; Widdig 2007). In these groups kin recognition is basically limited to relatives on the mother's side and further confined to those related through females. Uterine kinship is the most primitive and widespread form of kinship. It is also the most extensive component of kinship in nonhuman primates.

My aim here is to encapsulate the relevant information about uterine kinship in a small number of principles. I am concerned, in particular, with three aspects: the extent of kin recognition in terms of discriminated kin types, the developmental and cognitive processes underlying kin recognition, and the manner in which kinship affects social structure at the group level, creating various types of kinship structures. Needless to say, the corresponding principles are an integral part of any primatological theory about uterine kinship. I shall later argue that the underlying processes are generic and that they extend to kinship through males; in other words, that they are crucial for understanding the evolution of *bilateral* kinship, the reckoning of one's relatives on both the mother's and the father's side. Extensive bilateral kinship is possibly a uniquely human phenomenon.

Appraising Primate Kinship

To infer whether individuals recognize their kin among all their fellow group members and whether they further discriminate distinct kin types among all their kin, primatologists need to know the identity of ego's kin in the first place. Until recently, kinship along the paternal line was hardly known; hence the majority of kinship studies focused on uterine kinship. Uterine relatedness could be established only from birth records and knowledge of maternity based on long-term observational data on particular primate groups. Nowadays genetic maternity and paternity may be inferred from molecular genetic information in the absence of birth records and long-term studies of a given group. Noninvasive genotyping techniques make it possible to extract the relevant data from extremely small amounts of DNA in the hair and feces of free-ranging primates without capturing and manipulating the animals (Morin and Goldberg 2004; Woodruff 2004). Knowing how two individuals are genetically related to each other, whether from birth records or genetic data, makes it possible to assess whether they recognize each other as kin.

There is some confusion about the meaning of the expression "kin recognition." Primatologists (and ethologists in general) infer that animals recognize certain individuals as kin when they treat them in a discriminative manner compared to unrelated individuals. Thus the term does not imply that individuals recognize genetic relatedness and consanguinity per se. Nor does it mean that individuals have any conceptual understanding of kinship and kin categories. Kin recognition merely implies that individual primates perceive certain *correlates* of consanguineal kinship in others with whom they interact on this basis. From ego's perspective, a given relative might be no more than a long-term and disproportionately familiar associate whose behavior in various situations has become more or less predictable. For example, a kin might be the individual who is a consistent provider of comfort and security in certain circumstances and who acts as a reliable ally against certain opponents in certain social contexts. In other words, a kin is simply an individual who has a special significance to ego.

Genetic relatives are obviously not the only individuals that are special to ego. The group's alpha male and alpha female, for example, certainly enjoy a special status from ego's viewpoint, a status that they owe principally to their top position in the dominance order and to the level of power associated with it. Likewise two unrelated associates may have a special value to each other reflecting their particular relationship history as well as built-in factors such as age and sex—for example, same-sex peers who grew up together. Therefore, to assert that kinship has any reality as a factor affecting social life, one must have good reasons to believe that kin are special to each other by virtue of their kinship link. Put differently, ego's kin consist of that subset of individuals whose social significance stems from certain properties of the kinship bonds, properties that differ according to the kin type. The properties that characterize the mother–offspring bond, the grandmother–grandoffspring bond, or the sibling bond are a central aspect of primate kinship studies.

I use the word "kindred" to denote the subset of ego's consanguineal kin that ego actually recognizes. "Domain of kindred" and "domain of kin recognition" are thus synonymous. Primatologists circumscribe the domain of kindred by analyzing the distribution of nepotism. Etymologically, nepotism means favoritism directed to nephews. This is the meaning most social anthropologists have in mind when they use the word. But to primatologists and ethologists, nepotism refers to kin favoritism in general, irrespective of kin type. Nepotism takes the form of preferential bonds between relatives, manifest in interactions such as social grooming, tolerance at food sites, and aiding in conflicts. Primatologists conclude that kin recognition exists when the rates at which such positive interactions take place between individuals known to be related are significantly higher than rates of interactions between non-kin, which therefore stand as a comparative base line. Rates of behaviors provide quantitative criteria for kin recognition. But preferential bonds also have qualitative dimensions apparent in the exclusive occurrence of certain types of interactions. Among kin such interactions often involve risks for the individual performing them. Examples include protecting a relative against a higher-ranking and powerful aggressor or a predator. Interactions of this sort are often altruistic: they promote the beneficiary's well-being while providing no personal benefits to the performer, who takes risks and incurs potential fitness costs. The exclusive occurrence of altruistic behaviors among known consanguines is a particularly clear evidence of kin recognition.

Using nepotism to infer kin recognition carries with it one potential drawback. It may lead one to underestimate the domain of kin recognition. Primatologists might indeed fail to record kin recognition between two relatives if they happened to interact infrequently even though they recognized each other. Such a situation is especially likely between distant relatives for a number of reasons. First, individuals could satisfy most of their social needs—grooming, protection, alliances, and the like—by interacting with closer kin, for example their mother and sisters, especially in situations where close kin are numerous, as in groups with large matrilines (individuals related to a common female ancestor through females) (Chapais and Bélisle 2004). They might interact with their nieces or cousins infrequently for this reason alone. Second, it might be less profitable (in terms of fitness benefits) to interact with more distant kin than with closer kin. In theory, altruism is less profitable to the performer the lower the degree of relatedness between the two relatives. Therefore the absence of altruism between two distant kin might reflect the limit of the profitability of nepotism in this situation rather than the absence of kin recognition between them (Chapais, Girard, and Primi 2001; Chapais and Bélisle 2004). Third, nepotism might underestimate the extent of kin recognition if, from ego's perspective, cooperating with a competent but unrelated individual is consistently more advantageous than cooperating with a less competent relative. Competence refers to the partner's qualifications for a given task as these are affected by the partner's age, dominance status, skills, or past experience, among other things (Chapais 2006). For certain tasks, non-relatives may be more competent than relatives, whether close or distant. For all these reasons, individuals might fail to interact at significant rates with some relatives, especially the more distant ones, leading the researcher to underestimate the extent of kin recognition. It is thus important to keep in mind that inferences from nepotism are likely to yield a conservative assessment of the true domain of kin recognition.

Throughout the book I utilize George Peter Murdock's definitions of kin types, which are particularly useful in comparing domains of kin recognition across species (Murdock 1949, 94). Ego's primary consanguineal kin are its mother, father, brothers, sisters, sons, and daughters. Ego's secondary kin are the primary kin of each of its primary kin: grandparents, grandchildren, aunts, uncles, nieces, and nephews. Similarly, ego's tertiary relatives are the primary kin of it's secondary kin, that is, first cousins, great-aunts, and great-grandparents, among many others; and so on with quaternary kin and more distant categories.

The Domain of Uterine Kindred in Primates

Female philopatry—female localization coupled with male dispersal—is the most common "residence pattern" observed in multimale-multifemale primate groups. It characterizes species like macaques and baboons. It is also the pattern that provides the best context for studying the uterine component of kinship. Female philopatry generates kin groups comprising extensive, multigenerational matrilines, as shown in Figure 3.1. Because females stay in their birth group over their lifetime, any female coresides with her mother and her mother's consanguineal kin—her matrilateral kin. A female's matrilateral kin include mostly females because males leave their natal group at puberty. For example, a female coresides with her mother's sister and the sister's daughters, but not with her maternal uncle and his offspring. Females are part of a network of uterine kin that extends over three, rarely four, generations and includes females of all ages as well as young males that have not yet dispersed. The group's adult males come from other groups and therefore are unrelated or distantly related to the group's females. A female also lives with her father and her paternal siblings, but not with her father's kin unless some of these immigrated into her group. Patrilines are thus limited and fragmentary in female-philopatric groups.

In these species females have been found to approach, spend time with, groom, tolerate near food sources, ally with, protect, and sleep

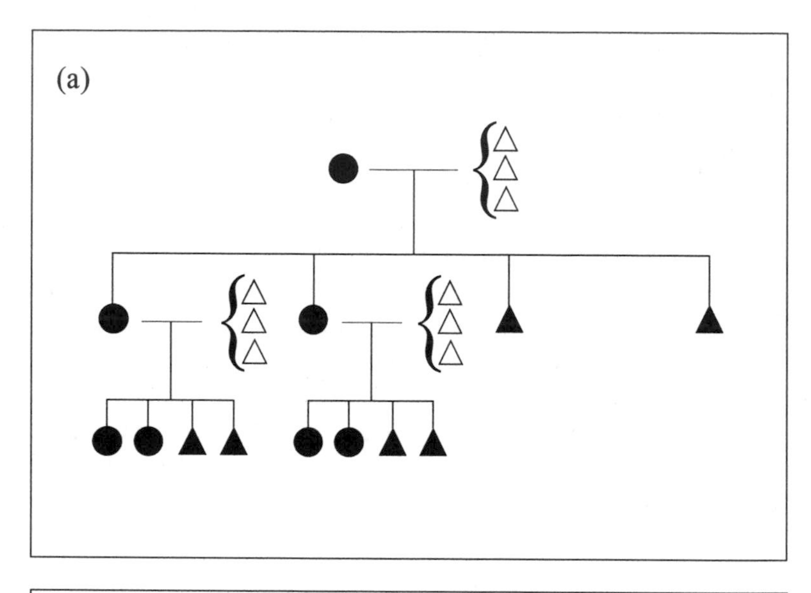

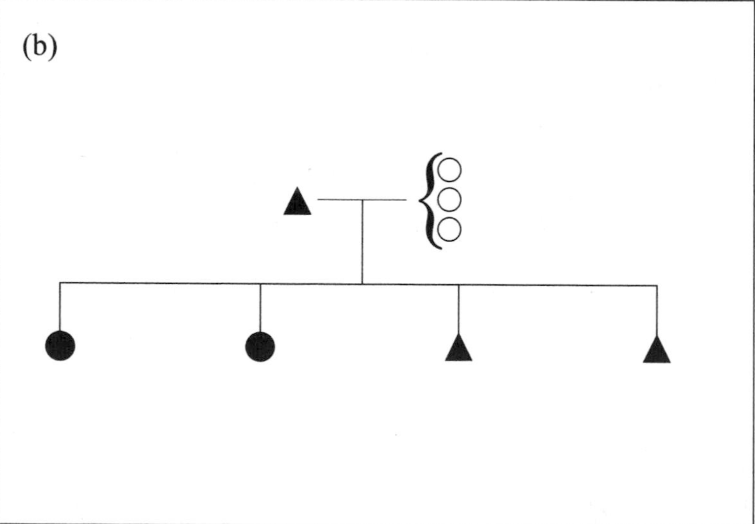

Figure 3.1. Comparison of a single matriline (a) and a single patriline (b) in a female-philopatric primate group. Circles represent females; triangles, males. Black symbols: members of the same matriline or patriline; empty symbols: nonmembers. Matrilines are much more extensive than patrilines because females breed in their natal groups whereas males breed elsewhere as a rule.

with their close uterine kin disproportionately more often than they do with nonkin (for reviews see Gouzoules 1984; Walters 1987; Bernstein 1991; Kapsalis and Berman 1996a, Silk 2001; Berman 2004; Kapsalis 2004). Kin biases in behavior among uterine relatives are thus particularly well documented. However, only a small fraction of the relevant studies are suitable for inferring kin recognition because only a few assessed nepotism according to category of kinship, by comparing rates of interactions of mother–daughter dyads, sister dyads, aunt–niece dyads, and so on. In most studies, individuals were classified in two broad categories, kin or nonkin, on the basis of predetermined criteria, the authors assessing whether behavior rates were higher among kin than nonkin. The criteria used to define the boundaries of kinship varied considerably across studies, from conservative definitions limiting kin to mother–daughter, grandmother–granddaughter, and sister dyads to extensive definitions such as "all dyads belonging to the same matriline." If this procedure certainly allows one to draw conclusions about the impact of kinship on behavior and the reality of nepotism, it is ill suited to infer the extent of kin recognition in terms of degree of kinship and kin types. As a result, much research remains to be done about the domain of kin recognition in nonhuman primates, even in species in which nepotism is best documented (for recent syntheses about primate kinship and behavior, see Chapais and Berman 2004b).

Among the studies in which nepotism was analyzed according to category of kinship, a few were experimental. For example, I carried out a series of experiments on Japanese macaques *(Macaca fuscata)* in which an adult female was given an opportunity to intervene on behalf of an immature kin, for example her daughter, against an unrelated dominant peer. By helping her young relative in conflicts, the adult female could easily induce her kin to outrank the dominant peer and assume a more advantageous social position. By creating various experimental subgroups of individuals, it was possible to analyze a female's propensity to aid a relative according to the degree of kinship between them (Chapais et al. 1997; Chapais, Savard, and Gauthier 2001). In a different type of experiment, a dominant female was given an opportunity to tolerate a lower-ranking relative near a highly prized food source that the dominant female could easily monopolize if she wished to. Again, by creating various experimental subgroups, one could measure a female's inclination to passively share food with a relative according to her degree of kinship with that kin (Bélisle and Chapais 2001). Overall, these experiments revealed that kin favoritism—aiding in conflict and tolerated cofeeding—was manifest between mothers and daughters, grandmothers and grandoffspring, great-grandmothers and great-grandoffspring, and sisters and sisters, but inconsistent between aunts and nieces/nephews and absent between grandaunts and grandnieces and between cousins. In sum, among *lineal* relatives, kin recognition encompassed four generations of individuals; in terms of degree of relatedness (r), it extended up to $r \leq 0.125$. Among *collateral* kin, it included sisters ($r = 0.25$) and, inconsistently, aunt–niece dyads and aunt–nephew dyads ($r = 0.125$).

In these experiments it is possible that a female failed to favor a distant kin not because she did not recognize the latter as a relative but because favoring that kin was not profitable from the female's perspective. Helping relatives in their conflicts with other individuals and sharing food with them were altruistic acts in the circumstances described, acts that do not benefit the performer personally. Thus the observed limit of kin favoritism in our experiments might reflect not so much the boundaries of kin recognition as the limit of the profitability of nepotism to the performer; kin recognition might, in effect, extend beyond the observed kinship limit. It is noteworthy, however, that similar results about the boundaries of nepotism were reported in another study, which employed a substantially different methodology. Ellen Kapsalis and Carol Berman (1996 a, b) analyzed the distribution of affiliative behaviors (proximity and grooming) among kin in free-ranging rhesus macaques *(Macaca mulatta)*. They found that a model postulating that kin recognition was limited to aunt–niece dyads and closer kin ($r \leq 0.125$) had a higher explanatory value than a model that also assumed recognition of more distant kin. In both sets of studies, rates of behavior diminished with decreasing degrees of kinship, suggesting that individuals discriminated distinct kin types rather than only larger categories such as kin versus nonkin. Recently Joan Silk and her colleagues (2006a) obtained similar results in a long-term study of wild baboons. Using an index of social affiliation, they found that mother–daughter dyads and maternal sisters had stronger bonds than both aunt–niece dyads and nonkin dyads, but that aunt–niece dyads did not have stronger bonds than cousins and nonkin dyads. Remarkably, then, three sets of studies carried out on three different species with different group compositions and living in different environments found a similar kinship threshold beyond which females interacted as nonkin.

Combining these results and keeping in mind that they are conservative, one may infer that a female's uterine kin include her mother, daughters, sons, brothers, sisters, grandmother, grandoffspring, great-grandmothers, and great-grandoffspring, and, less consistently, her aunts and nieces/nephews. Cousins and more distant relatives do not appear to be part of an individual's kindred. In terms of Murdock's categories of kin types, nonhuman primates recognize their primary uterine relatives (mother, offspring, siblings) and some of their secondary uterine relatives (the primary kin of their primary kin), namely their grandmother and grandoffspring. Other secondary kin, such as aunts and nieces, are part of the gray zone of kin recognition. Nonhuman primates also recognize some of their tertiary uterine kin (great-grandrelations) but not others, such as cousins. Figure 3.2 summarizes these results. It is noteworthy that among secondary and tertiary relatives, only lineal kin are recognized consistently.

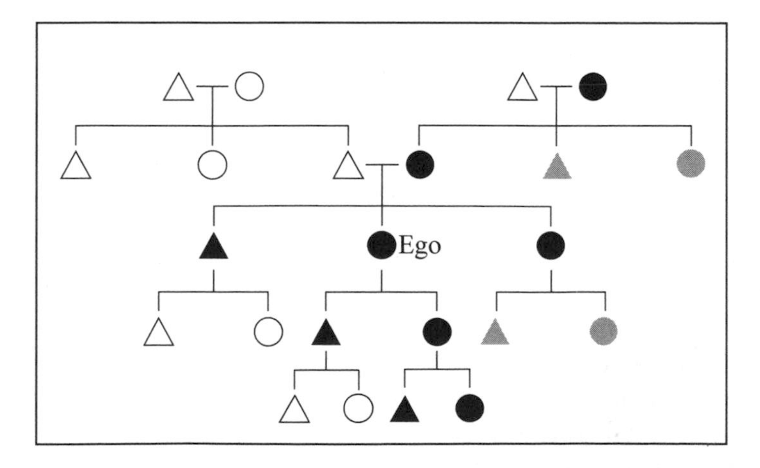

Figure 3.2. Ego's domain of uterine kindred (recognized kin) in a female-philopatric primate group. Only primary and secondary kin are pictured. Five generations are shown, but normally only three generations coexist. Black symbols: ego's kindred; gray symbols: ego's kin in the gray zone of kin recognition; empty symbols: ego's other kin.

How Are Uterine Kin Recognized?

The cornerstone of uterine kinship is the mother–offspring dyad. Maternal care in mammals translates into a uniquely intimate bond that provides both mother and offspring with a rich and enduring source of cues—auditory, visual, tactile, and olfactory—for recognizing each other (for a review, see Rendall 2004). Not surprisingly, both mother recognition by offspring and offspring recognition by mothers are well documented in nonhuman primates (Cheney and Seyfarth 1980; Symmes and Biben 1985; Nakamichi and Yoshida 1986; Pereira 1986; Hammerschmidt and Fisher 1998). It is the mother–offspring bond that in all likelihood generates kin recognition between other categories of uterine kin, for instance sisters. A basic principle is that kin recognition between sisters is conditional upon their being familiar with each other. Because the vast majority of nonhuman primate mothers give birth to a single offspring at a time, siblings do not share their mother's uterine environment as members of the same litter do, hence they cannot be familiar on this basis. Familiarity between siblings is postnatal and acquired in relation to the mother. This principle has been experimentally demonstrated among pigtail macaques *(Macaca nemestrina)*. Maternal half-siblings were raised apart from birth so as to be unfamiliar with each other. When the subjects were later given a choice between interacting with their sibling or with an unrelated individual, they did not select their sibling preferentially (Sackett and Fredrikson 1987; see also Welker et al. 1987). In theory, however, they might have done so. Sibling recognition might involve some form of *phenotype matching*, whereby females would be endowed with the ability to identify certain physical characteristics of their sisters (for example, their odor), match these traits with their own or their mothers', and recognize their sisters on this basis (Holmes and Sherman 1982; see also Rendall 2004). The fact is, however, that kin recognition between maternal sisters apparently depends on their sharing a common experience with the mother.

What then is the role played by the mother? Two different sets of cognitive processes are probably at work here, both involving the mother as the central mediator in kin recognition, but in two distinct ways. In the first process, the mother is merely a passive mediator of familiarity between her daughters. Accordingly, a female would recognize her sister as that particular individual she meets near her mother on a daily basis and with whom she has become disproportionately familiar as a result. In species in which females stay in their natal group with their maternal relatives, mother–daughter bonds often extend throughout the lifespan (Fairbanks 2000). And even in species in which females disperse away from their mothers, as in chimpanzees, bonds between mothers and daughters extend over several years, and those between mothers and their resident sons persist even longer. Now the existence of long-term bonds between mothers and daughters entails that two sisters, even though they are born a number of years apart, will spend time together, if only because they are attracted to the same mother. They are thereby in a privileged position to know each other intimately and to start developing their own relationship. The same process applies to other categories of uterine kin. Through proximity to her mother, a female is bound to regularly meet her mother's mother and her mother's sisters, that is, to become disproportionately familiar with her maternal grandmother and aunts. Moreover, the amount of time a female spends with a given kin through her mother's proximity to that kin should decrease in proportion to the decreasing degree of kinship between them. In theory, this provides females with a means, however approximate, to discriminate between broad categories of kin (Chapais 2001).

In the developmental scheme just outlined, it is through the mother that sons and daughters come in contact with their relatives. Based on this experience, the offspring themselves eventually play an active role in building relationships with their kin. Studies on the early development of kin networks in macaques provide empirical support for this view (Berman et al. 1997; Berman and Kapsalis 1999; Berman 2004). Nevertheless, the process envisioned here involves strictly the ego-centered dimension of kin recognition, and as such it captures only part of the whole picture. It focuses on the fact that when a female is placed in social situations involving her mother and a sister, she acquires information about her *own* relationship with her sister. This is certainly true, but it ignores the fact that the female simultaneously acquires information about the relationship *between* her mother and her sister. And it is that ability to learn about the relationships of others which provides a distinct cognitive basis for the operation of kin recognition. From a female's viewpoint, a sister may be not only that individual she is disproportionately familiar with but that particularly close associate of her mother, the one whom her mother protects against certain individuals in certain circumstances, grooms at certain rates, tolerates near food sources, and so on. Kin recognition in this case involves ego classifying others by using the mother as a point of reference. While the first process of kin recognition centers on ego's capacity to classify others from ego's own experience with them, the second process focuses on ego's capacity to classify others from knowledge about their relationships with the mother.

This cognitively more sophisticated process is not merely a theoretical possibility. The capacity of nonhuman primates to recognize the characteristics of the relationships of others is now well documented. Early indications of this ability in primates were anecdotal. It had been observed, for example, that potential aggressors sometimes refrained from threatening certain individuals when those individuals' close kin (and protectors) were nearby (Datta 1983). Such restraint implied that potential aggressors recognized associations between other individuals and their kin. We owe much of our present knowledge about this ability to the work of Dorothy Cheney and Robert Seyfarth on baboons and vervet monkeys. These researchers found that after a fight between two individuals, if one of the two combatants redirected aggression to a third individual, the latter was more likely than by chance to be a uterine relative of its opponent, a finding that pointed to the monkeys' ability to recognize associations between kin (Cheney and Seyfarth 1986; see also Aureli et al. 1992). In the same vein, they reported that in the aftermath of a fight, monkeys were more likely to reconcile with their opponent's uterine kin than with other individuals (Cheney and Seyfarth 1989). Finally, they conducted experimental studies revealing that upon hearing an individual scream, females looked at that individual's close relative, for example, the mother, further illustrating the monkeys' knowledge about the social networks of others (Cheney and Seyfarth 1980, 1999; see also Cheney and Seyfarth 1990, 2004).

In sum, it is most likely that primates learn the identity of their uterine kin by acquiring information about both the characteristics of their own relationships with their relatives and the characteristics of their mothers' relationships with these same kin. Crucially, the two processes of kin recognition just outlined depend on the *lasting character* of the mother-offspring bond. A female's capacity to recognize several uterine kin types besides her mother follows not from the mere existence of the mother-daughter bond but from its enduring nature. As already pointed out, for a newborn sister to be able to recognize an older sister, the latter must be maintaining a preferential bond with her mother so that the younger sister is biased toward interacting with her older sister. Thus, if the older sister is five years old upon her younger sister's birth, the older daughter's bond with her mother must last significantly longer than five years. The same principle applies to the recognition of grandrelations. For a granddaughter to recognize her maternal grandmother, the mother must herself be maintaining a preferential bond with her own mother when her daughter is born; that is, mother–daughter bonds must last significantly longer than the generation length. One may infer from this that prior to the evolution of lasting relationships between mothers and daughters, uterine kinship in primate societies was limited to mother–offspring dyads.

The Origin of Group-wide Kinship Structures

I have heretofore limited the discussion of uterine kinship to its effect on the content of dyadic, or interpersonal, relationships. But the impact of kinship on social behavior goes far beyond the patterning of dyadic bonds. Kinship has the potential to generate group-wide patterns of social relationships that map onto the group's entire genealogical structure and that are therefore predictable from knowledge of that structure. I call such patterns "kin-biased social structures," or *kinship structures*. These structures are a fundamental feature of human societies, and the evolutionary origin of some of the most conspicuous of these—unilineal descent groups—is discussed in detail later on. For now I am concerned with the conditions under which kinship structures arise in primate societies in which females are localized. The principles underlying the formation of uterine kinship structures in nonhuman primates will help us understand the emergence of human kinship structures, whether uterine or agnatic.

Let us carry out the thought experiment illustrated in Figure 3.3. Let us posit the existence of a female-philopatric multimale-multifemale group comprising two matrilines in which mother-daughter bonds terminate upon weaning; past that time, mothers and daughters maintain no preferential relationships. In such a situation and for the reasons just outlined, a female would not be able to use her mother as a mediator to recognize her sisters and other uterine kin; the domain of uterine

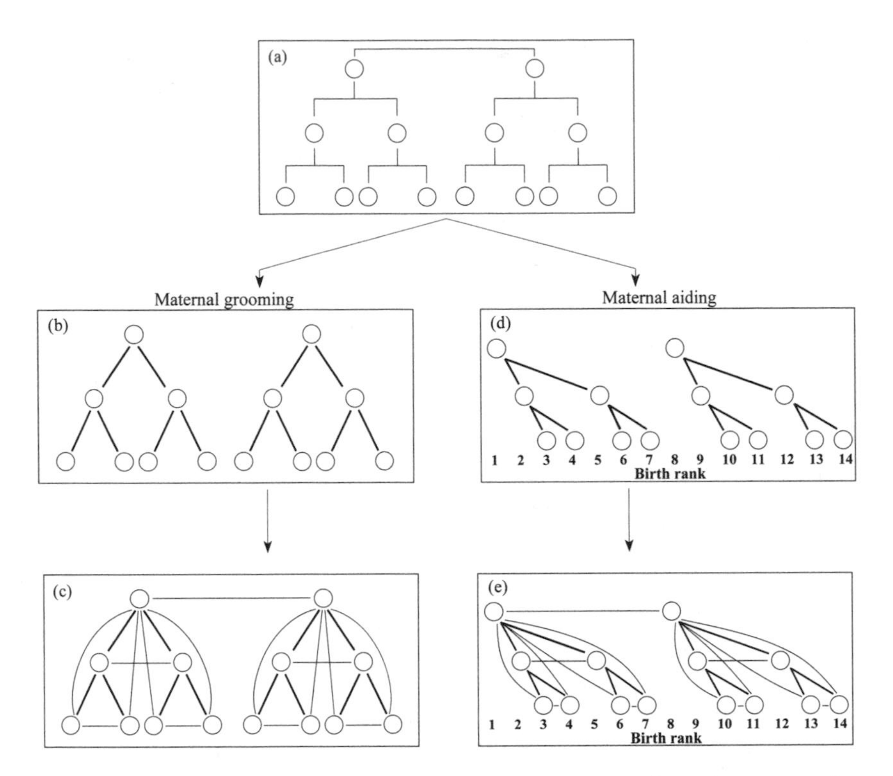

Figure 3.3. Two possible pathways leading to the formation of simple group-wide kinship structures in female-philopatric groups.

kindred would be limited to mother–daughter dyads. For the sake of clarity, let us further assume that there are no social bonds whatsoever between females. The genealogical structure is thus socially silent. Now suppose that mothers begin to exchange grooming with their daughters and that they do so over a number of years. Because all females have mothers, and because all females are related to each other through females, this *single bond* reveals the group's whole genealogical structure. It produces a kinship structure composed of a series of social cliques that map onto the genealogical structure, as shown in Figure 3.3b. From a socially dormant entity, the uterine pedigree has mutated into a conspicuous kinship structure—a grooming one—that connects all females. No kinship bond other than mother–offspring needs to be recognized for the genealogical structure to surface. However, assuming that mother–daughter bonds are maintained long enough, daughters are now in a position to use their mothers as mediators to recognize other uterine kin types, for example their sisters and grandmothers. The resulting "nepotistic cliques" become correspondingly more complex, as shown in Figure 3.3c. The important point here is that a strictly maternal pattern of favoritism, because it is reproduced in all mother-daughter dyads, reveals the whole group's hitherto silent genealogical structure. In short, lifetime maternal favoritism alone generates group-wide kinship structures.

Let us erase the preceding thought experiment and consider an alternative one taking place in the same initial group. Let us posit that adult females form a dominance order in which a female's rank is determined strictly on the basis of her size and physical strength. In such a situation, any newborn female is systematically defeated by larger females and ranks at the bottom of the dominance hierarchy. Only upon reaching adolescence or adulthood are females able to fight their way up the dominance order. Now suppose that mothers begin helping their immature daughters selectively against the females that rank below them (the mothers), and that they do so both consistently and efficiently. Eventually every daughter comes to rank immediately below her mother and above all the females that are subordinate to her mother. Mothers and their daughters now occupy adjacent positions in the dominance order, and that dominance structure maps upon the genealogical structure, as illustrated in Figure 3.3d. Compared to the previous kinship structure composed of a series of nepotistic cliques (3.3c), the present kinship structure integrates a further structural dimension: the cliques are ordered in relation to each other; the matrilines are ranked. Not only is the genealogical structure revealed, as in the previous situation, but it is organized along a group-wide dimension, one that regulates access to resources, among other things.

Such "matrilineal dominance structures" are well documented in macaques, baboons, and vervet monkeys. They are entirely predictable from knowledge of uterine kinship relations (for reviews, see Chapais 1992, 2004; Pereira 1992). In the classical system, a daughter socially inherits her mother's position in the linear order. Through her mother's active help, she eventually outranks all the females that rank below her mother, and she remains subordinate to her mother thereafter. She also outranks her older sisters, thanks to her mother's help. The fact that sisters ordered in relation to each other help their own daughters against lower-ranking females determines rank relations between aunts and nieces and those between first-degree cousins, second-degree cousins, and so on down generations. In other words, any female has a predetermined *birth rank*, which she attains upon reaching sexual maturity. Sons also inherit their mother's rank initially. But as they grow up, they begin to outrank higher-ranking females before leaving their natal group around puberty. Thus adult males are not part of the matrilineal dominance structure (Pereira 1989, 1992; Lee and Johnson 1992).

Note that no other kin type besides the mother-daughter dyad need be involved at this stage. In theory, matrilineal rank orders unfold from maternal favoritism alone. But assuming that mother–daughter bonds are long-lasting, other categories of uterine kin come to be in a position to recognize each other, as argued previously. At this point, evolutionary forces find new raw material to operate upon. For example, considering the dynamics of rank acquisition, other categories of allies besides the mother may now enter the picture. Empirical studies indicate that a female's grandmother and sisters are also actively involved in the female's acquisition of her birth rank (Chapais 2004). One thus obtains the pattern illustrated in Figure 3.2e. In the evolutionary scheme envisioned here, the involvement of these new actors constitutes a later development that came about only after the basic matrilineal dominance structure was put in place following changes in the mother–daughter bond.

To sum up, provided females are localized and the genealogical structure preexistent, the transformation of the mother-daughter bond from a short-term pattern of care into an enduring bond, whatever its content, readily generates social patterns that parallel the group's entire genealogical structure. Put differently, uterine kinship structures are byproducts of lasting maternal bonds; they are emergent properties of matrifilial relationships in certain social contexts. Remarkably, therefore, complex kinship structures arise through relatively simple processes. Later on I shall argue that the types of changes the transformation of mother–daughter bonds can accomplish in female kin groups could similarly be accomplished by the transformation of father–son bonds in male kin groups over the course of hominid evolution.

In stating a number of principles about the extent of uterine kin recognition in primates, the developmental processes involved, and the ways that group-wide kinship structures arise, I have outlined what amounts to a preliminary version of a partial "primatological theory" of uterine kinship. These principles have a wide taxonomic applicability, and taken together they provide a conservative assessment of the early hominids' cognitive potential for developing kinship bonds. They point to humankind's primate legacy, or phylogenetic heritage, in that area. As we shall see in the next chapter, however, human kinship goes well beyond biological kinship—so much so that what anthropologists call cultural kinship is considered by many to have little in common with genealogical kinship.

4

From Biological to Cultural Kinship

> A consanguine is someone who is defined by the society as a consanguine, and "blood" relationship in a genetic sense has not necessarily anything to do with it, although on the whole these tend to coincide in most societies of the world. . . . Quibbles aside, actual or putative genetic connexion, according to the local definition of "genetic" or "consanguineous," is usually the basis of kinship relations.
>
> *Robin Fox (1967, 34)*

> "Kinship" is an artifact of the anthropologists' analytic apparatus and has no concrete counterpart in the cultures of any of the societies we studied.
>
> *David Schneider ([1971] 1984, vii)*

The place and relative importance of kinship within anthropology has changed considerably since Lewis Henry Morgan established it as a fundamental branch of inquiry more than one hundred years ago. From its central place in the first half of the twentieth century, kinship had come close to extinction as a research domain in the 1980s before enjoying a true revival in the 1990s (Holy 1996; Stone 2001a). Among the critiques of the anthropology of kinship, one in particular, probably the most fundamental, is relevant here. Genealogical relations and consanguineal kinship networks stem, by definition, from the basic facts of procreation: males impregnate females and females bear offspring. The universality of reproduction, coupled with the observation that people all around the world have names for their consanguines—that kinship terminologies are universal—provided the early kinship theorists with their most basic assumption: namely, that kinship is universally based on people attributing social and symbolic meanings to consanguinity-based relations. In other words, kinship reflects genealogical relations.

But that basic assumption was hotly debated in the 1970s and 1980s, and eventually rejected by many. Anthropologists increasingly reported that kinship terminologies and the corresponding kin types encompassed much more than mere consanguinity, that kinship was often fictive from a biological viewpoint—in short, that human kinship categories were culturally defined. In its most extreme form, that view held that there was no such thing as a universal genealogical basis for kinship and that concepts such as motherhood, fatherhood, lineages, descent, marriage, and the like were no more than anthropological constructs existing only in the minds of ethnologists (Schneider 1984). But then, if human kinship is cultural, what happened to humankind's primate legacy of uterine kinship? How does one reconcile these two seemingly incompatible views?

Beyond Consanguineal Kinship

There is indeed a rather poor fit between the categories of relationships primatologists have in mind when they talk about kinship and those that anthropologists subsume under that term. Primatologists implicitly use three criteria to analyze the effect of kinship on social interaction: sex, generation, and genealogical distance (or degree of genetic relatedness). Primatologists use these three criteria because their study subjects interact differentially according to them. The effect of genealogical distance on social interactions is best evidenced when the two other factors, sex and generation, are held constant. It is manifest, for example, in ego's differential treatment of its sisters and female cousins, whose degrees of relatedness differ by a factor of four. Similarly, the effect of the generation factor stands out when sex and degree of relatedness are the same, as when one compares ego's interactions with its grandmother and its half-sister. And the effect of the sex criterion is best exemplified when generation and degree of relatedness are held constant. This is manifest in ego's differential treatment of its brothers and sisters, for example. Most often, the three primary factors combine in various ways to produce further differences. For example, the differential treatment of one's mother and grandmother would reflect the combined effect of generation and genealogical distance. The fact that the three criteria affect social relations in a wide array of primate species indicates that they are primitive aspects of kinship.

Ever since 1871, when Morgan published his pioneering study *Sys-*

tems of Consanguinity and Affinity of the Human Family, social anthropologists have been analyzing the way humans discriminate their relatives when verbally referring to them. Early in the first half of the twentieth century, the fast-growing ethnographical record allowed Alfred Louis Kroeber, Robert Lowie, and others to infer the distinctions made by humans when dealing with their kin and to identify the criteria they used (Murdock 1949, 101). The three categories of distinctions made by nonhuman primates—sex, generation, and genealogical distance (for which anthropologists use the term "collaterality")—are among the most important criteria used by humans to refer to their kin. From an evolutionary perspective, this basic similarity attests to the common biological background of primate kinship and human kinship; it points to the primate legacy within the human kinship realm. But although humans are able to differentiate their relatives according to the three criteria, in different cultures people create kin categories that belie these distinctions and that are artificial from a genealogical (biological) angle. *Classificatory kinship,* the grouping of persons belonging to different kin types in the same kinship category, is a universal phenomenon in human societies. For example, in societies that use what has been called a Hawaiian type of kinship terminology, cousins are called siblings and classified together under the same label; likewise, uncles are called fathers and classified together, and aunts are called mothers. Thus relatives differing by the criterion of genealogical distance—collateral and lineal relatives—are merged within the same class. In societies that exemplify the Crow and Omaha kinship terminologies, certain cousins are called aunts and classified with them. In this situation, relatives differing both by the criterion of generation and that of genealogical distance are lumped together.

In the same societies it is also the case that siblings are classified with so-called parallel cousins (children of same-sex siblings) and, significantly, both categories of relatives are subject to the same incest prohibitions; that is, both siblings and parallel cousins are proscribed partners. This last point brings us to a central issue. If classificatory kinship were merely a semantic affair with no impact on behavior, if culturally defined classes of consanguines did not translate into some patterns of social relationships common to all members of a class, kinship terminologies would be tantamount to interesting cultural curiosities. But as the assimilation of parallel cousins to siblings for the purpose of incest prohibitions illustrates, classificatory kinship does translate into behavior, here into the marriageability status of individuals. There is indeed a general correspondence between kinship terms and behavior patterns, so ego's attitudes and privileges when interacting with, for instance, his classificatory mothers—his mother's sister and his father's other wives—have more in common than ego's attitudes toward and privileges with other women, such as his paternal aunt, mother-in-law, or grandmother (Murdock 1949, 107). Thus cultural kinship categories are real social categories that often transcend genealogical categorization.

Equally important is the other side of the coin. Not only do humans lump together categories of consanguines differing in their degree of genetic relatedness, they may also treat in drastically different ways socially defined subcategories of relatives that are genealogically equivalent. A common distinction is that between parallel cousins and cross-cousins (children of different-sex siblings), which translates into profound differences in patterns of interactions. For example, cross-cousins are often prescribed marriage partners, whereas parallel cousins are often proscribed. In making these distinctions, humans demonstrate that they differentiate relatives not only according to sex, generation, and genealogical distance but also according to whether they are related through a male or a female relative. To Lévi-Strauss the differential treatment of cross-cousins and parallel cousins was so fundamental that it supposedly held the key to understanding incest prohibitions. Noting that the two types of cousins were "strictly interchangeable" in terms of biological relatedness, he wrote: "If we can understand why degrees of kinship which are equivalent from a biological point of view are nevertheless considered completely dissimilar from the social point of view, we can claim to have discovered the principle, not only of cross-cousin marriage, but of the incest prohibition itself" (Lévi-Strauss 1969, 122).

Both aspects of human kinship—merging certain kin categories and subdividing others—strikingly illustrate the impact of cultural constructs on consanguineal kinship. But humans not only juggle with genealogical categories, they may also negate the existence of real genealogical links, even the most basic ones. For example, some cultures deny that biological fathers have anything to do with procreation, while other cultures even negate the contribution of the mother to procreation. Finally, as alluded to earlier, the human primate commonly creates nongenealogical kinship categories, that is, categories that are fictive from a biological viewpoint. The prominent form of fictive kinship is probably affinity (bonds between in-laws). Marital alliances commonly create enduring bonds between the spouses' respective consanguineal kin, as between the husband and his brothers-in-law and father-in-law. Although humans differentiate between their consanguineal and affinal kin, it is common for in-laws to be labeled and treated as if they were blood relatives. To take just one example, in our own Western culture, uncles (and aunts), whether they are related to ego by blood or by marriage are classified together and treated the same way. Moreover, fictive kinship categories admit individuals that are neither consanguines nor affines. For example, coresident members of the same group may be labeled and treated like members of the same lineage, even though they are not descendants of a common ancestor. They are kin solely by coresidence (for a discussion of fictive kinship in an evolutionary perspective, see Rodseth and Wrangham 2004).

In sum, humans grossly manipulate genealogical relations: they merge different biological kin types, subdivide others, negate real blood relations, and create biologically artificial ones. As this type of evidence accumulated, Morgan's basic assumption that kinship all over the world reflects genealogical relations appeared untenable to many, a realization that led to an important shift in conceptions about kinship (Holy 1996; Stone 2001b). In the 1960s a large proportion of anthropologists probably agreed with Robin Fox's statement that even though kinship categories are defined by the society, "quibbles aside," cultural definitions and biological categories "tend to coincide in most societies of the world." But by the 1970s the domain of kinship and that of genealogical relations were conceived by many as basically unrelated. For example, referring to the collection of rights whose transmission is regulated by kinship, Rodney Needham could state that these rights were "all transmissible by modes which have nothing to do with the sex or genealogical status of transmitter or recipient. Certainly they have no intrinsic connexion with the facts, or the cultural idioms, of procreation" (1971, 4). Such views culminated in the mid-1980s in Schneider's denial of the existence of cross-culturally valid categories of kinship based on genealogical relations: fatherhood, motherhood, and lineages were seen as no more than the product of Western ethnocentrism. Schneider subsumed such concepts under the "Doctrine of the Genealogical Unity of Mankind," which he considered "one of the most important and explicit features of the conventional wisdom of kinship studies," but one that was basically "insupportable" (1984, 188, 198). For example, referring to motherhood, he remarked that "we must not translate or gloss every relationship between a woman and what appears to be the child she has borne as a mother-child relationship until that translation or gloss has been fully explored by examining in detail how the natives themselves conceptualize, define, or describe that relationship. . . . The same goes for the father–child and the marital and all other presumed kinship relations. To repeat, this means that the genealogical grid cannot be assumed but only held as a possible hypothesis" (200). According to this view, motherhood and all other components of genealogical kinship are mere hypotheses. Quite possibly, some cultures could have kinship systems lacking any genealogically based relations, kinship altogether without consanguinity.

While social anthropologists were debating the meaning of kinship among themselves and seriously questioning the genealogical foundation of human kinship, the discussion took a new turn as it was fed by the rise of sociobiology in the 1970s and 1980s. Armed with the recently born theory of kin selection (Hamilton 1964), sociobiologists took a diametrically opposed view, one that placed genealogical relations and genetic relatedness at the very heart of kinship studies. Emphasizing the importance of differentiating between what people say they do and what they actually do, they undertook to assess the extent to which genealogical relations per se explained patterns of human interactions. Strategically putting aside classificatory kinship categories and kinship terminologies in their analyses, they began looking for correlations between actual degrees of biological relatedness (r) and the differential treatment of kin categories (see, for example, Chagnon and Irons 1979; van den Berghe 1979; Shepher 1983). To take just one example, Richard Alexander suggested that the differential treatment of cross-cousins and parallel cousins, a theme that epitomized the impact of culture on kinship categories, might in fact reflect differences in degrees of genetic relatedness between these two categories. Alexander reasoned that in polygynous societies, the common practice of sororal polygyny (a man marrying sisters) entails that the offspring of sisters (matrilateral parallel cousins) are genetic paternal half-siblings. Another common practice, levirate (the rule by which a widow marries the brother of her deceased husband) similarly implies that the offspring of brothers (patrilateral parallel cousins) are genetic maternal half-siblings. Therefore, parallel cousins may be on average more closely related than cross-cousins. "This fact must somehow be explained away," he wrote, "before we can dismiss the differences in incest rules for cross- and parallel-cousin unions as genetically meaningless" (1979, 179). For recent overviews of the contribution of evolutionary psychology and Darwinian anthropology to kinship studies, see Barrett, Dunbar, and Lycett 2002; Burnstein 2005; and Kurland and Gaulin 2005.

Significantly, the debate about the biological content of human kinship going on within sociocultural anthropology proceeded independently of the rise of sociobiology. Anthropologists such as Needham and Schneider had begun questioning the genealogical basis of kinship as early as the 1960s and early 1970s. Somewhat ironically, barely a few years later, in the wake of the publication of Edward Wilson's *Sociobiology*, human sociobiologists were reducing human kinship to its genetic dimension for analytical purposes. Needless to say, sociocultural anthropologists were well primed to respond diligently to that external assault (Sahlins 1977). In retrospect, the timing of the two debates could hardly have been worse, with the views so polarized and the clash so severe. But that is another story.

The "Genealogical Unity of Mankind"

Both debates were certainly intrinsically important, but as far as the evolutionary history of human society is concerned, they are to all intents and purposes basically irrelevant. I am concerned here with the *origins* of human kinship. From this angle, the extent to which genealogical kinship categories actually fit with cultural kinship categories is pointless, as is the issue whether variations in degree of genetic relatedness explain variations in human behavior. My argument centers on two claims. The first is that even when culture negates or ignores the genealogical content of a kinship bond, for example that of motherhood, it does not necessarily preclude that bond from generating preferential relationships that do map onto genealogical kinship. Let us go back to Schneider's motherhood example. In some patricentric societies, the mother is seen as a kind of shelter for her child, an actor who does not contribute her own substance to her children (Fox 1967, 119; Godelier

1986). The denial of biological motherhood probably exemplifies the most intrusive incursion of culture into genealogical kinship. The fact that culture may negate even the most basic procreative bond supports particularly well the claim that human affairs are freed from the influence of genealogical kinship. But whether the mother's role in procreation is recognized or negated—whether in any particular society the "genealogical grid" applies to motherhood or does not—the fact remains that motherhood creates genealogical kinship. Whatever the ideology of procreation, the biological facts of pregnancy, parturition, lactation, and maternal care translate into matrifilial links. And regardless of cultural beliefs in the procreative role of mothers, matrifiliation generates preferential bonds between maternally related kin. Offspring recognize their mother, siblings recognize their common mother, and individuals recognize their matrilateral kin: their mother's parents, sisters, nieces, and so on. Motherhood cannot be removed from the genealogical grid. Motherhood creates kinship whether or not mothers are believed to be involved in procreation.

My point here is not that biological maternity transcends cultural practices about motherhood, invariably giving rise to maternal behavior and uterine kinship. It is certainly possible to eliminate maternity recognition by separating mothers from their children and having the children nursed and raised by other women. But this is exactly what ideologies that negate the mothers' role in procreation would need to achieve in order to prevent the formation of preferential bonds between mothers and children and thus the recognition of matrilateral kinship. A similar reasoning applies to situations wherein the institutionalized negation of paternity prevails. There is no question that culture intrudes on biological motherhood and fatherhood, and deeply so (Hrdy 1999), but even in those situations mother–child and father–child bonds are special bonds that produce specific outcomes; this is all that is needed for kinship, matrilateral and patrilateral, to be recognized and acted upon.

My second claim is the exact opposite of Schneider's: the genealogical unity of humankind does exist, and much of it comes from our primate legacy. Implied here is that all dimensions of cultural kinship—classificatory, social, and fictive—originated from the genealogical grid characterizing human society after it evolved its particular deep structure, the exogamy configuration. The genealogical unity of humankind includes, minimally, the following aspects: stable breeding bonds, motherhood, fatherhood, siblingship, intrafamilial incest avoidance, recognition of one's close matrilateral and patrilateral kin, recognition of one's in-laws and a propensity to treat them as allies, and certain kinship biases in mate selection. Taken together, these features define the *core kinship system of humankind*—itself an integral part of human society's deep structure—from which all known kinship systems have diversified. Depending on the circumstances, the system's basic elements may generate more complex phenomena, such as group-wide uterine or agnatic kinship structures and various patterns of descent.

It is true that from an anthropological viewpoint—from a strictly cross-cultural perspective—the existence of a universal set of kinship features is far from obvious. But this is because the unity is concealed beneath the extraordinary diversity of forms it has generated. Any regularity pointed out by anthropologists in this context is almost bound to be countered and swept away by exceptions, singularities, and contradictions. Motherhood, siblingship, descent, marital alliances, and the like are cross-culturally so diverse that they indeed defy any exception-free definitions. But it is precisely at this point that a change of perspective proves useful. Looking at the problem from an out-group's perspective, that of nonhuman primates, the genealogical unity of humankind is obvious. To make an analogy, the difference between anthropology and comparative primatology is like that between two groups of anthropologists seeking the structure and properties common to a large number of myths all derived from a single founding myth. The first group would carry out a classical comparative analysis of the variants without any knowledge about the founding myth. The second group would be given information about some major aspects of the original myth, which would no doubt facilitate the analysis of the resulting diversity and enlighten the multiple transformations the original myth went through. For example, features observed to characterize a large proportion of the variants but, significantly, not all of them could nevertheless be identified as "primeval." Similarly, features so variable as to defy any generalization could nonetheless be traced back to some primeval traits. In sum, exceptions and singularities would cease to provide sufficient grounds for rejecting principles that are general but not universal. The original myth would provide a sort of filter enabling one to retain certain principles as valid *despite* their not being universal. The difference between the two approaches is that between ignoring history and taking it into account.

The point, of course, is that comparative primatology may provide the original "myth," or at least a substantial part of the stem human kinship system; that is, primate studies help define the genealogical unity of humankind before it got blurred through cultural diversification. There is more in that statement than the mere acknowledgment of the primate origins of human kinship. What has been lacking in prior comparative studies is a systematic assessment of the domain of kin discrimination along both the maternal line and the paternal line in nonhuman primates and early hominids. Only through such an assessment can we hope to (1) characterize the maximal complexity that hominid consanguineal kinship networks reached prior to the evolution of language and symbolically mediated culture (as we shall see, that level of complexity was unprecedented in primate evolution), and (2) define the features and dimensions of genealogical kinship that are uniquely human and on this basis identify the factors responsible for the tremendous growth in complexity of human kinship networks over evolutionary time.

The Bilateral Character of Human Kinship

From an evolutionary outlook, the most important distinctive feature of consanguineal kinship in humans is its bilaterality. More specifically, human kinship is bifilial and thus bilateral. Kinship is traced through both the mother (maternal filiation) and the father (paternal filiation). Because kinship is bifilial, it includes ego's kin on both its mother's and father's sides and, on each side, kin related either through females or through males. For example, considering only the subcategory of lineal ascendants, ego recognizes its two parents and four grandparents and could recognize its eight great-grandparents if they were still alive. In marked contrast, if kin recognition were limited to maternal filiation, two consequences would ensue: ego's kinship network would be limited to its mother's side, and among its matrilateral relatives ego would recognize only those related through females. Hence, among its four grandparents, ego would recognize only its mother's mother, and among its eight great-grandparents, only its maternal grandmother's mother. As we saw, in many nonhuman primates kin recognition is limited to such a situation or something close to it. Thence the idea that bifilial kin recognition is the primary factor accounting for the unparalleled richness and complexity of human kinship networks. Indeed, well before the evolution of language and classificatory kinship, human consanguineal networks were already unequaled in their complexity simply because they stemmed from bifilial kin recognition. From an evolutionary perspective, therefore, the pertinent question concerns the origin of bifilial kin recognition rather than of its correlate, bilateral kinship.

Some definitions and clarifications of terminology are needed at this point. I could have used the words "matrilineal," "patrilineal," and "bilineal" to stand for maternal filiation, paternal filiation, and bifiliation, respectively. But this would be confusing. Primatologists commonly use the word "matrilineal" to refer to kinship relations through females only, and "patrilineal" to denote kinship relations through males only. Anthropologists, on the other hand, use these words to refer to types of *descent* groups, that is, to subgroups called lineages or clans whose members regard themselves as related to a common ancestor either through males only or through females only. Descent involves several dimensions that are not necessarily included in kinship, as we shall see in chapters 17–19. In order to avoid any confusion I have elected to stick to anthropological usage and use matrilineal, patrilineal, and bilineal in relation to descent only, never in relation to mere kinship. To refer to kin related through females only, to what primatologists call matrilineal kin, I use the term "uterine kin"; and to refer to kinship through males, I use the word "agnatic."

Throughout the book I utilize Murdock's definitions of kin types: primary, secondary, tertiary, and so on, as defined in chapter 3. I do not mean to imply here that all human societies differentiate these particular types. Classificatory kinship amply belies such a claim. My point is simply that human beings as a species have the potential to differentiate these various categories, and hence to make distinctions in terms of generation, sex, genealogical distance, and the sex of the relative who links them to a particular kin. Whether they make use of these criteria or not is a matter of cultural variation.

The fact that kinship is traced through both the mother and the father in all human societies does not mean that all kin categories have the same importance to ego. Lineage theory and the corresponding empirical data amply testify to the asymmetrical nature of kinship networks. *Unilineal descent* groups are prevalent in human societies. In such groups there is a profound asymmetry in the relative importance of genealogical relations according to kinship lines. For example, a patrilineal descent group is composed exclusively of the individuals descended from a given male ancestor, and through males only; the transmission of rights and property in these societies is strictly through the paternal line. Thus membership in a patrilineal descent group excludes a significant proportion of ego's genealogical relatives. But although the emphasis is placed on a single line of descent, and often strongly so, kinship is still reckoned bilaterally. Relatives outside ego's descent group are recognized, but as kin of a fundamentally different kind with whom ego has much less in common in terms of rights, property, and obligations. A classical example of bonds between relatives belonging to different descent groups is the avunculate, the set of special relationships existing between a man and his mother's brothers (between maternal uncles and their sororal nephews). The avunculate illustrates particularly well the principle that kinship, because it is intrinsically bilateral, transcends unilineal descent fundamentally.

5

The Incest Avoidance Legacy

> Strangely enough, it has not been suggested by any cultural anthropologist that before [the] incest taboo was institutionalized, there must have been a stage where pre-institutional prohibition or avoidance of it was already a social practice.
>
> *Kinji Imanishi ([1961] 1965, 117)*

If there is a single classic theme in the history of social anthropology and one that has been the object of a massive amount of debate, it is incest avoidance. Precisely because so much has been written on the topic by anthropologists, sociologists, psychologists, psychiatrists, ethologists, and primatologists, incest avoidance is ideally suited to further characterize the theoretical significance and implications of mankind's primate legacy.

Among the earliest statements about the primate origins of the incest taboo is an article written for a Japanese journal in 1961 by the primatologist Kinji Imanishi. Referring to Leslie White's belief that nonhuman primates have a strong inclination toward inbreeding and incest, Imanishi responded, "[White] seems to think that human beings should have an inherent desire for incest, because, he says, even monkeys have it; but he does not offer any evidence that monkeys have such an inclination toward inbreeding. As a matter of fact, there cannot be found such evidence. . . . It is to be regretted that this faulty Freudian hypothesis has been accepted widely in cultural anthropology without due consideration—for example by Murdock" (1965, 16). He then reported that in one population of free-ranging Japanese macaques observed over a number of mating seasons, "not a single case of incest between mother and son has been reported up to date" (119) despite the fact that mothers and sons had ample opportunities to engage in sexual activity. A few years later Donald Sade (1968) described similar findings on "inhibition of son–mother mating" in free-ranging rhesus macaques. Since then many other studies have confirmed that mating between close kin is avoided in nonhuman primates.

As with the topic of kinship in general, it is one thing to claim that incest avoidance is a generalized phenomenon in nonhuman primates, but it is another to argue that there is a phylogenetic continuity between incest avoidance in animals and incest prohibitions in humans, as done by Imanishi. The latter claim implies that some of the processes involved in the mere *behavioral* avoidance of incest in animals are also at work in the *normative* proscription of incest in humans. To prove the phylogenetic continuity hypothesis, one must first establish the full range of affinities between the two phenomena. That first step is by far the most important because the evolutionary-continuity argument rests entirely on the nature and extent of the similarities between the two phenomena. With this objective in mind, I describe here nine general principles that summarize our present knowledge about incest avoidance in nonhuman primates. I asserted earlier that the comparative study of primates is likely to generate domain-specific "theories" about various phenomena and that such primatological theories help characterize humankind's primate legacies in the corresponding areas. The nine principles amount to the core of such a primate theory of incest. At this stage in the evolution of our knowledge, my aim has more to do with identifying the themes and concepts that the theory will eventually encompass than with providing a mature theory.

Elements of a Primatological Theory of Incest Avoidance

Limiting the discussion of incest to the primate order may appear somewhat arbitrary. Why not extend it to other mammalian orders or even to nonmammals? After all, kinship is known to inhibit sexual activity between relatives in animals in general (Bateson 1983, 2004; Barnard and Aldhous 1991; Pusey and Wolf 1996). One reason for the limitation is that the basic biological and life-history characteristics of primates set important phylogenetic constraints on the types of mechanisms of incest avoidance at work in our species. Many primates are slow-maturing and long-lived organisms who commonly live in stable groups. Longevity and the age at sexual maturity are such that it is common for three generations of individuals to coreside in the same group, hence for the filial generation to live with the grandparental one. In the vast majority of primate species, females give birth to a single offspring at a time, often at intervals of at least one year, so siblings belong to different age classes. Maternal care extends well beyond weaning, and mother–offspring bonds are often enduring if not lifelong, providing kin with opportunities to recognize each other through association. On the other hand, because mating is often promiscuous, paternity recognition based on father–offspring association is difficult, if not impossible, depending on the species. All these characteristics and many others act as powerful constraints on the genealogical structure of groups, the mechanisms of kin recognition, and the patterns of incest avoidance; in other words, they set the stage for our primatological theory of incest.

PRINCIPLE 1: INBREEDING AVOIDANCE IS ADAPTIVE. This is the most general and least primate-specific of the nine principles. Animal (and plant) studies in general indicate that close inbreeding reduces an organism's reproductive success by lowering its fecundity and/or its offspring's chances of survival. Inbreeding depression results from increased homozygosity, which brings about the expression of deleterious recessive genes and reduces the levels of genetic variation among offspring (Charlesworth 1987; Hamilton 1993). The costs of inbreeding are variable across species and circumstances, and the strength of inbreeding depression has thus been questioned, but the balance of evidence indicates that too much inbreeding is negative (Ralls, Ballou, and Templeton 1988; Thornhill 1993; Pusey and Wolf 1996; Crnokrak and Roff 1999; Bittles 2004). If close inbreeding is costly, it follows that mechanisms of inbreeding avoidance would be favored by natural selection. In theory, such mechanisms could operate either by putting physical distance between kin or by preventing coresident relatives from mating.

PRINCIPLE 2: DISPERSAL LIMITS INBREEDING SUBSTANTIALLY. The most basic factor affecting mating opportunities among kin is the dispersal of individuals from their birthplace. In all primate species, individuals disperse away from their natal group to breed, which brings about the long-term physical separation of many categories of relatives (for reviews, see Greenwood 1980; Pusey 1987; Pusey and Packer 1987;

Clutton-Brock 1989a). Dispersal thus generates outbreeding and prevents incest among *some* kin categories. But this is not to say that dispersal patterns have evolved in animals solely to prevent inbreeding. Dispersal is also a means to reduce competition for food, territory, and sexual partners. However, a number of factors indicate that inbreeding avoidance is one major function of dispersal in the type of group we are most interested in here, the multimale-multifemale group. Most primate species with this type of group composition exhibit one of the two following patterns. Either the females remain in the group in which they were born while the males disperse and breed in other groups, or the males stay in their natal group while the females move out. Competition for food or territory can hardly explain such clear-cut sex-biased residence patterns in multimale-multifemale groups. The transfer of individuals between groups results in the replacement of natal individuals by nonnatal ones, so dispersal often does not reduce the number of food competitors in a group. Similarly, sexual competition between males may explain why males leave their natal groups in polygynous groups, but it hardly explains why all males leave a multimale-multifemale group and why they often seem to leave on their own. Inbreeding avoidance provides the best explanation of between-group transfer in multimale-multifemale groups (Packer 1979; Pusey and Packer 1987; Moore 1993).

The categories of relatives that have opportunities to commit incest depends on the exact dispersal pattern at work. If males stay put, they are in a position to commit incest with their mother, provided she is still alive when they reach maturity. But they are less likely to do so with their sisters and other female kin because they eventually emigrate. Reciprocally, if females stay put, they have opportunities to commit incest with their father if he is still present when his daughters reach maturity. However, their chances of mating with their brothers and other male relatives are much reduced owing to male dispersal. Thus primate dispersal patterns define some of the most basic constraints on incest opportunities. But given that not all kin are separated by dispersal patterns and that some sexually mature kin may coreside in the same group for some time before they emigrate, incestuous unions are theoretically possible.

PRINCIPLE 3: INCEST IS AVOIDED SYSTEMATICALLY ONLY BETWEEN RELATIVES WHO MAY RECOGNIZE EACH OTHER THROUGH ASSOCIATION. Any two kin, for example siblings, may be related through their mother only, their father only, or both. Primate studies indicate that incest is avoided systematically between certain categories of *maternally* related kin but not systematically between *paternally* related kin (reviewed in Paul and Kuester 2004). This reflects the fact that maternally related kin are in a position to recognize each other through association with their mother. In contrast, paternally related kin are often not in a position to recognize each other through association with their father because they often do not recognize their father in the first place. The latter situation prevails in primate groups composed of several adult males and females in which females mate with different males (Kuester, Paul, and Arnemann 1994). This is not to say that paternity recognition is not possible in primate groups: some social contexts are conducive to father-offspring recognition. For example, in groups composed of a single reproductive male, as in a majority of gorilla groups, there is indication that fathers and daughters avoid sexual activity. It cannot be excluded either that certain categories of paternally related individuals might recognize each other through mechanisms other than association (Widdig 2007). But the point remains that incest avoidance is markedly biased toward maternally related kin.

PRINCIPLE 4: INCEST AVOIDANCE DECREASES STEEPLY WITH DECREASING DEGREE OF GENETIC RELATEDNESS. Among maternally related kin that have opportunities to commit incest because they reside in the same group permanently or temporarily, mating is absent or rare between mothers and sons, brothers and sisters, and grandmothers and grandsons (Pusey 1990a, 2004; Paul and Kuester 2004). Depending on the species, mating may also be avoided between more distant categories of maternal relatives. For example, in one study on Barbary macaques *(Macaca sylvanus)* mating avoidance characterized not only mother–son, grandmother–grandson, and brother–sister pairs but also uncle–niece dyads, aunt–nephew dyads, and cousins (Paul and Kuester 2004, Table 12.1).

PRINCIPLE 5: THE WESTERMARCK EFFECT IS AT WORK IN PRIMATE SOCIETIES. In 1891 Edward Westermarck proposed that close intimacy between persons raised together from childhood bred “a remarkable absence of erotic feelings” between them. He reasoned that since close kinship is normally associated with high levels of familiarity during childhood, this factor, rather than the recognition of relatedness per se, explained incest avoidance. The primate data support the so-called Westermarck effect (Pusey 1990a, 2004; Paul and Kuester 2004). Among coresident, maternally related kin, incest appears to be prevented through a basic correlate of maternal kinship, which I call *developmental familiarity.* I use this expression to distinguish between familiarity levels that ego (say a female) cumulates with individuals who were present in her group at the time she was born and the levels of familiarity she shares with individuals that she met much later, for example as an adult. In many primate species, a female (or a male) cumulates developmental familiarity with several of her maternal relatives through her long-term association with her mother.

Primate studies make it possible to test the Westermack effect by dissociating the effect on sexual inhibition of mere developmental familiarity from that of genetic relatedness. As already pointed out, individuals that are closely related but unfamiliar to each other, as many paternal kin are, most often do not avoid incest, indicating that it is the familiarity correlate of kinship, not kinship per se, that accounts for incest avoidance. Interestingly, paternal relatives that are found to exhibit some degree of inhibition are typically members of the same birth cohort. Given that age peers share high levels of developmental familiarity, the inhibition probably reflects developmental familarity rather than kinship (Alberts 1999; Paul and Kuester 2004). Reciprocally, there is evidence that developmentally familiar but unrelated individuals exhibit mating inhibitions. Cross-fostered macaques were found to avoid mating with members of their foster matriline, reacting to one another as if they were kin (Smith 1995). Primate studies should make it possible to assess the relative importance of a number of components of developmental familiarity likely to affect mating inhibitions, such as degree of intimacy, nature of relationship, age difference, duration of the period of intimacy, and so on, but these questions remain to be fully investigated.

PRINCIPLE 6: DISPERSAL PATTERNS AND THE WESTERMARCK EFFECT MAY BE TWO ASPECTS OF THE SAME PROCESS. Dispersal patterns and the Westermarck affect are consistently treated as two separate and complementary mechanisms that reduce the likelihood of incest, the first by physically separating kin, the second by preventing coresident ones from mating. But I suggest that the two mechanisms may well be two aspects of the same phenomenon. Indeed, there is no a priori reason to think that the Westermarck effect is an all-or-none process, one that would produce, from ego's viewpoint, two classes of individuals: those that are nonattractive and those that are attractive. On the contrary, the inhibiting effect of developmental familiarity on sexual attraction is most probably a matter of degree, with the inhibition decreasing as the degree of developmental familiarity decreases. In a group in which females are resident (as in macaques), any natal male is, in effect, developmentally familiar with *all* the females in his group. The male is certainly most familiar with the members of his own matriline, but he is nonetheless familiar with all the other females he has grown up with over the years. Similarly, in a group in which the males are resident (as in chimpanzees), a natal female is familiar with all the male members of her natal group, though more so with the members of her matriline.

If the effect of developmental familiarity on sexual inhibition is indeed a matter of degree, it follows that an individual's (say a male's) level of sexual inhibition toward opposite-sex individuals should be (1) highest for his primary female kin and same-age female peers, (2) somewhat lower for the other female members of his matriline, (3) still lower for the remaining female members of his natal group, and (4) near zero for any females living in *distinct* social groups. Indeed, from the male's standpoint, all out-group (nonnatal) females, and only they, have a *zero level of developmental familiarity* with him. His level of sexual attraction should thus be highest for these females. From this perspective, the Westermarck effect would express the effect of developmental familiarity on sexual attraction at one end of a continuum of familiarity, the end at which levels of developmental familiarity are so high as to inhibit sexual activity almost completely. Below such levels sexual activity would be possible, but an individual's degree of sexual attraction toward the members of its natal group would be consistently lower than toward out-group individuals. This would explain why all males leave their natal group in species where females are resident and why all females leave their natal group when males are resident. In other words, the existence of a qualitative leap in sexual attraction between members of different local groups would account for the existence of clear-cut patterns of sex-biased dispersal.

PRINCIPLE 7: FEMALES SEEM TO HAVE A HIGHER MOTIVATION THAN MALES TO AVOID INCEST. Another aspect of a primatological theory of incest is the existence of sex differences in attitudes toward incest and incestuous attempts (Chapais and Mignault 1991; Manson and Perry 1993; Takahata, Huffman, and Bardi 2002; reviewed by Paul and Kuester 2004). When sexual activity does take place between close kin, it is typically initiated by the male and typically refused by the female, who often reacts vehemently (for data on chimpanzees, see Goodall 1986, 466; Pusey 2004). This sexual difference makes a great deal of sense from an evolutionary perspective, females having more to lose from consanguineal conceptions, in terms of reproductive success, compared to males (Clutton-Brock and Harvey 1976). In most cases, incestuous attempts are initiated by juvenile or young adolescent males, rarely by adult males. They are transitory in a male's life and apparently have to do primarily with experimentation and play (Paul and Kuester 2004; Pusey 2004).

PRINCIPLE 8. BOTH HETEROSEXUAL AND HOMOSEXUAL ACTIVITIES ARE AVOIDED BETWEEN CLOSE KIN. In theory, incest avoidance might be limited to sexual activity between opposite-sex partners. This would make sense given that the raison d'être of incest avoidance seems to relate to the biological costs of inbreeding. However, there is limited but clear evidence that incest avoidance is not restricted to heterosexual relations and that it extends to homosexual dyads as well, at least among female macaques. Japanese macaques are known for their high rates of female homosexual behavior in both captive and free-ranging populations (Eaton 1978; Baxter and Fedigan 1979; Takahata 1982; Vasey 1995; Wolfe 1984). Females who engage in homosexual activity also have heterosexual activity; no females are exclusively homosexual. Females who have homosexual interactions are in estrus and act like heterosexual pairs, performing series mounting with pelvic thrusts, sexual solicitations, and so forth. In one study on captive animals, homosexual activity was observed between unrelated females for over ten years. But it was never observed between mother and daughter, grandmother and granddaughter, or sister and sister. Interestingly, however, a large proportion of aunt–niece dyads engaged in homosexual activity, interacting as unrelated females in this respect (Chapais and Mignault 1991; Chapais et al. 1997). Remarkably, then, the boundaries of kin discrimination revealed by homosexual behavior were the same as those inferred from the distribution of nepotism (see chapter 3).

Homosexual incest avoidance appears to be a byproduct of the

Westermarck effect. Females refrain from sexual activity with developmentally familiar animals, be they males or females. These data further illustrate the rule-of-thumb character of the Westermarck effect. Developmental familiarity inhibits sexual attraction between individuals whether they are genetically related or not and whether sexual activity is potentially reproductive (heterosexual) or not.

PRINCIPLE 9: SEXUAL ACTIVITY BETWEEN KIN, WHEN IT OCCURS, IS UNLIKELY TO HAVE REPRODUCTIVE CONSEQUENCES. Relatives sometimes engage in sexual interactions, but they do so in a behaviorally atypical manner, so the activity is unlikely to have any reproductive consequence. It is characterized by the absence of a consortship, irregular mount series, little sexual interest by at least one of the two partners, and rare ejaculation; it rarely takes place during periods in which conception is most likely, and it is most often performed by prepubertal males (Paul and Kuester 2004; Pusey 2004). The fact that sexual activity between kin, when it occurs, has more to do with experimentation and play than with reproduction provides further support for the adaptive character of the Westermarck effect.

Humankind's Primate Heritage

The primate theory of incest avoidance just outlined is extremely fragmentary and provisional because it was derived from data on a relatively small number of species. Moreover, primatologists have barely begun investigating a number of crucial questions relating to incest avoidance, notably the issue of sexual differences in attitudes toward incest, the effect of type of kinship (maternal versus paternal), generation, and degree of kinship on sexual attraction, and the exact nature of the Westermarck effect. In relation to that topic, it is not an overstatement to say that we know almost nothing about how the Westermarck effect actually works. As pointed out earlier, primate studies make it possible to uncouple familiarity and relatedness and analyze the effect of familiarity per se. But much remains to be done in this area, considering that primates offer opportunities to carry out experimental studies on how differences in level and temporal patterning of developmental familiarity affect sexual attraction between various categories of kin. An equally important issue relates to the nature of familiarity itself. What does developmental familiarity mean exactly? One promising line of inquiry is suggested by studies of clinical cases of incest in humans (discussed in the next chapter). The quality of an individual's attachment experience with its primary kin during childhood appears to be a primary predictor of his/her incestuous tendencies. Attachment bonds—with their neurobiological, neuroendocrinological, emotional, and cognitive underpinnings—might well constitute one crucial component of developmental familiarity, through which individuals normally develop sexual indifference toward their closest kin and toward nonrelatives who play the same role. In fact, attachment might be a major key to our understanding of the Westermarck effect.

Notwithstanding the limitations of our primate theory of incest avoidance, let us come back to its significance for human evolution. In all likelihood, it would have applied equally well to early hominids immediately after the *Pan–Homo* divergence. This means that early in human evolution: (1) inbreeding avoidance was adaptive; (2) dispersal from one's birth group and between-group transfers limited incest opportunities substantially; (3) incest was avoided systematically only between relatives who could recognize each other through association, usually uterine kin; (4) incest avoidance decreased steeply with decreasing degree of genetic relatedness; (5) the Westermarck effect was at work; (6) if dispersal patterns and the Westermarck effect are aspects of the same process, both sexes were more sexually attracted to individuals born in groups other than their natal group; (7) females had a higher motivation than males to avoid sexual interactions with close kin; (8) incest was avoided in the context of both heterosexual and homosexual activity; and (9) sexual interactions between kin, when they took place, were unlikely to have significant reproductive consequences. Clearly, then, early hominids were far from Lévi-Strauss's blank slate with regard to incest. In fact, incest avoidance appears to be among the most primitive components of the exogamy configuration.

Early in the evolution of the human lineage, these general principles were at work in specific social contexts defined by the main characteristics of hominid populations. Of particular importance among such characteristics are three factors: group composition, mating system, and philopatry pattern. A group's composition and philopatry pattern largely determine its genealogical structure. For example, a multimale-multifemale group with a male philopatry pattern produces extensive patrilines but small matrilines (see chapter 12). In turn, the group's genealogical structure determines the exact categories of kin that are lifetime coresidents and hence have opportunities to commit incest. Finally, the mating system determines, among other things, whether paternity (father–offspring relation) is recognized or not. If it is, kinship is recognized on both the mother's and the father's side; it is bilateral. In this situation, one expects incest avoidance to extend well beyond the domain of uterine kinship and to include a male's daughters, a female's father, and patrilateral kin. I come back to this topic when I describe the genealogical environment of early hominids.

6

From Behavioral Regularities to Institutionalized Rules

> I have so far spoken of habits [of chimpanzees and primitive men], not of institutions. But there is an intimate connection between them. Social habits have a strong tendency to become customs, that is, rules of conduct in addition to their being habits.
>
> *Edward Westermarck (1926, 28)*

> The truth is that non-human primates lack any form of social organization or social structure in any sense comparable to that of humans. This is because the regularities that are observable in their modes of temporal and spatial association do not ensue from rules apprehended and conformed to.
>
> *Meyer Fortes (1983, 22)*

These two contradictory quotations embody particularly well the issue that lies at the heart of divergences about the import of primate studies for understanding human behavior. Mating avoidances in nonhuman primates pertain to the realm of interactional regularities. They are behavioral patterns shared by all group members, but their collective character results from what every individual does or does not do with others. By contrast, incest-related attitudes in humans belong to the realm of normative regularities, or rules. They are the object of a conceptually recognized consensus about what is good or bad, whether that consensus is framed in terms of prohibitions and legally enforced or reflects implicit norms of conduct. In part because they are symbolically encoded, rules about incest are extremely variable cross-culturally: the types of unions that are considered incestuous in a society vary greatly according to kinship terminologies, marriage rules, descent systems, and so forth. Moreover, incest prohibitions extend well beyond the domain of consanguineal kinship; they encompass categories of in-laws, for example. "It is notable," wrote Murdock, "that nearly all societies interpose an incest taboo between such artificial relatives as adoptive parents and adopted children, stepparents and stepchildren, godparents and godchildren, and persons who become brother and sister through the establishment of a bond of blood-brotherhood" (1949, 268; see also Godelier 2004a, 345–417).

For these reasons incest rules lend themselves particularly well to the widespread assumption that any phenomenon that is highly variable cross-culturally is entirely a cultural construct; that biology and cultural variation are incompatible. Accordingly, as pointed out earlier, a common view among social scientists is that the similarities between primate incest avoidance and human incest prohibitions are essentially metaphorical, hence meaningless; a somewhat weaker claim is that even granting that interactional regularities might be at the origin of human rules, they cannot explain their moral dimension. In what follows I critically examine both claims.

The Anthropologists' Treatment of the Primate Data

The first data on incest avoidance in nonhuman primates appeared in the 1960s and were limited to mother–son relationships. Nevertheless, in 1961 Imanishi was already arguing that "monkeys such as Japanese monkeys or rhesus monkeys, though distant from human beings in evolutionary level, have adumbrations of incest taboo and exogamy which have been institutionalized in human society" (1965, 120). But Imanishi's article had very little impact. It had been published in Japanese, and although it was translated into English in 1965, it appeared in a privately published edited volume with a narrow distribution. Moreover, the article presented no quantitative data. The next body of evidence about primate incest avoidance, and the first quantitative one to my knowledge, was published in 1968 by Donald Sade, and again it concentrated on mother–son relations. Sade concluded his analyses in the following terms: "The origin of at least the mother–son incest taboo may have been the elaboration of a phylogenetically older system," a system that would have become "invested with symbolic content during hominization" (1968, 37). Already at that time, the idea of a phylogenetic continuity between primate incest avoidance and the human incest taboo was beginning to circulate within sociocultural anthropology. For example, in 1971, referring to male gibbons who leave their parents or are expelled by their father at puberty, Kathleen Gough wrote: "Similar de facto, rudimentary 'incest prohibitions' may have been passed on to humans from their prehuman ancestors and later codified and elaborated through language, moral custom, and law" (1971, 762).

The first comprehensive essays on the "comparative ethology of incest avoidance" were written in the early 1970s by Norbert Bischof (1971, 1975), who concluded that human incest prohibitions had emerged from the biologically encoded mating regularities in animals. "The creation of cultural norms . . ." he wrote, "can be regarded as a cognitive achievement, an act of self-interpretation, and these norms will only remain satisfactory and stable if man is able to recognize his own natural image in this interpretation" (1975, 63). Put differently, cultural norms about incest would reflect natural tendencies and, as emphasized by Fox (1979), that which is cultural about incest would not be incest avoidance itself but humankind's collective *rules* about it.

The idea that primate behavioral regularities prefigured symbolically encoded rules of conduct was not born with or limited to the issue of incest avoidance. The principle had been clearly stated in relation to the origin of positive rules of conduct. In the early 1960s, referring to cooperative activities in nonhuman primates, Elman Service wrote:

> There is, of course, some prehuman basis for sharing and cooperation . . . Once symbolic thought and communication become possible new determinants of behavior can be invented on the basis of evidence or knowledge which is already present. . . . Thus sharing can be changed from mere situational expediency to a norm and a good "thing" . . . All that was necessary, then, was the symbolic ability to make some rules and values which would extend, intensify, and regularize tendencies which already existed . . . *Thus . . . social reciprocity (as an action or practice) and an appreciation of its positive results appeared first, the rules and values afterward.* (1962, 41–42; my emphasis)

In the mid-1970s Robin Fox further explored the theme of the phylogenetic continuity between primate behavioral regularities and normative rules, but in relation to kinship as a whole. The following quotation is particularly revealing:

> The contention here is that even in the absence of cultural rules and the logic of human imagination there would be kinship systems anyway, and that much of the rule-making and imaginative logic is simply

> (or complexly) playing games with a quite elaborate raw material. To those for whom kinship is a matter of categories this must be mystifying. But categories simply mean that linguistic labelling is possible, and that hence, with communication operating in another dimension than gesture and face-to-face contact, greater elaboration of this basic system is possible. (1975, 10)

In other words, as soon as hominids were in a position to label the "natural categories" of their social (and physical) environment, for example their kin, mere interactional recurrences between kin would eventually give rise to normative rules of conduct.

In marked contrast with such views, many contemporaries of Fox and Service held that the advent of the symbolic capacity had disconnected humans from their primate heritage. For example, in an early comparative analysis of human and nonhuman primate societies, Marshall Sahlins asserted that humans had invented cooperation and kinship from scratch, that thanks to culture and symbols "human society overcame or subordinated such primate propensities as selfishness, indiscriminate sexuality, dominance and brute competition. It substituted kinship and co-operation for conflict . . . In its early days it accomplished the greatest reform in history, the overthrow of human primate nature" (1960, 86). Sahlins was writing at a time when kin groups, kin discrimination, incest avoidance, and cooperative activities in nonhuman primates had barely been documented. But the same theme of the overthrow of man's primate heritage was expressed even more comprehensively in 1983 by Meyer Fortes, at a time when we knew much more about the interactional regularities of nonhuman primates, notably incest avoidance patterns. In an essay aptly titled *Rules and the Emergence of Society*—one of a very small number of comprehensive essays written by social anthropologists and essentially devoted to the significance of primate studies for social anthropology—Fortes argued forcefully in favor of the view that symbolic rule-making in humans had freed our species from its primate heritage. Following Lévi-Strauss, Fortes stated that the very emergence of human society coincided with the capacity to make and follow rules, and he emphasized the difference between rules and mere regularities:

> Rules convey norms for the conduct of social and personal life, norms which are socially and culturally authorised, sanctioned or prohibited

> . . . Rules stipulate regularities but not all regularities are rule-governed. The regularities of locomotion or peristaltis, for example, are not. Correspondingly, there are many regularities of behaviour among non-human primates, including the sexual avoidances that seem to be parallel to human incest avoidances, but they are not rule-governed. (1983, 10)

Referring to the primate data on incest avoidance, Fortes then asked the crucial question: What do these "striking parallels" with human incest prohibitions signify? His response was that to conceive of a phylogenetic continuity and common biological background between the two realms pertained to the "metaphorical fallacy." That is, incest avoidance and incest prohibitions may look alike, but they are unconnected, and the latter are not derived from the former. To illustrate his point Fortes gave the following example:

> Bischof and others argue that the occurrence of "incest" avoidances among nonhuman primates implies that they distinguish mothers and siblings from other conspecifics. There is also, it appears, some evidence that living mother's mother and some close collaterals are selectively recognized in some species. But this range of mutual recognition . . . is hardly comparable to a situation where a descendant of the same maternal ancestress five or more generations back is recognized and defined as a sexually prohibited sibling. (1983, 18)

In Fortes's view, the human capacity to extend incest prohibitions to distant relatives to whom one may be related only through dead ancestors, and above all the conceptualization and institutionalization of kinship that such extension implies, prove that primate incest avoidance and human incest prohibitions are two independent classes of phenomena. This argument is based on the cognitive discrepancy between humans and other primates: the level of cognitive sophistication manifest in incest prohibitions is such that incest-related rules cannot be derived from incest avoidance. But from an evolutionary perspective, such a qualitative gap between the two realms is precisely what one would expect. This argument therefore is not particularly convincing.

At this point another consideration helps us understand Fortes's position. In the same essay Fortes espoused Freud's theory that human beings are born with incestuous desires. "The indications are clear," he wrote, "that incestuous wishes are normally generated in the intergenerational and cross-sex relationships in the parental family and do not emerge in normal adult behavior because they are socially repudiated" (1983, 14). Obviously, if incest prohibitions serve to thwart natural tendencies for incest, they cannot be rules that codify and reinforce natural tendencies for incest avoidance, as suggested by Bischof, Fox, and others. Freud's theory is not easily compatible with the primate data, which indicate that our close relatives are born with familiarity-based mechanisms leading to incest inhibitions. Freud's theory directly contravenes the idea of an evolutionary continuity between incest avoidance and incest prohibitions through the Westermarck effect. Fortes therefore had at least two reasons for rejecting the relevance of the primate evidence: the cognitive discrepancy argument and his rejection of the Westermarck effect.

Still in the same essay, Fortes enlarged the discussion and argued that when ethologists and primatologists use terms such as "incest," "monogamy," "polygyny," "polyandry," and "matrilineages" and when they talk of "status" or "role" in their descriptions of animal interactions, they are making "no more than metaphorical attributions" and freely borrowing "anthropomorphic models." For example, criticizing descriptions of primate behavior made in terms of "an individual's age and sex status," Fortes asked: "Do non-human primates recognize age differences in any way, e.g. show respect to older animals? Do they show cognizance of sex status by, for example, mother-in-law avoidance?" (22). True, primatologists have not always been careful enough in their choice of words. But what Fortes is saying here goes beyond the semantic issue. He is arguing that because human behavior is rule-governed, incest prohibitions, polygyny, matrilineages, and the like are cultural constructs with no evolutionary ties whatsoever to similar phenomena in closely related species. One is therefore left to believe that the similarities, however striking, are merely coincidental.

Maurice Godelier's position differs substantially from Fortes's while agreeing with him on one central point. In a recent monograph on human kinship Godelier duly acknowledged the "great theoretical importance" of the primate data for understanding the origin of human incest prohibitions. "These mechanisms [those that prevent incest in nonhuman primates,]" he wrote, "would therefore constitute the background and initial material of what became, in the form of conscious proscriptions in the human species, the 'incest prohibition' (2004a, 473, my translation). Clearly this is far from Fortes's conception. But although

Godelier believed in the phylogenetic connection between the two phenomena, he expressed reservations about some important aspects of the primate data. He was notably skeptical about the role of the Westermarck effect. He was inclined to think, for example, that if a young male does not copulate with his mother, it is not because he is inhibited by a high level of familiarity, an explanation that Godelier ascribed to a biological mechanism. Rather, it is more likely that the male is inhibited by the presence of dominant males or that he is forced to disperse in order to minimize competition with other males, two explanations that Godelier attributed, in this case, to social mechanisms (467). Notwithstanding the fact that both explanations are biological *and* social, there are two problems with this conception. First, as discussed previously (principle 2), male sexual competition alone can hardly account for male dispersal in multimale-multifemale primate groups. Second, there is reason to believe that male dispersal and the Westermarck effect, far from being independent explanations of incest avoidance, are two aspects of the same phenomenon of lower sexual attraction to developmentally familiar individuals (principle 6).

Godelier's reluctance to accept a primate legacy of familiarity-based sexual inhibitions accords well with his assertion that there exists no biological mechanisms biasing the human child's sexual desires against or in favor of certain individuals: upon reaching sexual maturity, he wrote, "a child's sexual urges may spontaneously, that is, unconsciously, be directed to his mother or his sister if he is a male, or to her father or brother if she is a female" (481). This Freudian claim is not easily reconcilable with the primate data. This is apparently why Godelier, even if he accepts that the incest taboo is derived from primate incest avoidance, is not ready to accept one implication of the primate data, namely the support they provide for the Westermarck hypothesis. Thus, like Fortes, Godelier rejects the Westermarck effect. By removing this one big chunk from the primate legacy, Godelier is introducing a substantial discontinuity in the evolutionary history of the incest taboo. Indeed, his argument implies that for some reason our hominid forebears lost the biological underpinnings of incest avoidance, becoming atypical primates in this respect. It is difficult to see why such a widespread adaptation would have been selected against, or ceased to be selected for, specifically in the hominid lineage. Moreover, after freeing themselves from the Westermarck effect, hominids, in that view, eventually devised another means, cultural incest prohibitions, to achieve basically the same result. The continuity hypothesis is much more parsimonious.

The Westermarck Knot

Clearly, the Westermarck effect is one major obstacle that prevents many researchers from accepting the idea of a phylogenetic continuity between incest avoidance and incest prohibitions, either in whole, like Fortes, or in part, like Godelier. But there is a solid and now classic corpus of evidence about the inhibiting effect of childhood familiarity on sexual feelings in humans. At the risk of being somewhat redundant, it is worth summarizing that information. Briefly, one data set comes from studies of mate selection in Israeli communal villages (kibbutzim) where individuals were raised together from birth in peer groups of six to eight children. Building upon Yonina Talmon's (1964) pioneering analysis of mate selection in kibbutzim, Joseph Shepher (1971, 1983) carried out a detailed study on a larger sample of married couples and found that no marriages were contracted between individuals who had been raised continuously in the same peer group over their first six years, even though there was no prohibition to this effect. Shepher was further able to analyze premarital behavior in one kibbutz, and he reported not a single case of heterosexual activity between people, either as adolescents or as adults, raised in the same peer group as children, despite the occurrence of sexual play between some of them when they were children.

At about the same time, Arthur Wolf (1966, 1970, 1995) began a long-term study of the "minor" type of marriage in Taiwan, in which a girl is adopted by a family, often as an infant, and raised as a daughter-in-law *(sim-pua)* in close intimacy with her future husband. The girl and her future husband play, sleep, and take baths together. Wolf found that *sim-pua* couples were very reluctant to marry upon coming of age, that compared to women married in normal (major) marriages, women in minor marriages had a 40 percent lower fertility and a threefold higher chance of divorce, and that both wife and husband were significantly more likely to have extramarital affairs. More recently, Wolf (2004b) looked further into the Taiwanese minor marriage in an attempt to better characterize the relevant parameters of developmental familiarity that induced sexual aversion. Among other things, Wolf analyzed the effect of the wife's age at adoption, her husband's age at that time, and the number of years they had been in association prior to marriage. His results led him to conclude, paraphrasing Westermarck, that "there is a remarkable absence of erotic feelings between people who live together and play together before age ten. The absence is particularly marked among couples brought together before age three, and, for any given couple, largely depends on the age of the younger partner when they first met" (86).

These two sets of studies have in common the important characteristic that they are not about sex between consanguineal kin, hence not about incest avoidance as such. Precisely for this reason they constitute natural experiments that demonstrate the effect of developmental familiarity per se on sexual inhibition between individuals *reared as kin,* specifically as siblings. They thus provide particularly meaningful evidence in favor of Westermarck's explanation of how kinship breeds sexual indifference. To this ethnographical evidence one must add a growing body of evidence from evolutionary psychology. For example, Lieberman, Tooby, and Cosmides (2003) analyzed the moral judgment of American undergraduates about sibling incest by others but in relation to the subjects' experience with their *own* siblings. They found that an individual's length of coresidence with siblings of opposite sex predicted the strength of the negative sentiments expressed about sibling incest in general: the longer one's coresidence with siblings, the stronger the negative feelings about incestuous relations. Significantly, the effect of coresidence was independent of the actual degree of relatedness between siblings; adopted siblings had similar negative sentiments about sibling incest. These results suggest that the Westermarck effect operates in normal family settings (see, for example, Bevc and Silverman 1993, 2000), and they confirm the ethnographic evidence reported above about the mediating role of familiarity in the development of sexual indifference between close kin.

To many anthropologists, to argue that the Westermarck effect is at work in our species is simply incompatible with the fact that incestuous relations do occur in human societies. If the Westermarck effect were really in operation in humans, they reason, incestuous relations and incestuous marriage would be systematically avoided. That is to say, exceptions to incest avoidance rules prove that childhood familiarity does not breed sexual indifference and that close kin are not immune to sexual attraction. Fortes is particularly clear about this: "The fact," he wrote, "that incest occurs at all in our family system in spite of the sanctions of religion, the general climate of normative custom, and the definition of incest as a crime, must be taken as further confirmation of the validity in principle of the Freudian theory of its origins and nature" (1983, 14). This reasoning betrays a deterministic conception of the influence of biology on behavior, one holding that biological factors should produce no behavioral variation. This is erroneous for two distinct reasons.

First, the Westermarck effect is about the influence of a quantitative variable, developmental familiarity, on sexual inhibition; it is not an all-or-none phenomenon. Accordingly, one expects different levels of developmental familiarity between kin to generate varying levels of sexual inhibition between them. For example, one would expect siblings who have experienced low levels of developmental familiarity with each other to be less sexually inhibited compared to siblings who have experienced higher levels of familiarity. Among the factors likely to reduce sibling familiarity are (1) a larger age difference between them, (2) a shorter period of coresidence, and (3) less extensive physical contact during that period. In relation to the last factor, situations in which siblings have high rates of physical contact (play, touching, bathing, sleeping, sexual games) should breed sexual indifference, while situations in which they have much less physical intimacy would be expected to circumvent the Westermarck effect to some extent (see also Fox 1980, 27). Clearly, then, exceptions to incest rules do not necessarily negate the existence of the Westermarck effect; they may simply indicate that it could not operate fully in certain developmental situations.

A recent review of clinical cases of incest carried out by a psychiatrist is particularly revealing on this point. Mark Erickson (2004) considered separately two categories of incestuous relationships: those associated with an early separation of the incestuous kin—brothers and sisters, fathers and daughters, or mothers and sons—and those involving relatives who had not experienced an early separation. In the first category, early separation resulted most often from adoption, so in terms of developmental familiarity reunited kin were like unrelated individuals. Remarkably, reunited kin were reported to be fascinated by their physical and mental similarities, and many of them experienced strong sexual feelings toward each other even though they were aware of an incest taboo.

Some even married. These results indicate that in the absence of the Westermarck effect sexual attraction between close kin is not only possible but, as noted by Erickson, apparently exacerbated by phenotypic similarities.

The other category of incestuous relationships concerned kin who had not been separated and therefore should have been influenced by the Westermarck effect. The general pattern summarized by Erickson is clear: whether one considers father–child, mother–child, or sibling incest, "the most salient influence on incest behavior may be found in the childhood attachment experience" of the instigator, which is described as filled with rejection, neglect, physical abuse, emotional deprivation, hostility, and the like. In many cases the instigator had been sexually abused. Considering the primacy of intrafamilial relationships in a child's emotional and social development, it is not surprising to find that dysfunctional families produce behavior patterns that deviate from the species' norm. Erickson concluded his review in the following terms:

> As predicted by Westermarck's hypothesis, clinical studies show that incest is far more likely if kin are separated early in life. It is, virtually, only in this circumstance that incest may be mutually desired and eventuate in marriage. Taboos appear to have limited influence. By contrast given early association, incest is rarely, if ever mutually desired. It is perpetrated coercively, by fathers and brothers, and experienced as being intensely aversive by daughters or sisters. (2004, 170)

A deterministic conception of the Westermarck effect is erroneous for a second reason, namely that developmental familiarity is only one factor among others that affect sexual attraction and the likelihood of marital unions. It is certainly possible to oppose the Westermarck effect with cultural practices that for a number of reasons—for instance, the preservation of property and privileges within the family—encourage marriage between close kin. The well-known cases of institutionalized brother–sister incest in the Hawaiian, Incan, Persian, and Egyptian royal families (for example, see Godelier 2004a, 345–417) provide good examples. Again, the reasoning underlying the Westermarck effect in human societies is not that it should invariably produce incest avoidance between close kin independently of the circumstances but that it is a force consistently at work among close kin who experience high levels of developmental familiarity. Accordingly, the Westermarck effect would explain why, as long remarked by Lévi-Strauss himself, consanguineous marriages within royal families or the nobility were "either temporary and ritualistic, or, where permanent and official, nevertheless remain[ed] the privilege of a very limited social category" (1969, 9).

Significantly, however, not all sibling marriages are temporary, restricted to the elite, or socially coerced. Referring to Margaret Murray's (1934) findings about consanguineous marriages in ancient Egypt, Lévi-Strauss found that evidence "more disturbing," since it suggested "that consanguineous marriage, particularly between brother and sister, was perhaps a custom which extended to the petty officials and artisans, and was not, as formerly believed, limited to the reigning caste and to the later dynasties" (1969, 49). Census data preserved on papyrus that have since been published indicate that brother–sister marriage might have been the norm among ordinary residents in Roman Egypt (Hopkins 1980; Shaw 1992). This body of evidence has been brought forward, notably by Fortes and Edmund Leach, as constituting a decisive argument against the Westermarck effect and the claim that sibling incest avoidance is natural (Fortes 1983, 11; Leach 1991, 101–104). However, a recent analysis of the available evidence by Walter Scheidel has led him to conclude that as far as the correlation between early childhood association and sexual inhibition is concerned, Roman Egyptian sibling marriage and the Chinese data on minor marriages do not deviate significantly; both would obey the same general principles (2004, 105). This indicates that the most pertinent question is not whether sibling marriage was ever the norm in any human society but whether it was entirely immune to the Westermarck effect, as contended by many anthropologists. This appears highly unlikely.

The simplest way to summarize the whole argument presented in this chapter is as follows: the Westermarck effect is a *necessary* component of human incest prohibitions but not a *sufficient* one. It is a factor that must be taken into account in the "equation" of sexual attraction, along with others. And if it were not for humankind's primate legacy of incest avoidance, of which the Westermarck effect is an integral part, human societies would be free of any incest taboo because culture would have had no biological substrate to work upon. It is that very heritage (tentatively summarized under the nine principles in chapter 5) that set the biological foundation upon which culture developed a multitude of institutionalized variants, the sets of incest-related rules specific to every human society. Arguing in this sense is no different from saying that if it were not for our primate legacy of consanguineal kinship, the human primate would have no kinship, consanguineal, classificatory, or fictive.

The Morality Problem

That the Westermarck effect characterizes both human and nonhuman primates is certainly one important argument in favor of the phylogenetic continuity hypothesis. But recently the emphasis has shifted from the question of the existence of the Westermarck effect to its pertinence for understanding the moral dimension of the incest taboo. This issue refers to what Wolf called the "representation problem" (after Williams 1983) of incest: "Generally speaking, this is the problem of how the loves and hopes and fears and phobias of individuals give rise to *norms*, if they do" (Wolf 2004a, 10–11). Central to the representation problem is that incest is "a matter of public condemnation." If it is clear that the Westermarck effect provides an explanation for why people avoid sexual relations with their *own* close relatives, it is much less clear that the same developmental process could also account for the fact that people condemn *other* people for having sexual relations with their own close relatives and that "such condemnation elicits universal approbation." Indeed, the Westermarck effect is essentially a dyadic phenomenon: it is about what ego feels toward another individual, not about what ego thinks of what third parties do. Thus to state that the incest taboo evolved from primatelike patterns of behavior is to argue that the moral dimension of incest is no more than a further consequence of the primate legacy of incest avoidance merging with the symbolic capacity. Morality would be rooted in our biology.

The importance of Westermarck's reasoning in recent discussions about the evolutionary origins of morality, discussions fueled notably by Edward Wilson, can hardly be exaggerated. As noted by Larry Arnhart, it is significant that throughout his writings Wilson has used Westermarck's theory of the incest taboo as "the prime example of how biology can explain the moral sentiments" (Arnhart 2004, 191), Westermarck was indeed among the first to contend that "social habits" give rise to "customs and rules of conduct." Writing at a time when knowledge about primate behavior amounted to a few anecdotes, he hardly had the empirical data to understand how a primate legacy of interactional regularities might have given rise to humanlike rules of conduct. Nonetheless, his basic argument about the functional relation between dyadic sexual aversions and moral sentiments is clearly relevant to this issue. Westermarck's reasoning was in three steps: (1) "the absence of erotic feelings between persons living together very closely together from childhood" would produce (2) a "positive feeling of aversion when the [sexual] act [is] thought of" . . . and as a result, (3) the "aversion to sexual relations . . . displays itself in custom and law as a prohibition of intercourse between near kin" (1926, 80). In other words, a "social habit" of sexual indifference toward close kin, when cognitively apprehended, would generate a feeling of aversion, and this feeling would be the origin of the normative proscription against incest.

The major—and now classic—objection to Westermarck's hypothesis was that a natural aversion to incest did not have to be reinforced by a rule against incest. Therefore, the very existence of such a rule proved that it had originated as a cultural construct designed to thwart what was in fact a natural tendency to commit incest. In Westermack's words, "Sir James G. Frazer has argued against me that if exogamy had resulted from a natural instinct there would have been no need to reinforce that instinct by legal pains and penalties; . . . hence we may always safely assume that crimes forbidden by law are crimes which many men have a natural propensity to commit. This argument has been quoted with much appreciation by Dr. Freud" (1926, 90). Westermarck argued that this critique implied "a curious misconception of the origin of legal prohibitions." Referring to the "equally or even more stringent laws against other sexual offences, such as bestiality or sodomy," he asked: "Would they likewise be regarded as evidence of a general inclination or a strong temptation to commit these offences? Or would the exceptional severity with which parricide is treated by many law-books prove that a large number of men have a natural propensity to kill their parents?" (90). In sum, Westermarck reasoned that behaviors that are naturally avoided—mating with a close kin, having sex with another species, or killing one's father—should give rise to *proscriptive* rules against incest, bestiality, and parricide.

Let us examine on logical grounds what would be needed to transform a mere state of sexual indifference between opposite-sex siblings into a moral rule condemning incest between siblings in general. At least three conditions must be met. First, as stated by Westermarck, the thought of a sexual act with an opposite-sex sibling must elicit a feeling of aversion. Ego's conscious apperception of an incestuous act with its own sibling must be felt negatively. The Westermarck effect is precisely about such feelings. Second, to condemn sibling incest, ego must have a *concept* of siblingship in the first place. Ego must be able to recognize the properties common to sibling dyads in general and, on this basis, discriminate between siblings and nonsiblings. Some nonhuman primates display such a capacity even though they do not use symbols to label their concepts (Dasser 1988a, b). With the evolution of the symbolic capacity, hominids could label their concepts, both social and physical, and could manipulate them mentally to an unprecedented extent. Third, individuals not only need to have a concept of siblingship, they must be able to apply it to their own situation: they must realize that they, too, have siblings. To do so, ego must be able to make the following type of transitive reasoning: individual X stands in relation to individual Y (two siblings) the same way I stand in relation to my own sibling. Importantly, then, this kind of reasoning implies some degree of *self-awareness*. Although nonhuman primates exhibit some of the building blocks of self-awareness, this ability is uniquely human to a large extent (Parker et al. 1994; Keenan, Gallup, and Falk 2003).

Assuming that the three conditions are met, it follows that on witnessing a sexual act between two opposite-sex siblings, ego would be in a position to transpose it to its own situation. That is to say, ego would form a mental image of a sexual interaction with its own opposite-sex sibling and feel a sexual aversion. This kind of *empathy,* through which an individual vicariously experiences negative feelings by observing (or hearing about) interactions between others depends crucially on the third condition: ego must recognize the similarity between the bond linking the two incestuous partners (siblingship) and its own sibling bonds. This is quite demanding for a nonhuman primate but is part of common experience in humans. Now, given that all individuals who witness or hear about an incestuous act would go through the same empathy process, the aversion would be shared collectively. Moreover, the symbolic capacity would make it possible for group members to communicate with each other about their shared aversion, including using a symbolic label to refer to something like "sibling incest wrongness." A shared collective aversion may be expected to lead to the *differential treatment* of the "wrongdoers," from avoiding them to rejecting them—in other words, the implicit condemnation of sibling incest. At this stage one is quite close to a norm about sibling incest and its moral content.

I do not pretend to have demonstrated that moral rules derive from behavioral regularities or to have identified all the elements involved in the corresponding evolutionary transformation. My point is simply that to go from the Westermarck effect to a state of moral condemnation of incest, a number of major cognitive processes must be integrated, but unless it can be shown that the gap is not bridgeable, the moral content of rules is not an a priori argument for rejecting the derivation of the incest taboo from incest avoidance patterns. The same reasoning applies to any other proscription and, importantly, it applies equally well to social interactions that are naturally felt as *positive*. Examples include cooperative and altruistic interactions between close kin. In this situation one would expect corresponding *prescriptive* rules. For example, maternal caretaking is a major interactional regularity throughout the primate order and well beyond it. Applying Westermarck's argument, one would expect maternal caretaking to have given rise to normative rules of conduct—whether these are legally enforced or merely consensual—that prescribe maternal care and to other rules that prohibit behaviors contravening the principle of maternal care (for example, rules about child abuse).

Lessons from Comparative Anatomy

In closing this discussion about humankind's primate heritage, I want to come back to an issue alluded to earlier, that of the exact nature of the evidence supporting the contention that behavioral similarities between human and nonhuman primates reflect evolutionary relationships. My point is that whatever the topic considered, the relevant evidence is basically of the same nature as that employed by generations of comparative anatomists ever since cross-specific comparisons were performed. It is no different from the evidence supporting the claim, for example, that the human foot is homologous to other primate feet. That proof is based on detailed comparisons of the relevant structures and a demonstration that the similarities are too numerous and profound to be merely coincidental and shallow. This point is worth emphasizing. Detailed anatomical similarities between human and nonhuman primates have consistently been accepted as providing solid grounds to infer evolutionary connections. But behavioral—especially social—similarities between human and nonhuman primates are often considered suspicious and merely metaphorical. There is a double standard in the treatment of animal–human comparisons, one for anatomy, another for behavior. Why is that so? One important reason is certainly that behavioral similarities are more difficult to delineate and analyze than anatomical similarities, whether they concern homologies (similarities reflecting descent from a common ancestral species) or homoplasies (functional similarities independent of common descent). To illustrate my point here I focus on homologies.

Homologous anatomical structures are often plainly manifest—the hand in primates provides a good example. But other anatomical homologies may be extremely difficult to identify if the corresponding structures in different species have undergone such important evolutionary changes that they are hardly recognizable as homologous traits. A classic example is the homology between the bones forming the middle ear of mammals and those of the lower jaw of reptiles. To establish that a mammalian ear ossicle is phylogenetically derived from a certain reptilian jawbone, one can hardly rely on formal similarities between the two bones themselves—they have hardly anything in common in adult individuals. One powerful means to establish homologies in such a situation is through embryology. In the present example, the evolutionary connection between the two sets of bones can be detected in the embryonic development of mammals. One can literally see the embryonic precursors of jawbones "move into position" in the middle ear. That is how the homology was discovered by Carl Reichert in 1837 (Schmitt 2006, 297). One may also uncover such homologies by looking at the fossil record, which in the present case graphically reveals a number of intermediate stages (Futuyma 1998, 146–152). Finally, homologies between highly modified elements may be apparent in the way each element is physically connected to the parts surrounding it; that is, it may be seen in the overall structural pattern it belongs to. For example, a highly modified digit in one species may be recognized as homologous to a digit in a related species through its physical connections with other, less modified digits.

If anatomical homologies may be difficult to detect, the difficulties are only amplified in the case of behavioral homologies, especially complex behavioral systems such as those discussed in the present book. First, one cannot rely on fossils because complex behavioral systems do not fossilize. Second, compared to anatomical patterns, behavior patterns are environmentally much more variable. If only for this reason, the common biological foundation of potentially homologous behavioral patterns may be hidden underneath a thick layer of "background noise," so to speak. That foundation must first be delineated. Third, as with anatomical structures, homologous behavioral systems in different species may have changed a lot in the course of evolution, in which case the similarities would be difficult to recognize. As just pointed out, comparative anatomists facing that problem rely on two types of information: the overall structural pattern of which the structure is an integral part and its embryonic development. Needless to say, establishing such structural and ontogenetic connections with behavior patterns is considerably more arduous. But this is precisely, in principle, what needs to be done.

Consider incest avoidance, whose core process is, assumably, the Westermarck effect. The primary argument in support of the hypothesis that the Westermarck effect is homologous in human and nonhuman primates rests on the similarity of its overall manifestation in these species: mating is avoided between close kin that have undergone a period of developmental familiarity. But the homology hypothesis implies much more than that. First, the *processes* underlying the Westermarck effect—neurobiological, psychological, and developmental—should be similar in humans and other primates. This is at present unknown because we still do not understand how the Westermarck effect actually works. If, for example, it works through attachment bonds and involves some kind of critical period, the corresponding processes should be observable both in humans and in other primates. This remains to be ascertained. Second, the *structural correlates* of the Westermarck effect should also be similar between species. On this point the available evidence tends to further confirm the homology hypothesis. Indeed, in humans and other primates, sexual inhibition between close kin (1) is more strongly felt by females than by males (Erickson 2004); (2) affects both heterosexual and homosexual activity (Godelier 2004a, 360); (3) decreases with decreasing degree of kinship; and (4) generally prevents sexual intercourse between mature individuals but does not necessarily preclude sexual play between immatures; that is, the nature of the sexual inhibition varies according to absolute age (Shepher 1983; Bevc and Silverman 1993, 2000). These "secondary" similarities, and several others that need to be established, are significant. If incest avoidance obeyed different principles in different species, for example exhibiting a male bias in humans but not in other primates, this would somewhat reduce the amount of evidence in favor of the homology hypothesis.

In sum, behavioral homologies, like behavioral homoplasies, can be established only on the basis of detailed comparative analyses. If such analyses have been the hallmark of comparative anatomy over at least two centuries, they have not been the hallmark of comparative behavioral studies. Human–nonhuman comparisons, in particular, are often not sufficiently detailed. Sometimes the similarities are merely suggested through the use of a common label, a practice that raises legitimate doubts about the connection they suggest or purport to demonstrate. For example, between the primatologists' conception of a "matrilineage" and the sociocultural anthropologists' conceptualization of a "matrilineal descent group," also called a matrilineage or a matriclan, stand a number of important differences, structural and conceptual (see chapters 17–19). These need to be spelled out before inferences can be made regarding the exact evolutionary relationships between the two entities.

II

The Exogamy Configuration Decomposed

7

Lévi-Strauss and the Deep Structure of Human Society

> We have been careful to eliminate all historical speculation, all research into origins, and all attempts to reconstruct a hypothetical order in which institutions succeeded one another.
>
> *Claude Lévi-Strauss (1969, 84)*

Lévi-Strauss's words here convey particularly well the trademark of his whole approach: a strictly structural and synchronic—unchronological—analysis of human society. Accordingly, in Lévi-Strauss's alliance theory the contribution of humankind's evolutionary heritage to the emergence of human society is nil, essentially. Alliance theory thus amounts to some sort of "null hypothesis" about the phylogenetic history of human society. In retrospect, such a tabula rasa conception provides a particularly useful comparative base line from which we may better appreciate how human society was gradually put back into the evolutionary realm.

The key principle of reciprocal exogamy is the binding dimension of marriage, the fact that marital unions are not limited to the spouses themselves but extend to their respective kin and close associates, whether the bonded entities are nuclear families, extended families, small kin groups, or whole tribes. Although I concentrate on the contribution of Lévi-Strauss on this point, it is worth noting that the idea was very much circulating when he presented his in-depth exploration of it in his 1949 book. Some sixty years earlier, in 1889, Edward Tylor, referring to prior authors who had voiced it less formally, clearly articulated the principle that exogamy's fundamental and universal function was the "binding together [of] a whole community with ties of kinship and affinity" (1889b, 268)—his famous "marrying-out or being killed-out" aphorism:

> When tribes begin to adjoin and press on one another and quarrel, then the difference between marrying-in and marrying-out becomes patent. Endogamy is a policy of isolation . . . Among tribes of low culture there is but one means known of keeping up permanent alliance, and that means is intermarriage . . . Again and again in the world's history, savage tribes must have had plainly before their minds the simple practical alternative between marrying-out and being killed-out. (267)

Tylor's views were espoused by a number of scholars, but somewhat belatedly. Leslie White, in particular, adopted Tylor's interpretation of exogamy, further emphasizing that his explanation "provided the key to the understanding of the origin of the incest taboos" (1959, 88). White's formulation in 1949 of the functional linkage between exogamy and incest prohibitions strikingly parallels the cornerstone of Lévi-Strauss's reasoning published the same year. White wrote:

> Co-operation between families cannot be established if parent marries child; and brother, sister. A way must be found to overcome this centripetal tendency with a centrifugal force. This was found in the definition and prohibition of incest. If persons were forbidden to marry their parents or siblings they would be compelled to marry into some other family group . . . The leap was taken; a way was found to unite families with one another . . . Marriages came to be contracts first between families, later between even larger groups. (1949, 316)

White went on to discuss institutions like the betrothal of children, the levirate, the sororate, brideprice, and dowry as manifestations of the principle that marriage universally purported to establish and perpetuate alliances between groups, not individuals.

Clearly, therefore, the marriage-as-alliance principle was in the air. But Lévi-Strauss pushed the analysis much further, exploring its manifold ramifications, in particular the implications of the reciprocity dimension of marriage and the role of kinship in structuring marital alliances and societies. Lévi-Strauss's *Les Structures Élémentaires de la Parenté* was first published in 1949 and a revised edition in 1967. (The second edition was translated into English in 1969 under the supervision of Rodney Needham.) In this chapter I summarize Lévi-Strauss's thinking about the distinctive nature of human society and I address the temporal dimension of reciprocal exogamy. Lévi-Strauss repeatedly stated that he wanted to avoid any historical evolutionist interpretation of his analyses—for example, that simple forms of matrimonial exchange might antedate more complex ones. But from an evolutionary perspective, that which is simple and elementary may well be ancient and primitive.

Reciprocal Exogamy as a Deep Structuring Principle

Lévi-Strauss's lengthy and extremely detailed argument about reciprocal exogamy as the core organizational principle of human society is based essentially on the following four points. First, he conceived of reciprocity as a universal mental structure, a fundamental attribute of the human mind in the area of social life (1969, 84). As part of the human psyche, reciprocity and its underlying principles—duality, alternation, opposition, and symmetry—"are not so much matters to be explained, as basic and immediate data of mental and social reality which should be the starting point of any attempt at explanation" (136). In other words, reciprocity and exchange are part of humankind's neurobiological makeup. They are givens that organize social life the same way, one could argue today, humans are born with neurobiologically based linguistic abilities that need only be developed and expressed.

Second, of all the commodities that may be exchanged, "women are the most precious possession" (62). Accordingly, the exchange of women is the form of reciprocity whose social consequences are the most profound. Lévi-Strauss's reasoning here was founded on Marcel Mauss's (1923) conception of reciprocal gifts. Mauss argued that gifts and counter-gifts were much more than mere economic transactions and that exchanged commodities encompassed much more than their economic value, their significance being at once social, religious, utilitarian, sentimental, jural, and moral. Taking up the idea, Lévi-Strauss referred to the "synthetic nature of the gift," which he regarded, along with reciprocity, as a universal structure of the human mind: "The agreed transfer of a valuable from one individual to another makes these individuals into partners, and adds a new quality to the valuable transferred" (84). Thus the exchange of commodities between individuals would build social partnerships between them. Applying this principle to the most important of all commodities, women, Lévi-Strauss wrote: "Exchange—and consequently the rule of exogamy which expresses it—has in itself a social value. It provides the means of binding men together" (480). And so important are women-mediated partnerships between men that without matrimonial exchange there would be no social cohesion and no human groups as we know them: "Exogamy," Lévi-Strauss continued, "provides the only means of maintaining the group as a group, of avoiding the indefinite fission and segmentation which the practice of consanguineous marriages would bring about" (479).

The third point concerns the relationship of incest prohibitions to exogamy. Referring to Westermarck's explanation of the incest taboo, Lévi-Strauss stated that there was "nothing more dubious than this alleged instinctive repugnance, for although prohibited by law and morals, incest does exist." To him, Westermarck's hypothesis was all the more untenable because "an important modern school towards [*sic*] this problem runs completely counter to Havelock Ellis and Westermarck. Psychoanalysis," he continued, ". . . finds a universal phenomenon not in the repugnance towards incestuous relationships, but on the contrary in the pursuit of such relationships" (1969, 17). Lévi-Strauss thus endorsed the idea that human beings were sexually attracted to their close kin. This meant that to be in a position to exchange women with other groups, men first had to renounce marriage with their own daughters and sisters; they had to forbid marriage within the family. Practicing exogamy entailed the invention of incest prohibitions. To Lévi-Strauss incest prohibitions and exogamy were two sides of the same coin, incest prohibitions representing the negative and proscriptive aspect of matrimonial exchange, and exogamy rules, its positive and prescriptive aspect: "The prohibition on the sexual use of a daughter or a sister," he wrote, "compels them to be given in marriage to another man, and at the same time it establishes a right to the daughter or sister of this other man . . . The prohibition is tantamount to an obligation, and renunciation gives rise to a counter-claim" (51). In short, the prohibition of incest is itself a rule of reciprocity.

The fourth point is a correlate of the first three. If reciprocity is a universal mental structure and if men seek to build social partnerships by exchanging women, renouncing marriage with their close kin in the process, the exchange of women should be a universal property of the marriage institution. About this, Lévi-Strauss was repeatedly explicit: "No matter what form it takes, whether direct or indirect, general or special, immediate or deferred, explicit or implicit, closed or open, concrete or symbolic, it is exchange, always exchange, that emerges as the fundamental and common basis of the institution of marriage" (1969, 478–479). In sum, although marriage systems are cross-culturally extremely diverse, reciprocal exogamy would be universal; in all societies, marriage would bind groups.

Lévi-Strauss's alliance theory rests on three primary assumptions. First, primitive men lived in a social world that forced them to build alliances with other men. Second, men had a natural inclination to marry within their family. Third, the most efficient way for men to form alliances with other men was by exchanging their daughters and sisters. Because the second and third assumptions are contradictory, and because, presumably, the need for alliances outweighed incestuous inclinations, men solved the contradiction by inventing the incest taboo. When this happened, reciprocal exogamy was born and the nature-to-culture transition accomplished.

Of special interest from an evolutionary perspective are the simplest of all systems of reciprocal exogamy. In Lévi-Strauss's scheme, these belong to what he referred to as the *restricted exchange* category, typified by two exogamous groups (or moieties A and B) exchanging their daughters and sisters. In such a situation, reciprocity is direct and bilateral between the two moieties. A more complex category of matrimonial reciprocity incorporating more groups Lévi-Strauss labeled *generalized exchange*. In this situation, group A gives its women to group B, which gives its women to group C, which gives women to group A. The circulation of women is thus not bilateral, as in restricted exchange; it is unilateral, a group receiving its women from a group different from the one it gave its own daughters and sisters to. But considering the system as a whole, reciprocity is the rule here as well, provided the cycle of exchange is completed.

Lévi-Strauss subsumed these two categories of matrimonial exchange under the label of *elementary* structures. By that he meant that both restricted and generalized exchange systems specified positive rules about the choice of marriage partners—as opposed to *complex* matrimonial structures, as in Western societies, which define no such rules. Central to Lévi-Strauss's theory is that positive rules of mate assignment are based on kinship relations, that the overall genealogical structure embracing distinct exogamous units affects to a significant extent the formation of marital bonds between them. More specifically, elementary kinship structures hinge on cross-cousin marriage, which Lévi-Strauss conceived of as the "elementary formula for marriage by exchange" (1969, 129) and "the simplest conceivable form of reciprocity" (48).

Cross-cousins are the offspring of a brother and a sister; parallel cousins, the offspring of same-sex siblings (two brothers or two sisters). A widespread practice throughout the world is for a man to marry his cross-cousin on his mother's side (his mother's brother's daughter) or on his father's side (his father's sister's daughter), whereas marriage with a parallel cousin on either side is forbidden.

The importance of cross-cousin marriage in Lévi-Strauss's theory can hardly be overstated. But cross-cousin marriage, he emphasized, is only one manifestation of a more "general kinship structure," which includes various types of matrimonial privileges, matrimonial proscriptions, special relationships, and terminological classifications. Lévi-Strauss described that kinship structure as "being second only to the incest prohibition" in its universal character. All the phenomena he subsumed under it have in common the same asymmetry derived from the distinction between different-sex and same-sex siblings. Hereafter I refer to this kinship structure as the *brother–sister kinship complex.* Figure 7.1 illustrates the various types of relationships involved. Basically, the brother–sister complex is a two-generation system of relationships and corresponding terms founded on the brother–sister bond. As such it includes relationships between siblings, between cousins, and between uncle/aunts and nephews/nieces. Lévi-Strauss was particularly clear about the universal character of the brother–sister kinship complex and the underlying unity of all its manifestations, no matter how diverse:

> No doubt each of these features has its own particular history, and doubtless also its history can vary from group to group. At the same time, however, no feature is to be seen as an independent entity isolated from all others. On the contrary each appears as a variation on a basic theme, as a special modality outlined against a common backdrop . . . As systems without any of these features are far less common than systems with at least one of them and probably more, and as the latter systems are scattered all over the world, no region being completely devoid of them, *it is this general structure, of all the rules of kinship, which, second only to the incest prohibition, most nearly approaches universality.* (1969, 124; my emphasis)

For anyone interested in the evolutionary origins of human society, something identified as the simplest of all human social systems—restricted exchange and cross-cousin marriage—and stemming from a nearly universal kinship structure should be of cardinal importance. If

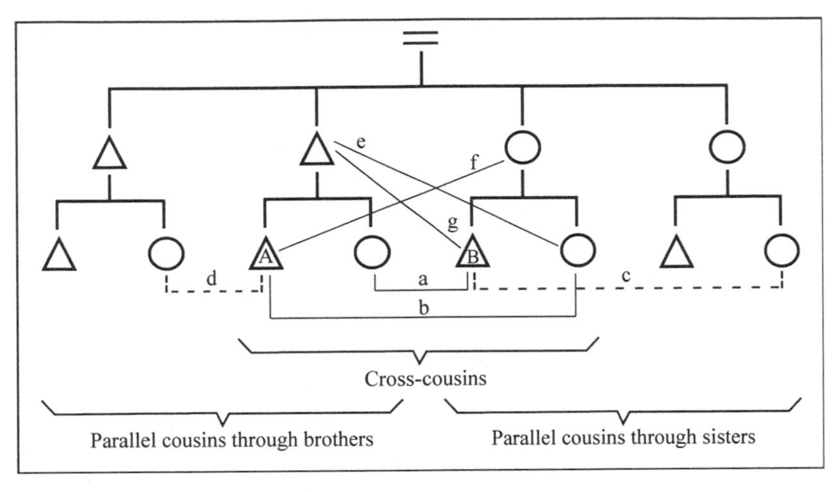

Figure 7.1. The components of the brother–sister kinship complex. The figure does not represent any existing social system. (a) Preferential marriage between male B and his matrilateral cross-cousin. (b) Preferential marriage between male A and his patrilateral cross-cousin. (c) Proscriptive marriage between male B and his matrilateral parallel cousin. Male B often calls his mother's sister (his parallel aunt) "mother," and her children "brothers" and "sisters." Parallel aunts often call their sisters' offspring "daughters" and "sons." (d) Proscriptive marriage between male A and his patrilateral parallel cousin. Male A often calls his father's brother (his parallel uncle) "father," and the uncle's children "brothers" and "sisters." Parallel uncles often call their brother's children "daughters" and "sons." (e) Matrimonial privilege of a maternal uncle over his sister's daughter. (f) Matrimonial privilege of a brother's son over his father's sister. (g) The avuncular relationship between maternal uncle and his sister's son, which is indulgent in patrilineal society, authoritarian in matrilineal society.

primatology has anything to contribute to our understanding of matrimonial exchange, it should be about the origins of cross-cousin marriage and the set of principles of which it is an integral part, the brother–sister kinship complex. But as one may readily observe, "simple" does not necessarily mean "archaic." This brings me to a consideration of the chronological/historical dimension of Lévi-Strauss's work.

Reciprocal Exogamy as Archaic

Lévi-Strauss consistently insisted that he was limiting himself to a structural analysis of human social systems whose relative complexity was not to be interpreted in terms of a temporal sequence. His concern to avoid any historical interpretation of his work—a fortiori, any connotations of historical evolutionism—is manifest in each of the three levels at which he discussed relationships between kinship structures, namely between elementary and complex kinship structures, between restricted and generalized exchange systems, and between different types of marriage within the category of restricted exchange. On logical and structural grounds, elementary kinship structures could antedate complex ones. Although Lévi-Strauss acknowledged that "this logical priority [may] correspond to an historical privilege," he excluded that issue from his concerns: "It is for the cultural historian to inquire into this" (1969, 465). Likewise, in relation to the connection between restricted and generalized forms of matrimonial exchange, he asserted that "we should resist to the very end any historico-geographical interpretation which would make restricted or generalized exchange the discovery of some particular culture, or of some stage in human development" (463). Finally, with respect to relationships between different types of matrimonial arrangements within the category of restricted exchange, Lévi-Strauss concluded his lengthy discussion of cross-cousin marriage—the crux of his reasoning on reciprocal exogamy—with the following particularly explicit cautionary remarks: "We have been careful to eliminate all historical speculation, all research into origins, and all attempts to reconstruct a hypothetical order in which institutions succeeded one another. By according cross-cousin marriage the premier position in our demonstration we have postulated neither its former universality nor its relative anteriority with respect to other forms of marriage" (142–143).

Lévi-Strauss's aversion to historical interpretations is clear. It is also easily understandable, considering the highly conjectural nature of earlier attempts to reconstruct the evolution of human societies, prominent among them those of the nineteenth-century historical evolutionists. But notwithstanding these reservations, Lévi-Strauss himself appeared to conceive of reciprocal exogamy not only as the key distinctive attribute of human societies but as the founding—in both senses of the term, basic and archaic—principle of human society. This interpretation may be inferred logically from his writings, notably from his repeated use of the phrase "the transition from nature to culture" in association with both exogamy and the incest taboo. Considering that exogamy and the incest taboo were two aspects of the same thing, when Lévi-Strauss writes that the incest taboo is "the fundamental step because of which, by which, but above all in which, the transition from nature to culture is accomplished" (1969, 24) he is in fact making a temporal statement: he is establishing a synchronicity between exogamy and the birth of human society. Clearly, then, in Lévi-Strauss's scheme, exogamy represents both the core organizational principle of human society and its archaic manifestation.

Lévi-Strauss's unchronological and strictly structuralist position has been endorsed by many. Fortes, for example, noted that "Lévi-Strauss's claim that the incest taboo marked the transition from nature to culture is not to be understood in a temporal or chronological sense. It is," he continued, "a structural generalization drawing attention to what most anthropologists regard as the crucial distinction between man and other mammals, notably the non-human primates" (Fortes 1983, 15). But again, granting that the passage from prehuman society to human society is tantamount to a deep structural transformation, that transition necessarily took place at a point in time. I intend to establish that Lévi-Strauss's structural analysis, however explicitly unhistorical and nonevolutionist it was meant to be, is fundamentally compatible with a phylogenetically comparative analysis, in particular with the conclusion that the exogamy configuration marked the beginning of human society.

The Convergence beyond the Critiques

Over the last fifty years, Lévi-Strauss's alliance theory has endured the well-articulated critiques and strictures of many scholars, anthropologists, and sociologists as well as nonanthropologists, including a few primatologists. Given the sheer theoretical scope and complexity of alliance theory, it would be impossible to review adequately all these critiques here. Fortunately this is no major handicap, because the point I want to make is that the convergence between alliance theory and the primate data, as far as the distinctive character of human society is concerned, rests on two extremely basic principles of Lévi-Strauss's theory. And these principles do appear to have survived the critiques.

The first principle concerns the nature of human society's deep structure. It states that the uniqueness of human society, the "transition from nature to culture," lies in reciprocal exogamy, in marital alliances linking distinct kin groups. One might question the validity of the claim that marriage universally creates bonds between kin groups through relations of reciprocity based on the direct exchange of wives or the exchange of wives for material goods. One might also question the idea that marriage is the main force ensuring the cohesion of intermarrying social units, including descent groups. These claims were indeed hotly disputed in the 1960s, most notably by the British descent theorists who held that social cohesion between descent groups reflected not marriage-based alliances but the unifying effect of kinship bonds between members of intermarrying descent groups (for a discussion, see Holy 1996, 131–137). In effect, repeated intermarriages between descent groups generate a complex intertwining of kinship bonds between them, such that any individual ends up having several consanguineal kin living in social units other than its own, including children, grandchildren, nieces, and cousins. Accordingly, descent theorists held that supragroup kinship networks, not marital bonds, were the main force unifying otherwise distinct descent groups.

The opposition between alliance theorists and descent theorists, however sound from a social anthropology perspective, is nonetheless secondary from an evolutionary outlook. Regardless of the relative role of marital bonds and kinship bonds in the social cohesion of intermarrying social units, the fact is that both categories of intergroup ties, those between affines and those between kin, are derived from intermarriage. Marital bonds create and reflect mutual obligations between affines. At the same time, marital unions produce children who have kin residing in social units other than their own, for example sororal nephews and their maternal uncles, or grandchildren and their maternal grandparents in patrilocal groups. Marriage thus consistently generates kinship bridges between groups in addition to affinity-based alliances. As we shall see, in this dual consequence of marital unions lies an important portion of the convergence between alliance theory and the primate data.

The second point of connection between alliance theory and the primate data is the idea that consanguineal kinship provides strong positive constraints on mate assignment, in other words that genealogical relations structure exogamous unions, a principle epitomized by cross-cousin marriage. Lévi-Strauss argued forcefully on this point and was particularly strongly criticized for it. It is indeed noteworthy that in many cases marriage does not take place between intermarrying kin groups, hence between relatives, and that even when it does, strict genealogical relations between intermarrying groups do not consistently determine the marriageability of individuals. For example, even in societies that prescribe cross-cousin marriage, spouses are often not real cross-cousins but classificatory ones, persons thus classified who are actually much more distantly related. In short, consanguineal kinship does not consistently govern marital arrangements, even in societies characterized by what Lévi-Strauss called elementary kinship structures. Interestingly, however, the present phylogenetically comparative analysis basically vindicates Lévi-Strauss's argument. The primate data suggest that genealogical relations did constrain the formation of marital unions in early hominids. Actual ethnographic exceptions would constitute more recent cultural elaborations all derived from the same stem pattern.

In sum, the convergence between alliance theory and the primate data centers on a limited number of aspects of Lévi-Strauss's theory, but they are the most fundamental ones: exogamy embodies the deep structure of human society, and in this context kinship acts as a strong constraint on marital unions. Several other aspects of Lévi-Strauss's theory have been the object of critiques. One of these concerns the relation between incest prohibitions and exogamy rules, which Lévi-Strauss and others, including Leslie White, considered as two sides of the same coin. Several social anthropologists have argued against this idea, noting that incest rules and exogamy rules are different and have their own distinct causes (for example, see Fox 1962, 1967, 1980). It is also the case that Lévi-Strauss's conception of the origin of incest prohibitions is plainly contradicted by the primate data, as already discussed. Notwithstanding this important discrepancy, the convergence between alliance theory and comparative primatology remains. Still another critique concerns Lévi-Strauss's conception of women as passive objects in the hands of men and his downplaying of the active role women play in the organization of marriage, not to mention the fact that men move between groups in a number of societies. Lévi-Strauss himself responded to that critique by noting that the important point is that marriage creates or perpetuates bonds between groups, whether it is the women who move between exogamous units or the men (1985: 60). But equally important, as we shall see, is that the primate data suggest it was women who initially moved between groups, rather than men, and hence that women were the *primeval* commodity of exchange between men.

Lévi-Strauss and the Primate Data

Before leaving Lévi-Strauss it is worth clarifying what he thought about the relevance of the primate evidence for understanding human society. In short, he basically ignored it. Initially he did so because the data were nonexistent, but later it was because he deemed the evidence irrelevant. To Lévi-Strauss reciprocal exogamy and its main components were cultural constructs that could only have appeared simultaneously: "As far as biology is concerned," he asserted, "hominization may have been a continuous process. To the sociologist, it could only happen all at once, as a synthesis of components which, considered in isolation, did not foretell the outcome of their union" (1958, 160; my translation). The conception of reciprocal exogamy as the outcome of a long evolutionary process rather than a cultural elaboration had to await the accumulation of primate data. In 1949, when Lévi-Strauss published *Les structures élémentaires de la parenté,* primatology was barely in its babbling phase. And when the revised edition was published in 1967, primate studies were well under way but just beginning to cross disciplinary barriers. In part for these reasons, Lévi-Strauss's conception of the social life of primates was deeply erroneous, even in the 1967 edition, which he did not revise on this point:

> The social life of monkeys does not lend itself to the formulation of any norm. Whether faced by male or female, the living or the dead, the young or the old, a relative or a stranger, the monkey's behavior is surprisingly changeable. Not only is the behaviour of a single subject inconsistent, but there is no regular pattern to be discerned in collective behaviour. . . . The relationships among members of a simian group . . . are left entirely to chance and accident and . . . the behaviour of an individual subject today teaches nothing about its congener's, nor guarantees anything about his own behaviour, tomorrow. (1969, 6–8)

Needless to say, such beliefs about the chaotic and unpredictable character of the social life of primates could only lead to a saltational conception of the transition from animal societies to human society. Reciprocal exogamy, or any other aspect of human social behavior, could hardly have any phylogenetic connection with the supposedly unpatterned behavior of other animals. But as ethologists and primatologists accumulated overwhelming evidence about the existence of regularities in the interactions of animals, they found that primate social relationships are highly structured, based on criteria such as sex, age, dominance status, generation, and degree of kinship. As we have seen, the pertinent question thus became: Do the interactional regularities of nonhuman primates have any bearing on our understanding of the origins of institutionalized rules in humans? In the preface to the 1967 edition of *Les Structures Élémentaires,* Lévi-Strauss briefly hinted at a conception of the relationship between nature and culture suggesting that his answer to that question would have been positive. The following quotation indeed departs significantly from his previous dichotomous conceptualization:

> To understand culture in its essence, we would have to trace it back to its source and run counter to its forward trend, to retie all the broken threads by seeking out their loose ends in other animals and even vegetable families. Ultimately we shall perhaps discover that the interrelationship between nature and culture does not favour culture to the extent of being hierarchically superimposed on nature and irreducible to it, rather it takes the form of a synthetic duplication of mechanisms already in existence but which the animal kingdom shows only in disjointed form and dispersed among its members. (1969, xxx)

Notwithstanding this promising statement, some fifteen years later Lévi-Strauss dismissed the relevance of recent findings on incest avoidance in primates for understanding the origins of incest prohibitions (1985, 53), on the ground that primate behavior is not rule-governed, as Fortes was arguing at about the same time. And another fifteen years later Lévi-Strauss still firmly held that nonhuman primates had little to contribute to our understanding of the origins of human society. Referring to the natural (biological) foundation of society, he stated:

> One often thinks that [this foundation] can be found in comparisons between animal behavior and human behavior, the former prefiguring the latter. This is illusory to a large extent. Although anthropologists have known for a long time that the relation between man and the chimpanzee is not a genealogical one, but one of cousinship, they have not drawn from this principle all the inferences they should have. Engaged in different specializations, man and the great apes are on the same horizontal line, at the same generation level. Knowledge about the great apes' present situation thus teaches us little about man's past, close or distant. (2000, 494; my translation)

By "genealogical relation" Lévi-Strauss presumably meant one of ancestorship. His argument that relationships of cousinship (relatedness through a common ancestor) between species are of no use for understanding the species' evolutionary history betrays his misunderstanding of the basic principles that underlie interspecific comparisons and phylogenetic reconstructions (the decomposition principle discussed in chapter 2). Evolutionary theory was born precisely from comparisons of species located on "the same horizontal line," and since then such analyses have continued to provide illuminating evidence about the evolutionary continuity between species and about "man's past, close or distant."

While Lévi-Strauss was dismissing somewhat peremptorily the significance of primate studies for understanding human society, other social anthropologists were pondering the points of articulation between the two realms, in particular kinship and exogamy, and I now turn to their contribution.

8

Human Society Out of the Evolutionary Vacuum

In the evolutionary process, whether it be social or biological, we almost always find the new growing out of, or based upon, the old.

Leslie White (1949, 315)

Even if we cannot deduce accurately the kinship systems of early man from those of the most primitive humans, we can do something better—we can distill the essence of kinship systems on the basis of comparative knowledge and find the elements of such systems that are logically, and hence in all probability chronologically, the "elementary forms of kinship."

Robin Fox (1975, 11)

In their attempts to characterize the society of the "primeval ape-man," the nineteenth-century evolutionists attempted to infer the past by doing what may be called "backward history." To Morgan, for example, "Systems of consanguinity . . . are found to remain substantially unchanged and in full vigor long after the marriage customs in which they originated have in part or wholly passed away" (Morgan [1877] 1974, 417). Thus one could deduce earlier marriage practices, for example consanguine group marriage, from present-day kinship terminologies, in this case the merging of collaterals. With the benefit of hindsight, the two assumptions underlying this reasoning—that marriage types determine kinship terminologies and that kinship terminologies "survive" more of less unchanged in new social settings—appear rather tenuous. In any case, even if they had proved to have some empirical validity, the evolutionists' retrograding method could not have gone far back in time.

The rise of the primate data by the middle of the last century afforded a whole new method for uncovering the roots of human society. It was now possible to reconstruct the past from knowledge of the evolutionary background from which that past had emerged; it was possible to do normal "forward history." A few social anthropologists were quick to avail themselves of the new body of evidence, notably Leslie White, Elman Service, and Robin Fox. In this chapter I explain how the ideas of these three researchers connect to one another and fit into a coherent whole, and how they paved the way for our present-day conceptions about the origin of human society. I also attempt to show that the history of ideas about the origin of human society is, to a large extent, that of the progressive decomposition of the exogamy configuration into more elementary elements that can be observed in living primates, even though this is not necessarily how the authors themselves conceived of their work.

White, writing in the 1950s, obviously was not the first to conceive of human society as emerging from the realm of primate societies. This was a view shared by most evolutionary biologists ever since Darwin's work. As a matter of fact, in 1891, Westermarck was already arguing "that we can no more stop within the limits of our own species, when trying to find the root of our psychical and social life, than we can understand the physical condition of the human race without taking into consideration that of the lower animals" (1891, 9). But Westermarck's proposition that exogamy had grown out of animal patterns of inbreeding avoidance, at a time when most sociologists and anthropologists believed that animals mated with their close kin, made him an outlier among his contemporaries—in retrospect, a prescient one. The reason I begin with White is that he was a *social* anthropologist interested in the origins of "immaterial" aspects of human social structure, such as kinship and marriage patterns, and that he was active when the primate data were just becoming available. In other words, he was well equipped, conceptually speaking, and at the right time.

Let us first briefly reckon what was known about the social life of monkeys and apes before 1960. Perhaps the most basic principle established by primatologists was that permanent societies and organized social life predated the evolution of the human lineage, that the human species had not invented society, contrary to Lévi-Strauss's belief at that time. Researchers had shown that many primate species form stable (year-long) bisexual groups and that such groups are not mere aggregations but real societies whose members interact differentially and predictably according to a number of built-in factors such as age and sex. They also knew that depending on the society, males and females could mate either promiscuously—each sex having several partners—or in the context of long-term associations between one male and one or more females. In other words, they knew that some of our close relatives form enduring breeding bonds (monogamous or polygynous) and live in "families." And most important for the topic of exogamy, primatologists had found that relations between local groups were generally unfriendly and that primates had no level of social organization above that of the local group. This had been stated particularly explicitly by Clarence Carpenter in the early 1940s:

> The characteristics of primate societies which cause them to react antagonistically to other groups of the same species and to defend a limited range prevent the *federation* or *combining* of organized groups. Therefore, not only does the lack of ability to use abstract symbols prevent super-group organization, but also the phenomena of territorialism and inter-group antagonism operate in the same direction. (1942, 38)

On the other hand, before the 1960s, primatologists and anthropologists were unaware of some fundamental aspects of primate social structures that are central to the exogamy configuration. They did not know that primates commonly disperse out of their natal group, form kin groups, are able to recognize a number of kin categories, and avoid incest with close kin. These aspects of human society anthropologists had to ascribe to the symbolic capacity and conceive of as cultural constructs. Remarkably, the issue of exogamy was a central theme running through all comparative analyses performed by social anthropologists.

Leslie White and the Primate Origins of Exogamy

White was writing about the origin of human society at about the same time Lévi-Strauss was arguing that matrimonial alliances lay at the heart of human social structure and marked "the transition from nature to culture." To White, as well, exogamy was a central aspect of the distinctiveness of human society. With the advent of incest prohibitions and exogamy, "social evolution as a *human* affair was launched upon its career," he asserted. "It would be difficult to exaggerate the significance of this step. Unless some way had been found to establish strong and enduring social ties between families, social evolution could have gone no further on the human level than among the anthropoids" (1949, 316). This is precisely what Lévi-Strauss was asserting at the time. But while Lévi-Strauss never phrased his argument in terms of phylogenetic history, White's analysis of the distinctiveness of human society was set entirely within an evolutionary framework. This is most explicit in *The Evolution of Culture,* published in 1959, in which he articulated the principles underlying a whole research program on the evolutionary origins of human society. His formulation is strikingly modern, even though some of his points are questionable in light of today's knowledge.

> We can construct a theory of the origins of human society—or more specifically, of the transformation of anthropoid society, the society of man's immediate prehuman ancestors, into human social organization—in the following way. (1) We know what the social organization of subhuman primates is like . . . (2) We are justified in believing that man's immediate prehuman ancestors conformed to this general primate pattern because of the fundamental similarity . . . between man and other primates. Therefore we have a valid conception of what the social organization of man's prehuman ancestors was like. (3) We know what distinguishes man from other primates from the standpoint of behavior: the use of symbols . . . (4) Therefore the transition from anthropoid society to human society . . . was effected by the operation of symbols. (1959, 72–73)

Of particular interest here is White's unequivocal position about the phylogenetic continuity between man and other primates in the *social sphere,* which contrasts dramatically with that of Lévi-Strauss. Equally notable is his belief that the society of "man's immediate prehuman ancestors" could be characterized using primate data. But notwithstanding the optimism conveyed by his first point, White and his contemporaries were still unaware of several basic aspects of primate social structures. White conceived of early human society as composed of a number of families living together in a group that defended its territory against other groups. Such communities, he surmised, were endogamous and the families incestuous; kinship relations were barely recognized, families were not tied together through social relationships, and links of affinity (bonds between in-laws) were nonexistent. This was, according to White, the social substrate upon which primitive humans had culturally constructed kinship networks, incest prohibitions, marital alliances, and affinal ties. These components of the exogamy configuration could enter the picture only after the symbolic capacity had evolved. Language, White argued, was the key for understanding human society, a theme that runs through every aspect of his reasoning.

For example, with respect to kinship, White acknowledged that nonhuman primates could discriminate their kin *within* their immediate family. Whether you are an anthropoid or a human being, he reasoned, "you see and you have dealings with your mate, your immature offspring, and, while you are still a child, with your parents." But, he argued, nonhuman primates are unable to recognize their *extrafamilial* relatives: "One cannot recognize his parents' parents and their siblings, for the simple reason that his parents have severed the *specific* social relationship which formerly existed between them and their parents and their parents' siblings. Consequently, on the anthropoid level, one cannot recognize and distinguish his grandparents, uncles, aunts, cousins, and so on" (1959, 82). The very principle underlying White's reasoning here—that kin recognition between certain relatives depends on lifetime bonds between kin—was to be amply confirmed by later primate studies: the recognition of secondary and tertiary relatives requires a mediator (the mother) who has not severed her relationships with her parents. Thus White was right on this point, but he mistakenly assumed that nonhuman primates did not meet this condition. Accordingly, the inferences he drew were somewhat erroneous. For extrafamilial kin to be recognized, he argued, language and the symbolic capacity were needed: "Specific genealogical relationships could not be defined, recognized, or expressed in the absence of articulate speech. How, for example, could 'cousin' be distinguished from 'uncle' or 'sibling' without language?" (81). Stated otherwise, to recognize their kin, humans must be able to label them. But as we have seen, kin recognition in nonhuman primates belies this statement. Familiarity differentials between kin and associative processes underly kin discrimination.

Before kin recognition came to encompass extrafamilial relatives, White continued, "families were in association with one another like marbles in a sack." Individuals had no reliable means to establish ties with members of other families. But with the extension of kin recognition brought about by language, families could develop cooperative bonds based on consanguinity. The social process envisioned by White here is extremely simple, and again the underlying principle bears striking parallels with what we know today about the development of kinship networks from ego's perspective among nonhuman primates. White reasoned that the primordial link between two families was ego, who stood as a child in one family and a parent in another:

> The first kind of union between families to become significant would undoubtedly be that between the family group in which one is a child and the family group in which he is a parent . . . Thus the social relations with my parents, established in infancy, would tend to persist after I had grown up, married and had children; and in turn my relations with my children would tend to persist beyond the time of their marriage and be extended to their children. (1959, 85)

In other words, thanks to language, ego could now act as a living link between his relatives of the ascending and descending generations; grandparents and grandchildren could recognize each other. Similarly, White continued, relationships between siblings would persist after they had grown up and reproduced, and these relations would extend to the children of these siblings (nieces, nephews, and cousins). In this manner extensive social networks embracing many families and even whole social groups could develop, which mapped onto lineal and collateral kinship relations.

Having dealt with the role of kinship ties in binding members of different families, White then turned to the role of marriage and affinity in relation to the same purpose. Because nonhuman primates were assumably endogamous and incestuous, White reasoned that these obstacles had to be overcome first. "The prehuman primate family was . . . marked by strong endogamous tendencies: attraction between father and daughter, between mother and son, and between brother and sister . . . The earliest of human families must have inherited many of these features from their anthropoid ancestors" (1959, 88). Therefore, if different families were to ally through marriage, they first had to subdue these centripetal forces; the institution of incest prohibitions was the solution. On this point White turned entirely and exclusively to Tylor's explanation about the origin of exogamy ("marrying-out or being killed-out"). He also held, like Lévi-Strauss, that incest taboos and exogamy were two sides of the same coin, but remarkably, he made no reference to Lévi-Strauss's lengthy argument on the subject, which had been published ten years earlier.

It is noteworthy that White conceived of exogamy as having originated *within* local groups: hominids initially would have been *family exogamous,* not local-group exogamous. Originally, White reasoned, matrimonial alliances purported to bind families belonging to the same local group; they were a means to favor within-group cooperation and solidarity. Only later would exogamy have extended beyond the local group, eventually linking distinct local groups within some higher-order social entity (clans or tribes). As we shall see later on, the primate data rather suggest that hominids were initially *local-group exogamous* and that primeval exogamy was a between-group phenomenon.

Finally, White's belief in the phylogenetic rooting of human society was particularly explicit in his discussion of marriage. He recognized the manifold and profound differences between human marriage and primate pair-bonding, emphasizing the uniquely human dimensions of marriage: institutional, legal, cooperative, economic, and political. But to him these real differences were no obstacle to the idea of a phylogenetic connection between primate pair-bonds and human marriage: "Marriage," he stated, "is the humanization, the institutionalization, the sociocultural expression of the relatively durable union between the sexes in subhuman primate society" (1959, 94). Clearly, then, to White, pair-bonding was part of humankind's primate legacy.

In sum, although White was unaware of the extent to which several fundamental aspects of exogamy were actually grounded in the behavior of nonhuman primates (especially outbreeding, kinship networks, and incest avoidance), he clearly perceived the phylogenetic origin of group living, territoriality, intrafamilial kinship, marriage, and exogamy. He also demonstrated a particularly insightful appraisal of some of the social processes involved in the transformation of a primatelike society into a humanlike society, notably in relation to the increasing complexity of kinship-based and affinity-based social networks.

Elman Service and the Primitive Exogamous Band

In his book *Primitive Social Organization: An Evolutionary Perspective,* published in 1962, Service acknowledged the "earliest and most profound . . . influences" of his teachers Leslie White and Julian Steward (viii). Like White, Service set his discussion of the origin of human society within an evolutionary framework, but he attributed to exogamy a more central role in defining the distinctive character of human society. After stressing the importance of dominance relations in regulating social relationships within groups, Service emphasized that although social integration prevails within primate groups, the situation is different *between* groups: "No consistent peace can be established [between groups] via physical dominance when there is no consistent contact." Consequently, Service reasoned, "relations among [prehuman] hordes were probably much more usually those of conflict and independence." If such a situation characterized early humans and was "to be altered by later human cultural innovations," he continued, "the most important attempts at rectification must have been made with respect to the antagonistic relation of groups to each other" (41). That is, one fundamental impact of culture on social relations would have been the *pacification* of intergroup relations.

This could not have been achieved prior to the evolution of the symbolic capacity, Service argued, because although cooperation and reciprocity exist in nonhuman primates, reciprocity is short-term, not based on agreements about future returns. Moreover, "once symbolic thought and communication become possible . . . sanctions, rules, proscriptions, and values can be created which inhibit conflict and strengthen solidarity" (1962, 41). Therefore, only when prehumans could avail themselves of this powerful cognitive tool were they in a position to engage in what Service considered the most basic form of reciprocity: female exchange. "It seems logical to assume that . . . the most significant of the early rules of reciprocity was related to the acquiescence of two social groups in the reciprocal giving of females" (43). Lévi-Strauss's influence is obvious here, and on the issue of exogamy, Service, contrary to White, referred both to Tylor and to Lévi-Strauss, arguing that humans had instituted exogamous marriage both to pacify intergroup relations through the exchange of daughters and to build alliances between groups.

Still referring to Lévi-Strauss, Service addressed the issue of why it should be women rather than men who are exchanged. If women are the "gifts," he asked, paraphrasing Lévi-Strauss, is this because in a hunting-gathering society women are the most precious possession a family can exchange? His answer was that Lévi-Strauss's proposition was "overlaid by more expedient reasons," which "have to do with the tendency toward virilocal marital residence" (1962, 47). Simply stated, it was because men were already "localized" that intergroup alliances were achieved through the exchange of women. And the reasons men were localized had to do with their hunting-gathering way of life and the specific constraints that this exploitative pattern imposed on relationships between males. Accordingly, Service concluded that the "early human structure in its simplest outlines" was probably the *exogamous patrilocal band of hunter-gatherers*. At this point it is Steward's influence that is most notable.

To sum up, Service adopted Lévi-Strauss's view that reciprocal exogamy was the distinctive trademark of human society, but he set exogamy within a phylogenetic framework, stating that it had emerged from primatelike behavioral patterns. Like White, he conceived of outbreeding, kin groups, kinship networks, and incest avoidance as cultural constructs. But contrary to White, Service saw exogamy as a means for pacifying intergroup relations, hence as originating in the realm of between-group relationships, a view that is supported by the primate evidence. Interestingly, Service's views about the primitive character of virilocality and female exchange were to feed to a considerable extent the idea of a phylogenetic continuity between the residence patterns of hominids and African apes, although Service himself did not argue in this sense.

By the 1960s the idea that reciprocal exogamy was a major divide between human and nonhuman societies had gained many adherents. In an article aimed at a large audience and published in 1960, Marshall Sahlins wrote:

> The primate horde is practically a closed social group . . . The typical relation between adjacent hordes is that of enmity . . . Territorial relations between neighboring human hunting-and-gathering bands . . . offer an instructive contrast. The band territory is never exclusive. Individuals and families may shift from group to group . . . These tendencies are powerfully reinforced by kinship and the cultural regulation of sex and marriage. Among all modern survivors of the Stone Age, marriage with close relatives is forbidden, while marriage outside the band is at least preferred and sometimes morally prescribed. The kin ties thereby created become social pathways of mutual aid and solidarity connecting band to band. (82)

As another example, in a textbook written in 1966 and otherwise devoted essentially to the anatomical aspects of human evolution, Bernard Campbell contrasted the absence of a supragroup level of social organization in nonhuman primates to its presence in humans and identified exogamy as the phenomenon bridging the gap:

> Primate intergroup behavior is unfriendly and antagonistic; there is, so far as we know at present, no supergroup social structure. Human intergroup behavior can be considerably more friendly, and a social structure generally exists that relates social groups. This broad structure exists as a result of exogamy and the political relationships developed between descent groups that exchange marriage partners. It seems clear that human social life is based on kinship and intermarriage . . . for marriage is historically a contract not only between individuals but between different descent groups. (283)

Clearly, then, the main principle was articulated, but the detailed analysis remained to be done. More data on primate behavior were needed, but, no less important, these data ideally would be interpreted by someone highly familiar with the structure of human society, in particular with human kinship structures.

Robin Fox and the Initial Deconstruction of Exogamy

The first systematic comparative analysis leading to the initial decomposition of the exogamy configuration was performed some twelve years after White's and Service's books were published. Armed with a solid background in the anthropology of kinship (Fox 1967), Robin Fox tackled the literature on kinship and behavior in nonhuman primates in a series of articles that appeared in the 1970s and in a book, *The Red Lamp of Incest,* published in 1980. His insights, thirty years later, appear remarkable. Fox began with what he called "a bold assertion" derived from Lévi-Strauss and his own work:

> The two elementary functions of human kinship are what I will call, borrowing somewhat recklessly from the jargon of social anthropology, descent and alliance . . . Kin are "grouped" and the grouping of people in this way, the deciding of who belongs with whom, etc., I am calling descent for convenience. Alliance refers to the allocation of mates. In

> all human kinship systems people have assigned mates, and the system determines who can mate with whom—crudely who gets whom, who is allied with whom. (1975, 11)

Thus, what Fox called descent "for convenience" amounted to a kin group, and what he called alliance was equivalent to marriage. Both elements, Fox argued, have analogues in nonhuman primate societies. The first is manifest in the form of enduring kin groups in primate societies composed of several males and females. Here lies the major difference between Fox and his predecessors White and Service. Writing in the 1970s, Fox knew about the existence of kin groups, kin recognition, and incest avoidance in multimale-multifemale primate groups. He knew that in these species females breed, give birth, and die in the group in which they are born, whereas males circulate between groups. He also knew about matrilineal dominance hierarchies, such that a female's position in the dominance order is determined by her mother's rank, and females belonging to the same matriline occupy close ranks in the hierarchy.

As emphasized by Fox, another important aspect of these societies is their mating system. Sexual activity occurs in the context of relatively brief associations. It is promiscuous in that any male mates with several females, and any female with different males; the two sexes do not form enduring breeding bonds. Accordingly, Fox argued, the second "elementary function of human kinship," the alliance component, was found in a different type of society. It was manifest in the form of long-term bonds between one male and one female (as in gibbons) or between one reproductive male and a number of females forming a polygynous unit (as in mountain gorillas). On the basis of data available at that time, Fox believed that females within polygynous units were not related to each other in most situations and therefore did not constitute a kin group. Hence, comparing the two types of societies, the multimale-multifemale group and the polygynous unit, he wrote:

> Each system has something that the other does not. The multi-male system has the enduring kinship groups, while the one-male system has the permanent polygamous families. If you like, one has kinship, but no marriage; the other has marriage but no kinship. This is just an analogy, but it may be an important one . . . And what intrigues me is

> this: Both these things—enduring ties based on relatedness and enduring mating relationships—exist in primates, but *not together in the same system*. Only in the human primate have they been put together. (1980, 104)

Or, stated differently, "the uniqueness of the human system, therefore, lies not in inventing something new, but in the combination of these two elements." (1975, 11–12).

The next step in the reasoning followed logically from the foregoing: it is only by combining and integrating descent and marriage (kin groups and pair-bonds) in the same social system that humans could invent exogamy, which is precisely "the use of kinship to define the boundaries of marriage" (1980, 104). How, then, did exogamy evolve, or, phrased in terms of Fox's two components, what led hominids to combine kinship and marriage in the same system? At this point Fox switched to a different level of explanation. Leaving the strictly comparative analysis, he proposed a scenario about the evolution of pair-bonds. The driving force behind the whole process, the "root of a truly human society" would lie in the transition to hunting and the ensuing division of labor between males and females. With hunting performed by the males and gathering by the females, the trading of these resources became possible and essential: "Men no longer needed women for sex only, and women no longer needed men for protection only, but each had a vested interest in the products of each other's labor" (144), the women providing vegetable food for the men, and the men providing meat for the women and their children.

Whatever the exact driving force behind the evolution of pair-bonds, the important point is that when this mating system entered the picture, new societies emerged that combined kin groups and pair-bonding. At this stage in human evolution, the last element needed to produce the exogamy configuration in Fox's scheme was the symbolic capacity, more specifically the capacity to label relationships and make normative rules, notably in relation to kinship. "When we first came to classify things," he wrote, "kin were high on the list. And if we can state with confidence one human universal, it is that there is not a single known society that does not classify kin" (1980, 149). Fox thus concurred with Lévi-Strauss, White, and Service about the importance of symbolic labeling and rule-making in defining human society, but with a most important nuance. He did not believe that the symbolic capacity had created kinship networks. He argued instead that the symbolic capacity was superimposed upon already existing primatelike kinship networks and that the two phenomena together had generated humankind's universal tendency to classify kin.

Now, with the three essential ingredients assembled in the same social system—kin groups, pair-bonding, and the ability to classify kin—exogamy could emerge, Fox thought:

> One of the effects of this [tendency to classify kin] was to divide kin into "marriageable" and "unmarriageable." We cannot have human kinship systems without this "marriageability component." This is the essence of exogamy . . . What our ancestors did, at some point, was to say, "Those people we classify as 'ours' on some kinship basis we may not have as assigned mates; those we classify as 'other' we may." Or less negatively, . . . "We will give you 'our' women—defined on some criterion of kinship—and you will give us 'your' women." (1980, 150)

That is to say, as soon as societies that had both kin groups and pair-bonding emerged, and inasmuch as hominids could label and classify their kin, they had, right from the outset, prescribed *rules of exogamy* as a means to exchange females between kin groups. But as one might justly remark, there is quite a gap—structural and cognitive—between a primatelike social system that merely integrates kin groups and pair-bonds and a humanlike system featuring, in addition, kin-group exogamy, female exchange, and marital alliances. Why could a man not marry *within* his own kin group and trade his kinswomen with men belonging to his own kin group? Why were a man's kinswomen nonmarriageable? In short, why systematic outmarriage—exogamy?

One might expect that Fox would have resorted here to the argument that men were exogamous because sex was prohibited within their kin group, as held by Tylor, Lévi-Strauss, White, and others. But Fox has consistently been a strong advocate of the idea that rules against incest and rules prescribing exogamy are two different things (1962, 1967, 1980). Along with others, he emphasized that "incest refers to sex, exogamy to marriage," (1980, 4) and that in theory societies could prohibit marriages within the group but still allow sex within it. The unmarriageability of members of one's kin group, in Fox's view, had little to do with incest prohibitions. It had to do instead with the advantages provided by female exchange. "As Lévi-Strauss saw so correctly . . . ," he wrote, "the impulse to exogamy was a positive, not a negative impulse. We did not just, as Freud pictured it, flee from tabooed women and so bump into others. We deliberately defined some females as 'our kinswomen' and made an exchange agreement with [other males] for their 'kinswomen'" (150). Thus if men married outside their kin group, the purpose was to build marital alliances with other kin groups. Clearly then, Fox adhered to Lévi-Strauss's views, and for this reason he needed to integrate female exchange and matrimonial alliances in his evolutionary scheme.

There are some important missing links in the proposed evolutionary sequence, and Fox's reasoning appears somewhat contrived at this point. As we shall see in more detail in the next chapter, the crucial consideration lacking in his reasoning, the fundamental missing link, is a discussion of the integration of primate *dispersal patterns* into the exogamy configuration. Fox knew that individual primates move between groups, he even knew that in chimpanzees it is the females who move, but he was apparently unaware of the systematic character of dispersal and its correlate, outbreeding. The following quotation is revealing of Fox's views on this point. He was referring here to a fission that took place in a large group of Japanese macaques comprising sixteen maternal lineages ranked from one to sixteen:

> When the group split, lineages one through seven remained, while eight through sixteen left to form another group. Let us call them A and B. The males began moving around, and after two years, all the males of A had moved to B and all those of B to A. This may or may not be unusual . . . but it is remarkable. For what we have here is only a step away from a human kinship system consisting of two matrilineal moieties with ranked matrilineages and a rule of moiety exogamy. The step is, of course, precisely the exogamy. *Before they moved, the males of A mated with the females of A. We would have to institute a rule forbidding this to get exogamy.* (1980, 104; my emphasis)

True, individuals may mate within their group before leaving, and Fox is right about humans inventing "a rule forbidding this." But within-group mating is limited in nonhuman primates; dispersal and outbreeding are the rule, which brings us one major step closer to human exogamy. Fox failed to see the evolutionary relationship between intergroup transfer in primates and marrying out of one's group in humans.

One last point is worth noting. Fox recognized that the integration of kin groups and pair-bonding in the same social system brought about another major component of exogamy, affinal kinship, a topic that he treated explicitly but rather briefly. The recognition of in-laws had become possible if only because individuals endowed with the ability to classify their kin were now in a position to recognize the relatives of their long-term mates ("spouses"). The new economic interdependence between men and women, Fox surmised, had allowed men to control their kinswomen. Males could now exchange their female kin to obtain sons-in-law, brothers-in-law, or even nephews-in-law. The human primate's contribution to kinship systems, Fox concluded, was "not the invention of kinship, but the invention of in-laws, affines, 'relatives by marriage'" (1980, 147), an understatement about an important distinction that was to stimulate further research.

Fox's comparative studies constituted a major leap toward an understanding of the origin of human society. First, more than any of his predecessors, Fox positioned exogamy at the very center of his analyses, identifying it as the cornerstone of the distinctiveness of human society. On that point Fox and Lévi-Strauss were fundamentally convergent, concurring that exogamy was the key for understanding the origin of human society. This meant, implicitly, that Fox's comparative (interspecific) analysis, framed in evolutionary terms, and Lévi-Strauss's comparative but strictly structural analysis of human societies were compatible. Second, Fox's comparative exercise led him naturally, that is, logically, to break down exogamy into more basic components and to assert that the originality of human society lay not so much in the building blocks themselves as in their combination. Similar deconstructions of human phenomena were going on in other areas, notably language. But in the 1970s Fox, a social anthropologist, was almost alone and well ahead of his time in his efforts to understand and extract the significance of the primate data for the origin of human society. This led him to conclude that humans had invented, not kinship and incest avoidance, as White and others believed, but other components of exogamy, like outbreeding, affinal kinship, and female exchange. More than ten years were to elapse before Fox's ideas were taken up in a comprehensive manner and the decomposition of exogamy significantly furthered.

9

The Building Blocks of Exogamy

> It would be of intense interest to find any species or population of non-human primates combining the two principles [kinship and marriage]. This would mean that even more was attributable to "nature" than has been envisaged here. The question would then be: on what basis was the combination made? And my own feeling is that it would not be on the basis of exogamy as we understand it in human kinship systems.
>
> *Robin Fox (1975, 29)*

Fox's hope of finding such a species was eventually fulfilled, and at the same time his "feeling" corroborated: other primate species have been found to combine kinship and enduring breeding bonds, but for all that, exogamy and its correlate, intergroup alliances, are not features of those species' social organization, a fact that in itself has interesting implications. In a seminal paper aptly titled "The human community as a primate society," published in 1991, Lars Rodseth, Richard Wrangham, Alisa Harrigan, and Barbara Smuts took up the comparative analysis of human and nonhuman primate societies where Fox had left off, using the primate data, which had grown exponentially over the preceding fifteen years. They further trimmed down the uniqueness of human exogamy and identified its distinctive traits more acutely.

Pinpointing the Distinctiveness of Exogamy

Rodseth and colleagues pointed out that the reason outbreeding had been identified with the onset of truly human society by several earlier theorists, including Tylor and Lévi-Strauss, was that all these scholars assumed, based on the data available at the time, that animals in general, hence including our ancestors, bred within their local groups: that they were both endogamous and incestuous. Ethnographic studies having revealed that human beings in all known cultures consistently marry out of their local group, outbreeding appeared as the fundamental divide between animal and human societies. But that view was by now untenable. Ethological and primatological studies had established that in all sexually reproducing species the members of at least one sex disperse and breed outside their natal group. The outbreeding component of exogamy thus ceased to be uniquely human, as noted by Rodseth and his colleagues: "The problem of exogamy, as Tylor, White and Lévi-Strauss saw it, was dislodging humans from their natal groups and that 'policy of isolation,' endogamy. But endogamy is an extremely unlikely candidate for the mating system of early hominids, given that dispersal is apparently universal in nonhuman primates" (1991, 234).

Although Fox knew that monkeys and apes were not incestuous or wholly endogamous, he nonetheless thought, as we saw, that systematic outbreeding was exclusive to humans. Even more to the point here, he emphasized that no primate species other than humans combined kin groups and enduring breeding bonds; hence, no primate was in a position to practice a form of kin-group exogamy wherein individuals would systematically establish breeding bonds outside their kin group. Contradicting this view, Rodseth and his colleagues pointed out that some primate species did combine kinship groups and exclusive breeding relationships and thus had opportunities to practice a form of kin-group exogamy. They focused in particular on the hamadryas baboon *(Papio hamadryas)*, one of two rather well-known species in which several breeding units are part of the same cohesive group, which is also a kin group. Each breeding unit is composed of a single male and a number of females and their dependent offspring. Several such units are part of a group called a clan. Although relatively little data were, and still are, available on the genealogical structure of hamadryas groups (Colmenares 2004), the adult males of the same clan appear to be related, and most of the females appear to breed outside their natal clan. Hence the hamadryas baboon combines male kin groups, outbreeding, and exclusive breeding bonds. It exemplifies a behavioral form of kin-group "exogamy" whereby a female born in a given clan transfers to another one in which she establishes a long-lasting, exclusive breeding bond with a male.

Species like the hamadryas baboon thus attested that human beings had invented neither kin-group outbreeding (breeding outside one's kin group) nor kin-group "exogamy" (pair-bonding outside one's kin group). This still left out one dimension of exogamy that is prevalent in humans but absent in nonhuman primates. This dimension is a dual one: the recognition of in-laws and its correlate, the formation of alliances between in-laws belonging to different local groups. In many simple human societies, marriages establish or reinforce relationships between the spouses' respective families or kin groups. This is possible because both the husband and his wife, regardless of whether they stay put or move into another group upon marriage, maintain long-term relationships with their respective kin. In a patrilocal society, for example, a wife lives with her husband's relatives (her in-laws); similarly, a husband, even if he does not live with his wife's relatives, meets with them and hears about them on a more or less regular basis, depending on the society. In human societies, therefore, exogamy translates into the bilateral recognition of affines.

The situation is very different in primate societies, as noted by Rodseth and colleagues. After transferring into a new group, individual primates are cut off from the relatives they left in their natal group. Dispersal severs kinship bonds. Thus a female who has transferred into a new group and established a breeding bond with a particular male may be in a position to recognize her mate's relatives (her "in-laws"). But the reciprocal is not true: the male cannot recognize his mate's relatives because the female does not come into contact with her kin after she has transferred into her new group. And this is true of all dispersing individuals, male or female, in all known primate species. "Affinity per se, then, is not uniquely human," concluded Rodseth and colleagues, but "affinity *between* groups, in contrast, is evidently unknown in nonhuman primates" (1991, 236). Put differently, the recognition of affines in nonhuman primates is unilateral at best; in humans it is consistently bilateral. Fox had concluded that humans had invented not kinship but affinity. Rodseth and his colleagues were qualifying that statement by noting that humans had invented not affinity as a whole but bilateral affinity.

It is worth reemphasizing that if affinal kinship is bilateral in humans, it is because the human primate is the only one in which the members of the dispersing sex maintain bonds with the relatives they have left behind after leaving their natal group. Stated from a different angle, "humans are the only primates that maintain lifelong relationships with dispersing offspring," which means that "both sexes therefore remain embedded in networks of consanguineal kin" (1991, 221). The adaptation that made this possible, Rodseth and colleagues continued, is the ability to maintain social relationships in the absence of actual physical proximity—the "capacity to sustain relationships in absentia." That capacity, they argued, would ultimately depend on language:

> Intergroup relationships similarly depend on the capacity for symbolic communication . . . In particular, symbolic communication is characterized by *displacement*—the ability to refer to things in other places and times (Hockett 1963 . . .). While displacement is widely recognized as one of the "design features" of human language, its social concomitant—the uncoupling of relationships from spatial proximity—has received little attention. Yet with this uncoupling, it may be argued, social evolution as a *human* affair was launched upon its career. (236)

Rodseth and colleagues were now in a position to pinpoint the unique character of human exogamy. Nonhuman primates may practice kin-group exogamy in the sense that they may establish enduring breeding bonds systematically with members of different kin groups, but such bonds do not give rise to *intergroup alliances* because the members of the dispersing sex lose contact with their natal kin, and therefore in-laws cannot engage in reciprocal bonding. "And *this* pattern," they insisted, "—the forging of alliances through systematic exchange of mates—seems indeed uniquely human" (236). In short, kin-group exogamy is not uniquely human, but *reciprocal exogamy* is.

Like Lévi-Strauss, White, Service, and Fox before them, Rodseth and his colleagues concurred in attributing to language a fundamental role in the elaboration of human exogamy. But these authors differed considerably on the exact role of language. To Lévi-Strauss human society as a whole was a cultural creation made possible by language, whereas to White and Service primeval human society was, basically, a primate-like society upon which the symbolic capacity had been superimposed, generating kinship networks, incest prohibitions, and female exchange in the process. Fox went much further in asserting that kinship networks and incest avoidance both predated the symbolic capacity. This still left out kin-group exogamy and female exchange as cultural products. Rodseth and colleagues reduced even further the creative part played by language, limiting it to female exchange and its correlates, bilateral affinity and intergroup alliances.

Reconstructing Human Society: The Task Ahead

This historical sketch shows that the phylogenetic decomposition of human society into more elementary building blocks is a natural outcome of comparisons between human and nonhuman primates. Although prior comparisons were not planned with this objective in mind, they have served to identify several important building blocks of the exogamy configuration, including kinship, marriage, outbreeding, affinity, and exchange. The list of components, however, is still incomplete. Some elements are all-embracing and too general. Kinship is a case in point; it could profitably be further broken down. And other important aspects of the exogamy configuration have not yet been discussed. The next question is, thus, Where are we today, and how exactly does the exogamy configuration break down? The danger here would be to define a vast number of traits more or less arbitrarily. This would be particularly easy because the differences between human society and other primate societies are innumerable. One solution, therefore, is to look for the smallest number of components whose combination generates the complete system. To achieve this, I used the following criterion: every building block of the exogamy configuration must be an autonomous entity, phylogenetically speaking. That is, it has to be a pattern that is structurally possible and duly observed in some species but not possible or present in other species. According to this criterion, a component of the exogamy configuration is a functional subsystem within the whole system, one that has its own set of determinants and its own set of properties. This eliminates a large number of aspects of lesser significance.

My own comparative analysis leads me to identify twelve such functionally distinct elements that are obligatory aspects of reciprocal exogamy—distinct elements but not independent ones. On the contrary, the twelve components are intimately interwoven, some of them even emerging out of the combination of others. For example, agnatic kinship (the recognition of kinship through males) is structurally connected to other components of the exogamy configuration, and its emergence was dependent on those components. But agnatic kinship is nonetheless a distinct component of the exogamy configuration: structurally it is either "feasible" or not, depending on the social system, and it possesses its own determinants and properties. Moreover, its emergence in the course of human evolution afforded a whole new range of opportunities for social evolution.

Of the twelve building blocks, a number are observed in nonhuman primate societies in forms that are more or less incomplete or adumbrated compared to their human analogues. Others are uniquely human, but even in this situation, some aspects or properties of the human form are found in the behavior of other primates. My aim at this stage is to provide an overall picture of the exogamy configuration through a brief description of each component in order to define the task ahead, that of explaining how the twelve blocks ultimately ended up in the same species. This is the object of Part III. Roughly speaking, I begin with the most primitive components and end with the most recent ones.

MULTIMALE-MULTIFEMALE GROUP COMPOSITION. The most basic building block of the exogamy configuration is the modal age-sex composition of human societies. Viewed from the perspective of the whole primate order, human societies are stable bisexual groups. This type of group is common in nonhuman primates, but it represents only one of several other equally frequent compositions. Most significantly, it characterizes our two closest relatives.

KIN-GROUP OUTBREEDING. A basic behavioral pattern in nonhuman primates is that a fraction of the individuals born in a given group stay and breed within their natal group, thereby producing a kin group, and that the other fraction leaves and breeds elsewhere, bringing about outbreeding. Nonhuman primates thus commonly practice kin-group outbreeding, and crucially, it is often sex-biased.

UTERINE KINSHIP. This building block is one of the most primitive aspects of the exogamy configuration, phylogenetically speaking. The importance of uterine kinship in the social life of primates stems from some basic biological facts: mothers bear offspring, mother–offspring recognition is a built-in aspect of maternal care, and female kin groups are widespread in primates. We shall see that some of the most basic processes involved in uterine kinship are relevant for understanding more recent aspects of the exogamy configuration, including agnatic kinship and affinal kinship.

INCEST AVOIDANCE. I refer here to the avoidance of sexual activity between kin that coreside in the same local group and are thereby in a position to commit incest, not to the avoidance of incest that results from dispersing away from one's natal group. Incest avoidance among uterine kin is probably as old as uterine kinship itself.

STABLE BREEDING BONDS. The pivotal component of the exogamy configuration is a mating system based on long-term breeding bonds between particular males and females. Such bonds occur in various types of primate groups. If enduring breeding bonds are not a human invention, several dimensions of marital unions are uniquely human, from the sexual division of labor to the institutionalization of marriage. But the point is that well before these dimensions evolved, primitive breeding bonds had already produced several of the most important transformations that were to bring about the exogamy configuration.

AGNATIC KINSHIP. Kinship in humans is recognized both matrilaterally and patrilaterally; it is bilateral. Bilateral kin recognition is probably uniquely human, at least in its broad extent. For kinship to become truly bilateral in the course of hominid evolution, systematic paternity recognition had to evolve. Group-wide agnatic kinship structures are particularly original aspects of human society.

BILATERAL AFFINITY. In humans, both spouses recognize their respective in-laws; affinity is symmetric or bilateral. In addition, relationships between affines commonly translate into positive bonds or alliances. Bilateral affinity is thus uniquely human. However, some basic processes involved in the recognition of affines are found in the behavior of nonhuman primates. Moreover, human affines are often allies, a situation that has deep phylogenetic roots.

THE TRIBE. Human kinship networks commonly encompass more than one local group. Any human group is typically embedded in a larger social structure, which is itself part of a still larger social entity. This situation contrasts markedly with that in nonhuman primates, for whom, as a rule, the local group constitutes the highest level of social structure. I use the term "tribe" in a generic sense to refer to levels of social structure above that of the local group. I shall argue that in its most primitive version the tribe amounted merely to a state of mutual tolerance between specific local groups brought about by the evolution of stable breeding bonds.

POSTMARITAL RESIDENCE PATTERNS. Human postmarital residence patterns—patrilocality, matrilocality, bilocality, and so forth—are diverse and highly flexible. They are causally connected to some basic aspects of social structure, including genealogical composition, form of descent, and exogamy rules. In all likelihood postmarital residence patterns grew out of their counterparts in nonhuman primates, so-called philopatry patterns. The discrepancies between the two sets of phenomena are important, but their evolutionary connection is no less clear.

THE BROTHER–SISTER KINSHIP COMPLEX. This building block is perhaps the single most original trait of human kinship from an evolutionary outlook. It is manifest in a wide array of phenomena, including avuncular relationships and rules about marriage between cousins. It is thus intimately related to exogamy rules. Although the manifestations of the brother–sister complex are diverse, all aspects are ultimately derived from the strengthening of the brother–sister relationship in the course of hominid evolution. I shall be concerned in particular with the factors that contributed to that change, and with the evolutionary precursors of exogamy rules.

DESCENT. Unilineal descent groups—patrilineal, matrilineal, or bilineal—provide the most conspicuous illustration of kinship's potency in organizing human affairs. Descent is a fundamental principle regulating exogamy in many human societies, namely those that prescribe descent-group exogamy. Embryonic forms of descent groups are present in a number of primate species. More specifically, some primate societies display several major properties of unisexual descent groups, an observation that provides realistic hypotheses about the origins of unisexual descent groups in hominids.

MATRIMONIAL EXCHANGE. Reciprocal exogamy as defined by Lévi-Strauss hinged on alliances between kin groups founded on the direct exchange of women or the exchange of women for material goods. All of the aforementioned building blocks of the exogamy configuration are an integral part of matrimonial exchange, but altogether they still do not amount to the complete phenomenon. What is lacking is the reciprocity, or trading dimension, of female exchange, which constitutes a further refinement of human exogamy and a building block of its own.

A Once Irreducible System

The phylogenetic decomposition of the exogamy configuration has major implications for the structural and functional explanations of exogamy that have been proposed by social anthropologists. Before the primate data were available, exogamy appeared to be an irreducible whole. All aspects of the exogamy configuration—kinship, incest prohibitions, marriage, female exchange, intergroup alliances, and so on—are, in effect, inextricably linked in any human society. Accordingly, the irreducibility of exogamy is a conception that has consistently guided social anthropologists in their attempts to explain the phenomenon. Based on the assumption that exogamy was a cultural creation in its entirety, the anthropologists' explanations necessarily implied structural, functional, and causal links among all of its aspects. Lévi-Strauss's alliance theory is probably the best illustration of this. Because he assumed that reciprocal exogamy had arisen, right from the onset, as a system of matrimonial exchange between kin groups, he had to conceive of it as a self-explanatory system—one whose parts were born simultaneously with the complete system and that made sense only as elements of that system. Accordingly, in his scheme, it is the men's need for alliances, in combination with the preexistence of reciprocity as a universal mental structure, that acts as the prime mover of female exchange. It is the same driving force that (1) generates incest prohibitions as a means to compel men to marry out of their group, (2) brings about the circulation of women between groups, (3) creates marital unions, (4) engenders kinship-based rules of mate assignment, and (5) forges bilateral bonds between affines and alliances between groups. The system is irreducible and self-explanatory in that each of its constituent elements finds its sense and purpose within the system itself.

The primate data plainly contradict this view. They indicate that the exogamy configuration is eminently reducible and that it is not a self-explanatory system. They show that most of its components have their own taxonomic distributions and evolutionary histories. The primate evidence thus belies several of the causal and functional relations posited by Lévi-Strauss and others. It shows, for example, that incest avoidance long predated the evolution of pair-bonding, and therefore that incest prohibitions were not invented to make one's sons and daughters available for marital unions outside the family; they were available well before pair-bonds evolved. Similarly, the primate data strongly suggest that female circulation between local groups long predated the evolution of pair-bonds and the exchange of women between groups. In other words, exogamy existed in bits and pieces well before it existed as a whole. Hence the phylogenetic decomposition of exogamy, besides forcing one to rethink the origins of human society, prompts a reappraisal of the validity of some basic assumptions of anthropological theories about human social structure. To paraphrase Lévi-Strauss, who was referring to culture in general, to make sense of the exogamy configuration one needs to "trace it back to its source and . . . retie all the broken threads by seeking out their loose ends in other animals" (1969, xxx). Lévi-Strauss was reluctant to abide by his own precept, but we now have the necessary data to identify exogamy's loose ends in nonhuman primates and to understand, at least partially, how they came together in the human species.

III

The Exogamy Configuration Reconstructed

10

The Ancestral Male Kin Group Hypothesis

> Early human social structure in its simplest outlines was probably that of a pre-human primate group altered and subdivided in ways directly related to reciprocal, virilocal marriage modes.
>
> *Elman Service (1962, 50)*

Determining the starting point of the evolutionary sequence that led to the exogamy configuration can only be arbitrary. Nevertheless, it is convenient to start at the point in time when the line leading to the human species, the *Homo* line, separated from the line leading to the two chimpanzee species, or *Pan* line (see Figure 2.1). Starting right after the *Pan–Homo* split makes it possible to describe the general features of the last common ancestor that we shared with another living species. What then was "human" society like at the beginning of hominid evolution? From a comparative perspective, the modal social system of human societies is the *multifamily community*. As explained earlier, this is a rare system that combines a multimale-multifemale group composition with a mating system featuring stable breeding bonds. Accordingly, the multifamily system may have evolved through two different pathways, depending on the social system that characterized the earliest hominids. According to the first possibility, the ancestral hominid species had a multimale-multifemale group composition and a promiscuous mating system, in which case the humanlike multifamily system evolved through a change from sexual promiscuity to stable breeding bonds. This hypothesis assumes that the society of ancestral hominids was chimpanzee-like in its broad outline; I refer to it as the *ancestral male kin group hypothesis*. The second possibility propounds the reverse evolutionary sequence:

stable breeding bonds appeared first, followed by the multimale-multifemale group composition. Accordingly, the ancestral hominid system was gorilla-like, with a unimale-multifemale group composition, and the multifamily system evolved through the amalgamation of several such autonomous polygynous units into a cohesive multifamily group. I come back to this hypothesis later on. For now I focus on the first possibility.

The hypothesis that early hominids initially formed male kin groups was born with the realization that some basic aspects of chimpanzee/bonobo society, notably their philopatry pattern, had much in common with some important features of the "patrilocal band model" of human hunter-gatherers. The ancestral male kin group hypothesis was proposed in the late 1980s and endorsed by many, but since then various objections have been raised and new data on both humans and nonhuman primates have accumulated. In this chapter I review the empirical evidence relating both to the philopatry patterns of our closest relatives and to the patrilocal band model; I spell out the assumptions, problems, and implications of the ancestral male kin group hypothesis; and I end with a somewhat updated version of it.

The Patrilocal Band Model

As has often been reported, patrilocality is the prevalent postmarital residence pattern worldwide, with roughly 70 percent of the 1,153 societies coded in Murdock's (1967) *Ethnographic Atlas* classified as patrilocal or virilocal. Of particular interest for the present discussion are the residence patterns of the simplest of all human societies: foraging, or hunter-gatherer, societies. In his influential book *Theory of Culture Change,* published in 1955, Julian Steward developed the "concept and method of cultural ecology" and applied it to what he called the hunter-gatherer *patrilineal band,* a "cultural type whose essential features—patrilineality, patrilocality, exogamy, land ownership, and lineage composition—constituted a cultural core which recurred cross-culturally with great regularity" (122). Steward aimed to provide a cultural ecological explanation of the patrilineal band—what primatologists would call a socioecological explanation. The similarities between geographically distant and culturally different patrilineal bands resulted, he argued, from the identity of the exploitative patterns at work in similar, but not necessarily identical, environments. Central among the ecological bases of the patrilineal band were two factors: a reliance on limited and scattered resources that "restricted population to a low density" and prevented "large, permanent aggregates" and a dependence on nonmigratory game rather than large migratory herds, because "this kind of game can support only small aggregates of people, who remain on a restricted territory" (123–124).

According to Steward these bands were patrilineal *because they were patrilocal*. "In these small bands," he wrote, "patrilocal residence will produce the fact or fiction that all members of the band are patrilineally related" (125). Stated differently, the residence pattern determines the group's genealogical structure and the ensuing kinship network. But, then, why are these bands patrilocal? Steward mentioned four reasons, about which he was notably succinct. First, patrilocality might result from "innate male dominance," which "would give men a commanding position." Second, it might reflect men's greater economic importance in a hunting culture. Third, in a note in which he referred to Radcliffe-Brown (1931), Steward briefly hinted at the importance for hunters of remaining in a country they knew from childhood and were highly familiar with (125). Fourth, patrilocality would necessarily follow from the existence of competition between groups of hunters for territories holding scarce resources: "As the men tend to remain more or less in the territory in which they have been reared and with which they are familiar, patrilineally related families would tend to band together to protect their game resources. The territory would then become divided among these patrilineal bands" (135). In short, patrilocality would be an adaptation to territoriality.

A few years later Service (1962) argued that the term "patrilineal band" was misleading because it carried "the connotation that the group conceptualizes the patrilineal descent line, and further, that membership in the group is essentially a matter of reckoning descent" (52). Service pointed out that if virilocal residence in association with outmarriage caused the members of a local group to be patrilineally related, patrilineal kinship did not imply patrilineal *descent*, which, he emphasized, "is a much later development in the evolution of social organization," with "social consequences not found in band society" (53). For this reason he proposed the term *patrilocal band*. Compared to Steward, Service was notably more explicit about the causes of patrilocality. He rejected the male dominance argument as a sufficient explanation, noting that male dominance is actually compatible with a variety of residence patterns. Likewise he discarded the argument that virilocality reflected the greater economic importance of men, asserting that bands in which women have an important economic role are not necessarily uxorilocal. He was equally skeptical about effective hunting requiring long residence and high levels of familiarity with the hunting territory, noting that hunters typically range far beyond the band's "country."

Service favored explanations of patrilocality based on male cooperation. "Male cooperation," he wrote, is "most efficiently practiced among men who have grown up together in the same locality and know each other's habits and capabilities and who trust each other" (48). He considered two different goals of male cooperation. The first was success in hunting, not only in catching prey animals but in finding them, bringing them home, and, "most important of all," sharing them with unsuccessful hunters. The second goal of cooperation was territorial defense, or "competition among societies," which, he contended, "could have been the most important cause of virilocality, for if offense-defense requirements are important, then the trusting cooperation among brothers and other closely linked male relatives would be more important than anything else—depending of course on the severity of competition" (49). Patrilocality, therefore, would be an adaptation to male cooperative activities pertaining to hunting: cooperation in the hunt itself and cooperation in defense of the hunting territories.

Service further argued that other types of "rather amorphous" social structures observed in hunter-gatherers, those that are neither patrilocal nor matrilocal—which he and Julian Steward called composite bands—reflected the effect of contact between hunter-gatherers and modern civilization and the destruction of the original band organization. That is, prior to acculturation, patrilocality was the norm among hunter-gatherers. The ideas of Steward, Service, and others—notably Radcliffe-Brown, who had earlier defined the *patrilineal horde* in reference to Australian tribes—led to the widespread view that "the patrilocal band . . . is perhaps the most primeval of truly human groups and was probably the social unit of our paleolithic hunting and foraging ancestors," as Fox (1967, 93) put it.

But the patrilocal band model was soon challenged. In their introduction to *Man the Hunter,* which came out of a symposium of the same name, Richard Lee and Irven DeVore (1968) summarized the problems with the model. They noted that the patrilocal band did exist, that it was "not an empty category," but that it was "certainly not the universal form of hunter group structure that Service thought it was" (8) They reported that several researchers who had worked among hunter-gatherers failed to confirm the patrilocal band model, observing instead flexible local groups, fluid social organizations, bilateral kinship bonds, and bilocality. These features, Lee and DeVore further claimed, were independent of the influence of acculturation and modern civilization. Similar arguments were brought forward by many others (for example, Gough 1971).

A few years later, however, Carol Ember (1978) provided new arguments in favor of the patrilocal band model. After noting that the prevalent view among researchers about the postmarital residence patterns of hunter-gatherers had apparently shifted from patrilocality to bilocality, she disputed the claim that hunter-gatherers were typically bilocal, on the grounds of a lack of statistical analyses supporting it. She presented her own analyses based on tabulations of hunter-gatherer residence patterns coded in Murdock's *Ethnographic Atlas*. Out of 179 societies, Ember argued, 62 percent were patrilocal/virilocal, 16 percent matrilocal/uxorilocal, and 16 percent bilocal (the remaining societies were avunculocal or neolocal). Even after removing the societies whose hunter-gatherer status might be disputed—those that used horses and those that were largely dependent on fishing and aquatic resources—she still found that patrilocality (56 percent) outnumbered both matrilocality (20 percent) and bilocality (22 percent). The prevalence of patrilocality further held true in each of the five large geographical areas with hunting-gathering societies. Clearly, then, Ember's analyses provided strong support to the patrilocal band model.

Further major critiques were leveled at the patrilocal band model, but before examining these, I turn to the second body of evidence that gave rise to the ancestral male kin group hypothesis.

Male Philopatry in Apes

By far the most common philopatry pattern found in nonhuman primates is female philopatry. This pattern characterizes most species of Old World primates (Pusey and Packer 1987). From that angle, the residence pattern of our two closest relatives, chimpanzees and bonobos, appears truly unusual, for both species exhibit male philopatry. In all known populations of chimpanzees and bonobos, males stay in their natal group (Goodall 1986; Furuichi 1989; Nishida 1990; Kano 1992; Boesch and Boesch-Achermann 2000, Nishida et al. 2003). Female dispersal is the rule as well: in three different populations of chimpanzees (Mahale, Taï, and Ngogo) the proportion of females emigrating was around 90 percent (Nishida et al. 2003; see also Gerloff et al. 1999). However, in the Gombe population of chimpanzees, 50 percent of the females did not emigrate and bred instead in their natal group. This may reflect the fact that the main study group at Gombe had only two neighbor communities, whereas the three populations in which the majority of females were observed to transfer had several (Boesch and Boesch-Achermann 2000, 46; Nishida et al. 2003).

Diane Doran and her colleagues (2002) carried out a quantitative analysis of the behavioral variation of chimpanzees and bonobos in an attempt to understand the respective roles of phylogeny, ecology, and demography in the behavioral variation observed both between populations of the same species and between the two species. They compared the distribution of fifty-seven behavioral characters across four chimpanzee populations and two bonobo populations. They found that although chimpanzees and bonobos differed substantially in several important characters, such as the degree of male sociality, the degree of female sociality, the extent to which males were dominant to females, the frequency of hunting, and so forth, they consistently shared the same pattern of male philopatry. Moreover, they did so independently of significant differences in their behavioral ecology.

This raises the question of the adaptive function of male philopatry in *Pan*. The current explanation has two components, one that pertains to male localization, the other to female dispersal. Basically, male localization would reflect the advantages for males raised in the same group and highly familiar with each other to stick together and cooperate. Chimpanzees are highly territorial, and they cooperate actively in the defense of their community's range. Intercommunity relationships are tense, generally hostile. Border patrols and aggressive displays are common (Goodall 1986; Pusey 2001; Watts and Mitani 2001; Williams et al. 2002, 2004). Aggressive encounters between members of different communities, some of them leading to lethal aggressions, have been observed in all known chimpanzee populations. Males combine forces in these activities. Territorial displays and border patrols are conducted by parties of male chimpanzees, and attacks on or killings of non-community individuals are typically coalitionary in nature (Manson and Wrangham 1991; Wrangham 1999; Boesch and Boesch-Achermann 2000; Wilson and Wrangham 2003). But what exactly do male chimpanzees cooperatively defend? This has been the object of some debate (Wrangham 1979; Ghiglieri 1987; Boesch and Boesch-Achermann 2000; Pusey 2001; Wilson and Wrangham 2003). Paternity analyses indicate that the offspring born in a given community are fathered by the community's resident males (Constable et al. 2001; Vigilant et al. 2001), supporting the view that males effectively prevent extragroup paternity. Such data are compatible with the hypothesis that male cooperation aims at the defense of the females themselves, hence that it is part of the male *mating* effort (Wilson and Wrangham 2003). Complementary data support one other major function of male cooperation: the defense of feeding territories. A recent long-term study by Jennifer Williams and colleagues (2004) suggests that males defend a feeding territory both for their own sake and for that of the females of their community. By excluding other male and female feeding competitors, males are helping the females of their community to reproduce and their offspring to survive. Male cooperation would thus also be part of a male's *parenting* effort.

The second part of the explanation for male philopatry is that because males benefit by staying in their natal community, the females have no choice, evolutionarily speaking, but to leave their birth group to avoid inbreeding and its associated costs (Clutton-Brock and Harvey 1976; Pusey and Packer 1987; Clutton-Brock 1989a; Isbell 2004). Importantly, female dispersal in *Pan* translates into outbreeding: although females may mate in their natal group before dispersing, they do not normally breed in it (Constable et al. 2001). Females play an active role here. They avoid sexual relationships with their maternal brothers—the single category of male relatives systematically recognized in a male-philopatric society (see chapter 12). When they reach adolescence, the females voluntarily leave their natal community, commonly visit other communities, show sexual attraction to foreign males (typically, females visit other groups while in estrus), and eventually establish themselves in a new group (Pusey 1980, 2001). In sum, male philopatry in chimpanzees would reflect the combination of two elements: the benefits for males of cooperatively defending the territory and the "outbreeding pressure" operating on females.

Interestingly, our third closest relative, the gorilla, also displays a male philopatric pattern, though it is not as complete as that of chimpanzees and bonobos. Most groups of mountain gorillas *(Gorilla gorilla beringei)* have a single adult male, but a large proportion (around 40 percent) have more than one adult male (Robbins 1995, 2001; Watts 1996). The majority of sexually mature males born in a given group (about 64 percent) remain in their natal group. Those that disperse either join an all-male group (a group of bachelors) or become solitary. Importantly, dispersing males do not transfer to other bisexual groups. In contrast, the majority of sexually mature females born in a given group (72 percent) disperse away from their natal group and join other bisexual groups (Doran and McNeilage 1998; Robbins 2001). Thus while females commonly transfer between bisexual social groups, males do not. Both aspects, the higher rates of male residence compared to female residence and the fact that males never transfer between bisexual groups, indicate that mountain gorillas have a partial pattern of male philopatry, in that not all males stay in their natal group.

Recent genetic studies on the closely related but comparatively little known western gorilla *(G. gorilla gorilla)* provide further evidence for a male philopatry pattern in gorillas. Brenda Bradley and her colleagues (2004) reported that silverback males leading nearby groups were often genetically related to each other (father–son or half-brothers) and that males dispersed less far than females. They suggested the existence of "dispersed male networks" based on kinship among silverback males and manifest in the occurrence of peaceful encounters between neighboring groups.

The Homology Hypothesis

In 1987 Richard Wrangham proposed that male philopatry in apes and patrilocality in hunter-gatherers had a common origin. The idea that primate philopatry patterns and human residence patterns are evolutionarily connected had been in the air for some time, even among nonprimatologists. For example, in an article on the origin of the human family written in 1971, Kathleen Gough, a sociocultural anthropologist, criticized the patrilocal band model. After noting that "among apes and monkeys, it is almost always males who leave the troop" while "females stay closer to their mothers and their original site," she argued that this type of evidence supported the claim that "the earlier hunters had matrilocal rather than patrilocal families" (769). Gough was thus using the argument of phylogenetic continuity between primate philopatry patterns and human residence patterns in an attempt to prove her point about the ancestral hominid residence pattern. Owing to the relative paucity of data on apes at that time, she had it wrong with regard to the specifics, the African apes being "patrilocal" rather than "matrilocal." Had she known this, she would have had to argue the other way around and conclude that the primate data supported the ancestral patrilocality hypothesis. As another example, Claude Masset, an archaeologist writing in 1986, remarked that in both chimpanzees and most hunter-gatherer societies it is the females who move. This led him to conclude that "whether it is the man or the woman who moves, we are faced with one of the most ancient traits of human family systems" (108, my translation).

What was lacking at that time was a formal analysis of the similarities between the two categories of phenomena. Wrangham (1987) carried out just such an analysis by comparing humans to the three African apes (chimpanzees, bonobos, and gorillas) in an effort to identify the behavioral traits they share and thereby characterize the last common ancestor of all four species. As discussed earlier, the traits shared by species descended from a common ancestral species may represent homologies, and thus it might be possible to reconstruct the behavioral profile of the stem species—in the present case, the traits of the common ancestor of humans and the three African apes. But there is an alternative hypothesis. Some of the similarities among the four species might be the result of similar selective pressures acting separately on each of them; they might be homoplasies. This explanation is not easily discarded, but it appears less parsimonious than the common-descent hypothesis, considering that the four species differ importantly in their diet and behavioral ecology and that their respective ancestors, accordingly, must have undergone distinct selective pressures.

Endorsing the view that patrilocality was the preponderant residence pattern in humans, Wrangham found a number of basic similarities in the social structures of humans and the three African apes, including female dispersal. With respect to its complement, male localization, Wrangham had to make do with a paucity of empirical data on bonobos and gorillas in the 1980s. He concluded that male residence would be "the likeliest finding" of future studies, which eventually proved right. The traits characterizing the common ancestor of African apes and humans implied the existence of an "ancestral suite" of similar traits in the populations of hominids that evolved immediately after the *Pan–Homo* split, thus right from the onset of human evolution. Female dispersal and male localization would therefore exemplify phylogenetically conservative traits in humans. Implicit in Wrangham's reasoning was that human patrilocality and male philopatry in African apes were homologous phenomena.

At about the same time, Michael Ghiglieri was carrying out a similar comparative analysis. But, citing molecular evidence indicating that chimpanzees, bonobos, and humans shared an exclusive common ancestor (that is, gorillas had split off earlier), he limited his characterization of the stem species to the last common ancestor of these three species. The exclusion of gorillas, with their unimale group composition, allowed him to produce a larger list of shared traits, and hence a more comprehensive characterization of the common ancestor. Central to that list were the multimale-multifemale community, female dispersal, male residence, male kinship groups, and cooperative defense of the territory by males. Ghiglieri concluded explicitly that the behavioral similarities between *Pan* and *Homo* constituted homologies: "Because communal breeding strategies by male kin are so extremely rare among nonhuman primates—and among mammals in general," he wrote, "the chances that each of the three most recent species from the common ape-human ancestor evolved them independently seem extremely small. The behaviors appear to be homologies" (1987, 347). Robert Foley, also using a cladistic approach and citing evidence that chimpanzees and humans formed a clade relative to the gorilla, similarly argued that "their last common ancestor [of chimpanzees and humans] was probably similar in social organization to the modern common chimpanzee" and that the "phylogenetically inherited socioecology" of chimpanzees and the earliest hominids consisted of multimale-multifemale groups, male philopatry, and strong bonds between males (1989, 485). All three researchers, Wrangham, Ghiglieri, and Foley, assumed that patrilocality was the majority residence pattern in hunter-gatherer societies, a view that received empirical support, notably through Ember's studies, but was questioned by others. The hypothesis that male philopatry in *Pan* and patrilocality in contemporary hunter-gatherers are homologies has since been endorsed by many.

Before examining recent analyses of ethnographic data that dispute the ancestral male kin group hypothesis, I should note that the hypothesis suffered from internal problems, notably a significant contradiction. If male philopatry and human patrilocality are traits similar by descent from a common ancestor, it follows that both systems initially had the same adaptive function, namely, the one it served in the common ancestor. Yet the adaptive function of male localization differs depending on whether one is concerned with ape philopatry along the *Pan* lineage or with ancestral patrilocality along the human lineage. Anthropology, through the patrilocal band model, and primatology, through philopatry theory, provide distinct explanations of the origin of male philopatry. But if the homology hypothesis is correct, the two explanations must be concordant. Steward and Service did not know about male philopatry in apes. Accordingly, they assumed that patrilocality had evolved after the *Pan–Homo* split for reasons specific to hominids. They proposed, as we saw, that patrilocality was an adaptation to the hunting-gathering mode of subsistence. Implicit in their reasoning was that hunting had driven the evolution of patrilocality. But the primate data indicate that male philopatry long antedated the evolution of systematic hunting in hominids, and furthermore that male philopatry in apes did not evolve in relation to hunting.

Ever since the pioneering work of Jane Goodall on chimpanzee hunting, a number of studies have confirmed that males in all known populations of chimpanzees hunt and, when doing so, may join forces and even coordinate their actions in sophisticated ways (Goodall 1986; Boesch and Boesch 1989; Boesch 1994; Stanford et al. 1994; Stanford 1999; Mitani and Watts 1999, 2001; Watts and Mitani 2002). But hunting accounts for only a small percentage of the daily caloric intake of chimpanzees. Hillard Kaplan and his colleagues computed data showing that while meat represents between 30 percent and 80 percent of the diet of hunter-gatherers (n = 10 societies, excluding high-latitude foraging societies that would have further increased the proportion of meat in the diet), it accounts for only about 2 percent of the diet of chimpanzees

(Kaplan et al. 2000, Table 3). They estimated the daily meat intake of chimpanzees at about 10 to 40 grams per day and that of humans at about 270 to 1,400 grams per day. The difference is substantial, and the reasons are clear. Hunting in chimpanzees is an opportunistic activity; males and females are vegetarian essentially. By contrast, hunter-gatherers hunt systematically and males *specialize* in hunting. The reason they are able to do so is that females specialize in gathering, their production compensating for the highly unpredictable and variable outcomes of hunting. In allowing males to specialize in hunting, the sexual division of labor transformed hunting from an opportunistic activity to a systematic one with high caloric returns.

For their part, bonobos also practice opportunistic hunting, but at substantially lower rates than chimpanzees (Fruth and Hohmann 2002). Thus bonobos and chimpanzees are vegetarians, yet they are male philopatric, which implies that male localization in these species cannot be an adaptation to hunting. If, as argued above, the cooperative defense of feeding territories by localized males is the driving force behind male philopatry in *Pan,* it is fruits and other vegetable resources that are at stake, not game. For that matter, gorillas do not hunt at all, and they too are male philopatric. What these data suggest, then, is that hominids were already localized and forming male kin groups well before systematic hunting evolved. Therefore, none of Service's arguments about the links between hunting and cooperation explain the very *origin* of patrilocality, and in this sense the primate data contradict the original patrilocal band model.

One last remark on this point. To state that hunting and its cooperative aspects have not driven the evolution of ancestral patrilocality is not to deny that hunting and patrilocality get along well. Patrilocality may facilitate cooperation between males, including cooperative activities relating to hunting, as asserted by early theorists as well as contemporary ones. For example, Keith Otterbein (2005) argued that 75 percent of the societies in which hunting and fishing dominate (that is, whose combined hunting and fishing score is 70 percent or above) are virilocal (but see Marlowe 2005). If hunting, cooperation, and patrilocality are in effect causally related, primitive male philopatry in the *Pan–Homo* ancestor would have created conditions favorable to the evolution of cooperative hunting among coresident males. Stated otherwise, ancestral male philopatry and its correlate, male cooperation, would have acted as a *preadaptation* for cooperative hunting.

Updating the Ancestral Male Kin Group Hypothesis

A long-standing assumption underlying the hypothesis that male philopatry and patrilocality are homologies is that patrilocality was the majority pattern among hunter-gatherers. This assumption stemmed from two distinct sources. First, it was a basic tenet of the patrilocal band model, which stated it not as an assumption but as a fact. Second, for male localization in apes and humans to be homologous traits, they must characterize both groups of species, by definition. But the assumption that patrilocality is the preponderant residential pattern of hunter-gatherers has been challenged by a number of authors (for example, Barnard 1983; Knauft 1991) since *Man the Hunter* (ed. Lee and DeVore) was published in 1968. More recently it has been the object of two quantitative analyses that converge in their conclusion that bilocality, not patrilocality, is the majority residence pattern among modern hunter-gatherer societies.

The first study was carried out by Helen Alvarez (2004), who reexamined Ember's classification of postmarital residence schemes among hunter-gatherers. Alvarez reviewed all the ethnographies cited by Murdock (1967), which Ember had used in the first (1975) of her two studies on the subject. Alvarez examined forty-eight of the fifty original sources cited by Murdock. As she noted,

> With a few notable exceptions, the ethnographies from which the atlas is coded are totally devoid of the actual number of cases for any described behavior. At their best, the early ethnographies are good descriptions of behavior observed by ethnographers who spent time living with their informants . . . Others contain simple declarative statements with no supporting material. In salvage ethnographies assembled from the recollections of one or a few informants, nothing was measured. (426–427)

Alvarez obtained proportions different from those reported by Ember. She found that 25 percent of the societies were patrilocal or predominantly patrilocal (versus 56 percent in Ember's study), 39.6 percent were bilocal (versus 28 percent), and 22.9 percent were matrilocal or predominantly matrilocal (versus 16 percent) (Alvarez 2004, Table 18.2). On this basis she concluded that bilocality was the preponderant postmarital residence pattern and further inferred that patrilocality was unlikely to be the ancestral residence pattern of hominids.

The second study was conducted by Frank Marlowe, who analyzed the residence pattern of thirty-six foraging societies taken from the *Standard Cross-Cultural Sample,* which "includes 186 societies with good ethnographic coverage that have been chosen to create an unbiased sample of the world's societies with respect to geographic region, language, family, and cultural areas" (2004, 278). Noting that Ember had not taken into account changes in residence throughout the duration of a marriage, Marlowe defined "multilocal residence" to refer to situations in which a couple transferred to different places either in the first years after marriage or in later years. Multilocal residence thus includes various combinations of virilocality, uxorilocality, and neolocality. Marlowe found that 34 percent of the societies were virilocal throughout marriage, 43 percent were multilocal, and 23 percent uxorilocal, and he concluded that "contrary to the orthodox view, most foragers are not virilocal." Marlowe also discussed the reasons underlying the prevalence of multilocality among hunter-gatherers, which, like others before him, he attributed to a foraging diet whose elements are spatially dispersed, hence hardly defensible, and which requires day ranges that are much larger than those of apes. In a response to Marlowe, Otterbein (2005) further stressed that the exact nature of subsistence activities in hunter-gatherer societies markedly affected the nature of residence patterns. Specifically, he argued that in societies in which gathering rather than hunting and fishing dominates, multilocalily or uxorilocality are the favored residence patterns. Given that subsistence patterns vary importantly among hunter-gatherers, residence patterns should as well.

Let us assume that Alvarez's and Marlowe's figures are closer to the reality of hunter-gatherers than those of Murdock and Ember, which appears likely, and that patrilocality is indeed not the majority pattern among modern hunter-gatherers. Does that throw away the ancestral patrilocality hypothesis? It might, if hominids had been hunter-gatherers ever since the *Pan–Homo* split, for thus one could argue that residential flexibility rather than male philopatry characterized the very first hominids. Indeed, if we knew that for sure, there would be much less ground to assert that primitive hominids were male philopatric and to infer on that basis that our last common ancestor with *Pan* was male philopatric as well. But all we know is that modern hunter-gatherers are often residentially flexible and that the exact residence pattern depends in part on their subsistence pattern. Now there is every reason to believe that hunting evolved much later in the course of hominid evolution. The earliest recorded zooarchaeological evidence that hominids butchered mammals is dated 2.5 million years ago and concerns *Australopithecus garhi* from Ethiopia (Asfaw et al. 1999; de Heinzelin et al. 1999). This means that hominids were primarily vegetarians for some 3.5–4 million years. The next question thus becomes: Do we have reasons to believe that male philopatry was the most likely residence pattern of hominids before they adopted hunting as a primary source of subsistence? Based on cladistic considerations, the answer is yes. Our three closest primate relatives and a majority of human societies being male philopatric, an ancestral pattern of female philopatry, as in macaques, is unlikely. As for mixed patterns of residence such as bilocality or multilocality, they are unlikely as well, for reasons discussed in chapter 15. Briefly, residential diversity (the occurrence of several residence patterns in the same species) and residential flexibility (changes in residence patterns over the lifespan) could hardly evolve before males were able to move between local groups with relative impunity. This stage probably coincided with the emergence of supragroup social structures and between-group pacification—the primitive tribe—which occurred much later in human evolution.

The fact that patrilocality is not ubiquitous in human societies, and especially that it is not the norm in hunter-gatherers, may appear incompatible with the idea that patrilocality is homologous to ape philopatry. A similar argument was made by Knauft (1991) in relation to the issue of male violence. Knauft compared intragroup and intergroup patterns of violence in three categories of societies: ape societies, "simple human societies" (mostly nomadic foraging societies), and "middle-range" (or prestate) human societies. Knauft rightly noted that as far as patterns of aggressive competition are concerned, apes have much more in common with middle-range societies than with hunter-gatherer societies. On this basis he concluded that "correspondences between great ape and middle-range human societies are not homologous based on phylogenetic continuity through human evolution" (407). In other words, the egalitarian nature of hunter-gatherers would reveal the phylogenetic gap separating apes and middle-range societies. On similar grounds, Knauft argued that the fact that patrilocality is not the majority pattern in foraging societies while it is common in middle-range societies proves that ape philopatry and human patrilocality are phylogenetically disconnected and hence not homologous.

This reasoning stems from a somewhat erroneous conception of the influence of biology on behavior, one that posits that evolved patterns of behavior should be manifest independently of circumstances. Assuming that chimpanzee and human patterns of violence have some homologous basis, the human predisposition to compete aggressively for resources should manifest itself only in certain situations, specifically when resources deemed intrinsically valuable are limited in quantity and spatially distributed in such a way that they may be aggressively defended against competitors at reasonable costs to the aggressor. Implicit in this statement is that human violence represents a potential that may or may not be expressed, depending on the context. If middle-range societies often meet the above conditions, simple societies do so much less often. As a result, one obtains a phylogenetic "discontinuity" in patterns of violence between apes and middle-range societies. Likewise, assuming that ape philopatry and human patrilocality are homologous, the males' propensity to stick together and cooperate should be realized only in situations where this is advantageous. Now it might well be the case that such situations occur infrequently among foragers scattered across vast territories and much more often in sedentary populations with fixed property and higher population densities. Again one obtains an apparent phylogenetic discontinuity between ape philopatry and human patrilocality.

Although the homology hypothesis cannot be discarded, it is not the only possibility. Similarities between apes and middle-range human societies might have evolved through convergent evolution; that is, they might reflect homoplasies. Whether patterns of intergroup aggression seen in humans and chimpanzees result from homology or homoplasy "is currently unclear," noted Wilson and Wrangham (2003, 385). Favoring the homoplasy explanation, Laura Betzig argued that because chimpanzees and humans are separated by much evolutionary time, "it seems reasonable, then, to suspect that similarities between them exist because similar behaviors have been adaptive under similar conditions rather than because phylogenetic legacies made them persist against selective force" (Betzig 1991, 410). Homoplasies are certainly possible. But contrasting homologies and homoplasies as if they were alternative explanations should not obscure the fact that any homoplasy between two closely related species such as chimpanzees and humans is likely to have developed out of homologous structures present in their last common ancestor. In other words, homoplasious similarities between chimpanzees and humans are likely to have evolved through *parallel* evolution, in which case the observed similarities would reflect *both* shared ancestry and similar selective pressures.

An equally important point is that phenomena like "violence" and "philopatry" are not unitary; they do not have a single cause or a single function. Some aspects of the patterns of violence shared by chimpanzees and humans might be homologous while others are homoplasious. For example, the potential for territorial defense might be a legacy of the last common ancestor of *Pan* and *Homo,* whereas lethal raiding might be the product of convergent evolution. A broad, multicomposite behavioral category such as violence certainly breaks down into a large number of contextual, motivational, and functional subcategories, each with its own evolutionary history. On this basis alone, it would be simplistic to assert that ape violence and human violence have no homologous components whatsoever or, on the contrary, that they are entirely homologous.

Likewise, some aspects of the similarities between ape philopatry and human patrilocality might represent homologies and others, homoplasies. But in this case the homology hypothesis appears more likely. This is because philopatry patterns appear to be phylogenetically conservative traits (chapter 15), more so than patterns of violence taken as a whole. The fact that male philopatry characterizes our three closest relatives makes it likely that our last common ancestor with *Pan* was male philopatric and hence that patrilocality evolved from male philopatry.

The Gorilla Alternative

As pointed out earlier, structurally speaking, the ancestral male kin group hypothesis is one of two main possibilities regarding the evolution of the human social system, the multifamily community. This hypothesis propounds that the multimale-multifemale composition characterized the last common ancestor of chimpanzees and humans and that stable breeding bonds evolved at some point after the *Pan–Homo* split, as illustrated in Figure 10.1a. Another possibility is the reverse sequence: autonomous families existed first, and the human social system arose from the combination of several such families into cohesive multifamily groups. This hypothesis is not merely a structural possibility. It rests on the fact that the modal social unit of our closest primate kin next to chimpanzees and bonobos, the gorilla, is the unimale-multifemale unit (or autonomous polygynous family). In accordance with this view, stable breeding bonds would be much more primitive than posited under the ancestral male kin group hypothesis.

There are two possible evolutionary sequences here. The first was recently suggested by David Geary (2005). Not only would stable breeding bonds have antedated the *Pan–Homo* split, but the multi–family system would have been in place by that time; that is, the last common ancestor of chimpanzees, bonobos, and hominids would have had a humanlike multifamily structure, as illustrated in Figure 10.1b. There are two problems with the proposed sequence. Chimpanzees and bonobos mate promiscuously. Positing that the common ancestor of the two species had a multifamily structure implies, as noted by Geary, that both species lost the long-term breeding-bond trait at some point in their evolution (the H → C sequence in 10.1b). Apart from having to account for such a major evolutionary reversal, one finds no vestige of long-term breeding bonds in chimpanzees and bonobos, for example in the form of long-term bonds between particular males and females.

Another problem is the assumption that the multifamily type of group would have evolved through the grouping together of several independent polygynous groups (G → H sequence in 10.1b). This is unlikely for two reasons. The first has to do with what we know about the origin of multifamily groups in the few primate species that exhibit this type of structure. Robert Barton (1999) carried out a cladistic comparison of the social structures and mating systems of cercopithecine primates, a family that includes a large number of species with promiscuous multimale-multifemale groups and four species with a multifamily group structure (hamadryas baboons, gelada baboons, mandrills, and drills). His results suggest that the multifamily structure did not evolve through the amalgamation of independent polygynous groups but from an ancestral multimale group and the *subsequent* conversion of the mating system from promiscuity to stable breeding bonds. On this basis alone, the evo-

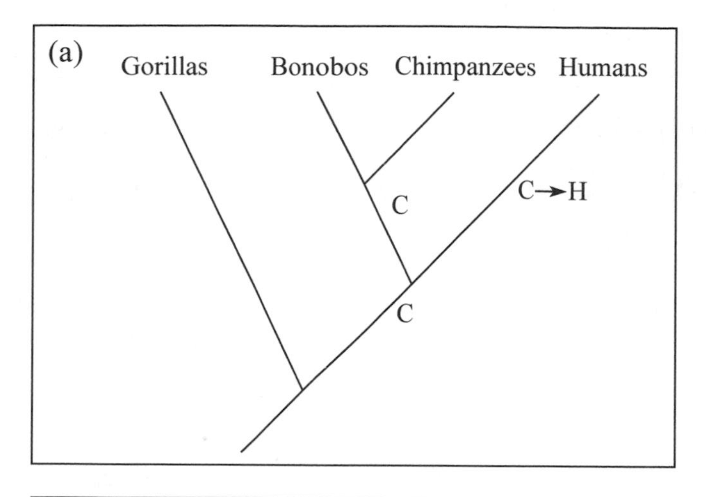

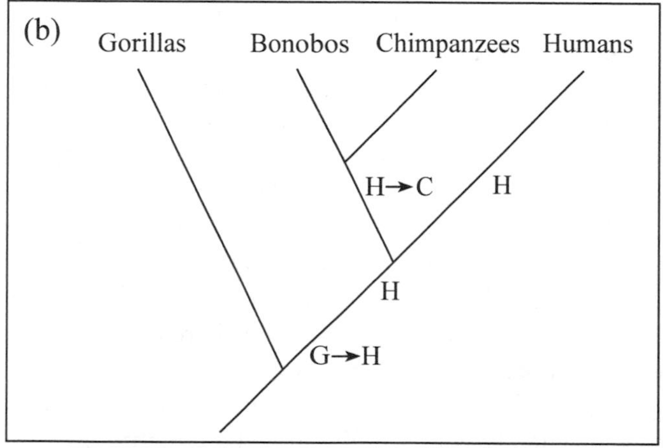

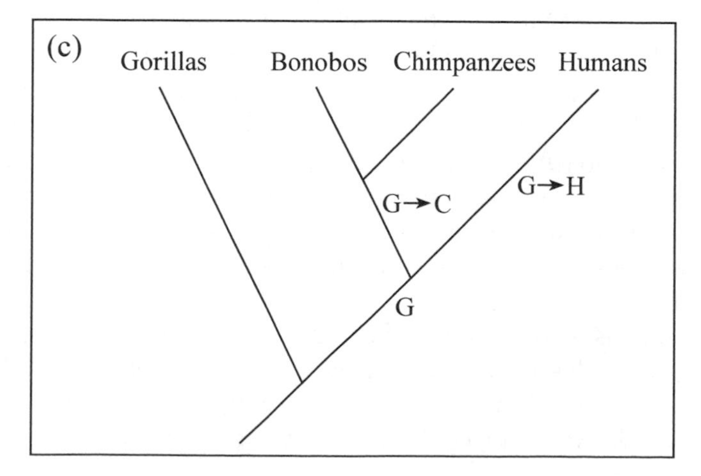

Figure 10.1. Three hypotheses about the evolutionary history of human society. C: chimpanzee-like society (multimale-multifemale composition, sexual promiscuity). H: humanlike society (multifamily groups). G: gorilla-like society (independent polygynous units).

lution of the hominid multifamily structure through the merging of gorilla-like polygynous families appears unlikely.

The second reason has to do with the proximate and developmental processes posited. Geary's hypothesis assumes that lower levels of competition and higher levels of cooperation between fathers and sons would lead formerly independent polygynous units to mingle peacefully and form a cohesive group. But the primary effect of reduced levels of sexual competition between males would be to increase the adult male's level of tolerance toward his sons as they grow up in their natal group. The outcome would be the extension of male philopatry to natal males in general, that is, the formation of a multimale-multifemale group. Put simply, if adult males were to grow more tolerant of one another, they would tolerate the presence of their sons in the natal group rather than evicting them out of it only to tolerate them later on. One also expects the mating system to evolve toward some sort of dominance-regulated sexual promiscuity of the kind one observes in chimpanzees. This reasoning finds support in theoretical considerations and empirical data relating to the evolution of female kin groups. As argued by Lynne Isbell (2004), female kin groups form when mothers are in a position, ecologically and reproductively, to tolerate their daughters in the natal home range. Maternal tolerance translates into female retention and the extension of female philopatry (see chapter 19).

The other possible evolutionary sequence is a more parsimonious variant of the latter hypothesis, illustrated in Figure 10.1c. It posits that the last common ancestor of chimpanzees and humans prior to the *Pan–Homo* split displayed the gorilla-like polygynous group. Along the *Pan* line the ancestral polygynous group would have generated the multimale chimpanzee group through male retention (G → C sequence). This presents no particular problem from a structural viewpoint, as just discussed. But along the hominid line, the multifamily group would have evolved through the grouping of polygynous units. This sequence (G → H) is unlikely for the same two reasons it was unlikely in the previous model.

Kinji Imanishi was perhaps the first primatologist to propose this third variant in his 1961 article on the origin of the human family. But the same variant was proposed more recently by Sillén-Tullberg and Møller (1993), who mapped the mating systems of anthropoid primates

(New World monkeys, Old World monkeys, and apes) onto a phylogenetic tree. These authors were mainly concerned with the relationship between concealed ovulation and mating systems. Nevertheless, their analysis led them to conclude that the gorilla-like polygynous family characterized the *Pan–Homo* stem species, and therefore that chimpanzees had evolved the multimale-multifemale composition on their own, as shown in Figure 10.1c. But they reached that conclusion by attributing the same mating system to orangutans, gorillas, and humans (1993, Fig. 3). Indeed, for the sake of their analyses they defined only three types of mating systems: the monogamous unit, the polygynous unit, and the multimale-multifemale group, ignoring a fourth type, the multifamily structure. Instead they lumped that system with polygynous units and "refrained from a finer subdivision of mating systems" for methodological reasons. With the ascription of the gorilla polygynous system to humans and orangutans, that system necessarily became the primitive or stem pattern, and the chimpanzee multimale pattern became the derived one. But clearly, orangutans, gorillas, and humans have very distinct mating systems; moreover, chimpanzees and humans share a most basic aspect of group composition: the multimale-multifemale component. This brings us back to variant 1 (Figure 10.1a).

All in all, therefore, the ancestral male kin group hypothesis appears the most parsimonious of the three evolutionary sequences. It is the sequence that minimizes the number of evolutionary changes required to arrive at the multifamily structure—one instead of two changes—and it does not involve the unlikely G → H sequence, the grouping together of independent families. One might object that the ancestral male kin group hypothesis has its own weakness in that it also posits an evolutionary change, namely, the C → H sequence: the transformation of a promiscuous multimale system into a multifamily one. But as we shall see in the next chapter, this sequence is supported both by empirical evidence and theoretical considerations.

To these phylogenetic arguments in favor of the ancestral male kin group hypothesis, one may add a different type of evidence, paleoanthropological and anatomical. The fossil record indicates that early hominids (australopithecines) were anatomically more similar to chimpanzees and bonobos than to gorillas in terms of absolute body size and relative brain size (McHenry 1992). The fact that early hominids had a chimpanzee-like body mass suggests that their diet had more in common with the chimpanzee diet than with the low-quality diet of gorillas. This in turn suggests that the behavioral ecology of early hominids, in terms of group size, group composition, ranging pattern, and so forth, more closely resembled that of chimpanzees.

11

The Evolutionary History of Pair-Bonding

Marriage is the humanization, the institutionalization, the sociocultural expression of the relatively durable union between the sexes in subhuman primate society. In the transformation of anthropoid society into human society mating became marriage.

Leslie White (1959, 94)

The transition from sexual promiscuity to stable breeding bonds in the ancestral male kin group is the key event that launched the exogamy configuration on its evolutionary path. Many scenarios about the origin of pair-bonding have been proposed, and my aim here is not to present an additional one. Evolutionary scenarios are educated stories about the unfolding of events leading to important adaptations of the human species, from bipedalism or brain expansion to concealed ovulation and language. Scenarios have several of the following features. First, they postulate the advent of certain climatic, ecological, demographic, or social transitions as prime movers of evolutionary change. Second, they posit the operation of specific selective pressures responsible for the evolution of certain traits. In doing so, they specify the exact functional consequences, among all those possible, that supposedly led to the traits being selected. Third, they often ascribe evolutionary changes to specific hominid species or time frames. Fourth, they commonly attempt to causally link various categories of adaptations—anatomical, physiological, mental, behavioral, and social. It is thus in the very nature of evolutionary scenarios to be riddled with untested assumptions and high levels of speculation. In spite of this, scenarios have proven remarkably useful as sources of testable hypotheses about human evolution, and therein lies most of their interest.

Be that as it may, my approach in this chapter has none of the main characteristics of scenarios. No specific environmental changes, selective pressures, or adaptive consequences are hypothesized. A single major assumption underlies the whole reasoning, namely, that models, explanations, and theories that are good for nonhuman primates and mammals in general are equally good for hominids. Sticking to such models appears to be the best way to avoid the pitfall of anthropomorphic explanations, which are particularly tempting when dealing with the evolution of the family. In keeping with this objective, I present what amounts, I hope, to a minimally speculative and maximally parsimonious account of the evolutionary history of human pair-bonds. It centers on five points. (1) Stable breeding bonds in hominids have a biological basis, just as they have in other species, and they constituted the evolutionary precursors of marital unions. (2) The fact that the costs of maternity are disproportionately high in humans and the observation that the father helps alleviate those costs are compatible with the hypothesis that parental collaboration is a major adaptive function of pair-bonding. (3) Notwithstanding this, the primate data do not support the classic view that pair-bonds originated as parental partnerships. (4) Stable breeding bonds initially evolved among hominids as a mate-guarding strategy, and they followed a two-step evolutionary sequence, from sexual promiscuity to generalized polygyny, and from polygyny to generalized monogamy. (5) The core feature of cooperative childcare in humans, the sexual division of labor, evolved in a number of steps that were fortuitous byproducts of the combination of other evolutionary events.

The "Invariant Core of the Family"

Depending on their specialty, anthropologists hold diametrically opposed views about the nature and origins of the marital bond and its correlate, the human family. To many if not a majority of biological anthropologists and human behavioral ecologists, stable breeding bonds pertain to the realm of biological adaptations, like bipedalism, language, or kin favoritism, for that matter. One implication of this view, albeit one that is less often voiced explicitly, is that pair-bonds were the evolutionary precursors of marital unions. On the other hand, many sociocultural anthropologists take it for granted that marriage is a cultural construct with no biological underpinnings whatsoever. Whether the discrepancy between the two views reflects a real disagreement or merely a lack of background information or even of concern for the issue, it is still very much alive, and for this reason it must be addressed before one tackles the evolutionary history of marriage and the family. In the first place, how could one prove that stable breeding bonds in humans have a biological foundation? The most direct type of evidence in support of the biological hypothesis would be that long-term bonds between men and women are sustained by specific physiological, neurobiological, and psychological processes. I return to this topic shortly. A completely different type of proof would be that some transculturally "invariant social core" is present in all types of marital bonds and families. But that is precisely the argument invoked against the claim that marital bonds arose from "natural constraints." The culturalist interpretation of marriage and the family is indeed predicated on the absence of an invariant core of the family. It relies on the observation that all aspects of the human family vary to a considerable extent across cultures, from the preponderant type of marital union and the exact composition of domestic groups to the precise division of tasks between wives and husbands. No one would dispute this fact, and therefore some clarifications are needed.

Not only do marital unions vary across societies among the three basic types—monogamy, polygyny, and polyandry—but they also vary *within* the same society, a remarkable fact from an evolutionary standpoint. In polygynous societies, for example, a fraction of men are polygynously married and a majority monogamously involved. Moreover, to these basic types of unions one must add many other, more exotic variants. To take a single example, Evans-Pritchard (1940) described marriage between women among the Nuer. If a woman was sterile, she could marry a woman and have children with her through a male genitor who visited the woman's wife from time to time. The sterile woman was considered a man and called father by her children, and as such she could inherit cattle from her lineage. The composition of human families is no less variable cross-culturally. Kathleen Gough (1971) listed five main forms of "kin-based households": (1) nuclear families; (2) "compound families," polygynous or polyandrous; (3) "extended families" composed of three generations of married brothers or sisters; (4) "grandfamilies" descended from a single pair of grandparents; and (5) "matrilineal households" consisting of brothers living with their sisters and the sisters' children, with the husbands visiting their wives in other homes.

So many cultural variants of marital unions, domestic groups, and the division of labor exist that to many anthropologists this alone testifies to their cultural nature. "The sexual distribution of tasks . . . ," stated Françoise Zonabend, "has no physiological character, it is in no way embedded in the nature of men or women, *given that it varies across cultures, even across groups*" (1986, 95; my translation and emphasis). The assumption underlying that reasoning is widespread: an entity that is symbolically elaborated and cross-culturally variable must be intrinsically cultural. From this it follows that searching for the universal core of that entity is pointless. In that vein, Sylvia Yanagisako ended her cross-cultural review of the family and domestic groups in the following terms: "Our emphasis on reproduction as the core of the family's activities all betray our Malinowskian heritage. The belief that the facts of procreation and the intense emotional bonds that grow out of it generate an invariant core to the family is what sustains our search for universals. But the units we label as families are undeniably about more than procreation and socialization. They are as much about production, exchange, power, inequality, and status." Therefore, she concluded, "we must . . . abandon our search for the irreducible core of the family and its universal definition" (1979, 199–200). We dealt with that same reasoning in relation to the incest issue. The fact that incest rules vary extensively across cultures in no way precludes the possibility that all incest rules are cultural offshoots derived from the same stem pattern of incest avoidance observable in nonhuman primates. Similarly, all types of marital unions could be cultural variants ultimately derived from the same ancestral mating pattern. What, then, was that stem pattern like?

In all likelihood its essence lay in the single major feature that transcends all types of marital arrangements: their selective and relatively stable character. Regardless of the exact nature of marital unions, men and women form enduring *pair-bonds*, a term that encompasses more than strictly dyadic, monogamous unions. In polygynous unions two or more women are paired with a single man, who is in fact maintaining two or more distinct pair-bonds simultaneously. The same applies to polyandrous unions, with a woman maintaining several pair-bonds concurrently. This is precisely why the expression "pair-bonding" is commonly used to characterize the human mating system even though a large number of marital unions are not limited to a single dyad. Pair-bonding is, essentially and simply, the opposite of sexual promiscuity—short-term mating relations with several partners. It includes all types of enduring reproductive relations between particular males and females. Pair-bonding thus refers to a specific property of mating arrangements: their duration.

True, not all human sexual activity takes place in the context of lasting relationships. At any point in life, a man or woman may have short-term sexual interactions with a number of sexual partners. But in no human society is sexual promiscuity the sole or the main form of mating arrangement, as it is in other primate societies categorized as sexually promiscuous, such as chimpanzees and macaques. Promiscuity consistently accompanies various forms of stable breeding bonds. As a matter of fact, much of the sexual promiscuity in humans takes place premaritally, and the permissive character of premarital sex only serves to emphasize the social importance of marital unions. Moreover, sexual promiscuity by married individuals is universally disapproved of. Thus not only are stable breeding bonds consistently present across human societies, but they prevail over promiscuity. For that matter, procreation in humans is expected to take place in the context of marital bonds. All human societies differentiate between legitimate and illegitimate children, a distinction that further illustrates the centrality of pair-bonding in the human mating system. Logically, therefore, the stem ancestral pattern out of which emerged all known human mating arrangements, whether polygynous, polyandrous, or even homosexual, is the enduring breeding association itself. In theory, ancestral pair-bonding could have been of the monogamous or the polygynous type, and I address this question later on.

Significantly, the human species does not have a monopoly on stable breeding bonds. In nonhuman primates they are observed in various types of groups: in single monogamous pairs, single polygynous units, single polyandrous units (a rare form), and multiharem groups (several polygynous units assembled together). But an important difference from the human situation is that in other primates they exhibit relatively little intraspecific variation. For example, in species that form polygynous units—either independent units, as in gorillas, or aggregated ones, as in hamadryas baboons—polygyny is the rule across groups of the same species; there are no groups or populations in which all individuals practice monogamy or in which all individuals practice polyandry as this is observed in humans. Similarly, within a given group, adult males are either polygynously mated or they have no females. Monogamous bonds may be observed, but they are temporary arrangements characterizing young males in the initial phase of harem building; monogamy is not a regular adult pattern found alongside polygyny, as it often is in our species. Compared to humans, therefore, other primates have relatively rigid mating systems. The remarkable flexibility of human mating systems is obviously a matter of cultural variation. We have an unusual capacity to adjust mating arrangements in relation to subsistance patterns, resource distribution, degree of economic stratification, residence patterns, forms of descent, and so forth. But however culturally variable, human mating arrangements always build upon the same core feature: the pair-bond. This brings us to a simple but consequential principle. Given that pair-bonding evolved in a large number of animal species in which it is a biological phenomenon, human pair-bonding is also a biological phenomenon. From this parsimonious inference it follows, no less parsimoniously, that marital unions, however culturally embedded and institutionalized they are, grew out of that biological substrate. This is precisely what Leslie White was arguing in 1959 (see the chapter epigraph).

If marital bonds are biologically grounded, this should be manifest in various physiological, neurobiological, and emotional processes involved in the formation and maintenance of human pair-bonds. In her book *Why We Love,* Helen Fisher (2006) looks into the "chemistry of love" and argues that human pair-bonds involve a mixture of three distinct categories of emotions: (1) lust, the craving for sexual gratification; (2) attachment, the feelings of comfort and security felt with a long-term partner; and (3) romantic love, the passion directed to a preferred individual. Each system would have its own functional raison d'être and neurochemical bases. Research on the biological underpinnings of human pair-bonding is only beginning (see also Flinn, Ward, and Noone 2005). It remains to be seen how specific to humans some of the mechanisms discussed by Fisher are. When the data become available, it will be interesting to compare the biological foundation of pair-bonds in humans and other pair-bonded species, primate and nonprimate.

Pair-Bonds as Parental Partnerships

Why and how did pair-bonding evolve? The classic explanation here is what I refer to as the parental collaboration hypothesis. It conceives of human pair-bonds as parental partnerships, or cooperative breeding units, based on a sexual division of labor between the mother and the father. It states that the family came into being when the father joined the mother to help her raise their children. I examine separately two major correlates of this hypothesis. The first is that the costs of maternity are disproportionately high in our species compared to other primates. This correlate is an implicit assumption of the parental collaboration hypothesis, because nonhuman primate mothers commonly raise their offspring without cooperation from third parties. The second correlate is that the father does alleviate the costs of maternity to a significant extent.

The Costs of Maternity in a Comparative Perspective

The costs of maternity are best appraised by differentiating three periods: pregnancy, lactation, and postweaning care. The costs of pregnancy are the sum of the metabolic costs relating to fetal growth, maternal tissue growth, maternal fat deposit, increased basic metabolism, and higher levels of physical activity. These costs are known to increase a mother's daily expenditure of energy by about 20–30 percent in mammals in general (Aiello and Key 2002), and the question is whether the costs of gestation incurred by human mothers are higher than those incurred by other primate mothers of similar size (body size must be taken into account because larger animals have higher daily energy requirements). The length of gestation is not a significant factor in these costs because it is similar in humans and the great apes. Yet the human neonate is relatively larger than most other primate neonates. This does not seem to relate to the human infant's larger brain, because its relative weight is similar to that of great apes. The difference may in part be attributable to the greater amount of fat in human neonates (nearly four times higher than expected from body size), a pattern "expected to place energy demands on human mothers that are not faced by nonhuman primates" (Dufour and Sauther 2002, 586). In sum, the daily costs of gestation incurred by human mothers may be somewhat higher than that in other primates of similar size, but not by a particularly big amount.

The energy costs of lactation—basically, the volume of milk produced per day and its energy content—are known to be at least twice as high as the energy costs of pregnancy in mammals in general (Clutton-Brock 1991), including humans (Dufour and Sauther 2002). Human milk does not appear to be more costly to produce than the milk of other primates: its composition in terms of proteins, carbohydrates, and fat is similar (Dufour and Sauther 2002; Robson 2004). Nor are the *daily* costs of lactation in human mothers obviously and disproportionately higher than those of other primates of similar size. The *total* costs of lactation in humans are actually lower than those of chimpanzees because the duration of lactation in humans is only about 60 percent (Kaplan et al. 2000) or 50 percent (Aiello and Key 2002) that of chimpanzees. However, by weaning their children at an earlier age, human mothers must start provisioning them with solid food. It is difficult to compare the costs incurred by a chimpanzee mother suckling a child over four years with the costs incurred by a human mother suckling her child over the first two years and provisioning it over the next two years. But this does not matter too much for the present discussion because provisioning activities extend well beyond four years, so the bulk of mothering costs in humans is incurred well after weaning.

The cost of provisioning is the difference between the quantity of food a child produces and the quantity it consumes. For chimpanzees this difference is practically nil because weaned chimpanzees are largely self-sufficient in terms of food acquisition. The situation is drastically different in humans, owing to important differences in life-history parameters. The human brain is three times bigger than the brain of a primate of comparable size. Yet, as just pointed out, the relative weight of the human neonate's brain is similar to that in great apes. This means that the human baby is born at a relatively earlier stage in its development and also that it will grow over a longer period to attain adult size. Humans, indeed, have the lowest growth rates of all primates. While the chimpanzee brain stops growing at three to four years of age, the human brain continues to grow until a little past ten years. Using data on human hunter-gatherers and wild chimpanzees, Hillard Kaplan and colleagues calculated that humans have their first offspring about five years later than chimpanzees and have a prereproductive (juvenile) period 1.4 times longer than that of chimpanzees. As a result, human children are dependent on their mother over a longer period. Remarkably, they eat more than they gather until they reach their mid to late teens (Kaplan et al. 2000). We touch here on the most original and important component of the human parental load: human mothers must feed more than one child at a time. For example, they may be suckling an infant while provisioning two or three other children. Nonhuman primate mothers never have to feed more than one offspring at a time (except in the rare case of twinning). Maternal feeding of offspring in nonhuman primates is a sequential activity along a female's reproductive career; in humans it is a cumulative activity to a large extent. The first correlate of the parental collaboration hypothesis is thus substantiated: the costs of maternity are disproportionately high in our species.

Hunting and the Father's Contribution

The very nature of provisioning means that individuals other than the mother may help her meet the postweaning costs of maternity. And the same helpers are also in a position to help mothers alleviate the costs of gestation and lactation. Nonhuman primate mothers meet the costs of pregnancy and lactation by spending more time eating. Human mothers also eat more while pregnant or lactating, but much less than one would predict based on primate trends. This is because they deal with these costs by reducing their levels of physical activity and by resting more (Dufour and Sauther 2002), which they can do because others provision them. Food sharing is thus the single most important factor allowing human mothers to alleviate the costs of maternity at all three stages of maternal investment. Based on primate patterns of alloparental care, a number of candidates could help with provisioning, including the mother's mother, sisters, and older offspring (Nicolson 1987; Fairbanks 1990; Chism 2000). This is confirmed by actual patterns of provisioning in hunter-gatherers (Hawkes et al. 1998; O'Connell, Hawkes, and Blurton-Jones 1999; Hrdy 2005). But according to the parental collaboration hypothesis, the father's contribution should also be important, if not predominant.

The best data to address this problem come from foraging societies. The intuitively sound idea that in hunter-gatherer societies men hunt *in order* to provision their wives and children has come under severe criticism from behavioral ecologists studying food acquisition and distribution in these societies. In a series of articles that began to appear in the early 1990s, Kristen Hawkes and her colleagues opposed on several grounds the idea that hunting by males and gathering by females were integral to cooperative partnerships (Hawkes 1991, 1993, 2004; Bird 1999; Hawkes, O'Connell, and Blurton-Jones 2001). First, they noted that in the societies they studied successful hunters did not necessarily favor their own families but frequently gave away a large proportion of the carcass so that "women gain little extra meat for themselves by marrying better hunters." (Hawkes, O'Connell, and Blurton-Jones 2001, 682). Second, they emphasized that hunters most often did not receive meat from others in proportion to what they had given and that dyadic reciprocity was either nonexistent or extremely asymmetrical (Bird 1999; Hawkes, O'Connell, and Blurton-Jones 2001). Hawkes and her coworkers inferred from this that successful hunters had no control over the distribution of big game; in other words, meat was a public good, not a private property. The upshot was that most of the meat a wife obtained came from hunters other than her husband. Third, they asserted that meat yielded lower return rates in terms of calories per hour of work compared to certain types of vegetables gathered by women, even assuming that a calorie of meat had a nutritional value several times higher than a calorie of plant food. They concluded that "men could earn higher rates of nutrient gain for their families by gathering plant foods than they earn as big-game hunters." (Hawkes, O'Connell, and Blurton-Jones 2001, 686) They also calculated that big-game hunting alone was less productive than a combination of big-game and small-game hunting but that hunters often ignored small game, as if strict nutritional gains were not a priority. Another way of saying the same thing is that if men's activities are less productive than women's activities, men seeking to maximize their family's income would be better off doing what women do. If they don't, this is presumably because they have other objectives in mind. Fourth, Hawkes and colleagues emphasized that big-game hunting is a subsistence activity with highly variable payoffs, owing to its unpredictable character and extremely high daily rates of failure, and hence is not a particularly reliable means to ensure that one's children are fed on a regular basis; that it is, in fact, "an ineffective strategy for provisioning a family."

If men do not hunt to feed their families, they might do so to obtain other types of benefits. Successful hunters are popular, enjoy high status and high levels of prestige, have more allies, gain the attention of others, have a greater influence on group decisions, and may have more wives and obtain more sexual favors. This led Hawkes (1991) to propose the "showing off hypothesis," according to which hunting would be part of a male's mating effort rather than his paternal effort. Males would hunt big game, ultimately, to increase their reproductive success. Any benefits to the hunter/father's family would be ancillary. This particularly clear position was bound to stir up heated discussion and stimulate much further research. More than fifteen years have now passed since the "hunting as mating effort" idea was first put forward, and the present overall picture does not appear to support the view that male hunting is *only* a mating effort. Briefly, the claim that large game is a common good over which successful hunters have no control is difficult to reconcile with the abundance of ethnographic accounts stressing the importance of reciprocity or with reported negative attitudes toward freeloading (see, for example, comments of Altman, Beckerman, and Smith in Hawkes 1993; and see Gurven 2004). In many societies some levels of reciprocity *are* observed. Quantitative data on meat transfers in hunter-gatherers are few, but Gurven (2004) calculated "the proportion of receiving that is contingent on giving" in eight societies and found significant correlations in most cases. Significantly, reciprocity does not need to be symmetrical to be advantageous; it can be substantially asymmetrical and still provide each partner with payoffs superior to what they would get by not sharing (Boyd 1992). Regarding the claim that the families of successful hunters did not receive larger shares, Gurven's extensive cross-cultural analysis (2004) led him to conclude that in all cases except one, the hunter's nuclear family got considerably more than any other family (see also Wood 2006). As to the claim that big-game hunting yields lower return rates compared to certain types of plant food, Gurven and Hill (n.d.) reported that in the three populations for which data on nutrient rates were available, big-game hunting provided substantially more calories per unit time of work compared to gathering or to small-game hunting. They also stressed that the unpredictability of big-game hunting is compensated by mechanisms that reduce the daily variation in food availability, notably food sharing and storage.

What this combined evidence suggests is not that hunting has nothing to do with the male mating effort—there is much evidence that hunting *is* mating effort (van Schaik and Paul 1996)—but that it may also be a parental effort. Perhaps in part for heuristic purposes, Hawkes and coworkers have consistently discussed the mating and parental functions of male hunting in terms of mutually exclusive alternatives, as if the two could hardly coexist. But successful hunters appear to provision their families *and* obtain other social benefits, and one aspect need not be merely an ancillary byproduct of the other; both might be adaptive. This is precisely what one would expect if a trait's evolutionary history was marked by different functions evolving sequentially. To take a simple example, primate studies indicate that the smile is derived from the bared-teeth grimace of nonhuman primates, a ritualized facial display that signals submissiveness and appeasement. Apparently this original function has survived until today. But smiling encompasses more than submissiveness; it may also convey guilt, reassurance, and friendliness, among other things. It serves different functions depending on the context (van Hooff 1972; Preuschoft 1997). Hunting too might well serve more than one function, regardless of the exact evolutionary history of these functions. For example, long after pair-bonding had evolved, possibly as a mate-guarding strategy, as discussed below, the dietary shift to hunting might have initially served a family-provisioning function in response to increasing costs of maternity, and only subsequently served the reproductive functions subsumed under the showing-off label. Alternatively, hunting might have initially served nonnutritional goals—as in chimpanzees who use meat to reinforce alliances with other males and/or to obtain sexually receptive females—and only subsequently evolved a provisioning function. In either case one expects hunting to serve a dual function among contemporary hunter-gatherers.

To sum up, two major correlates of the parental collaboration hypothesis receive empirical support: the costs of raising human children are disproportionate in a comparative perspective, and the father does contribute to reducing these costs. That is to say, pair-bonds are, in effect, parental partnerships. At this point it would be tempting to conclude that pair-bonds emerged as cooperative alliances right from the beginning. But inferring the primeval function of the human family from its present-day working and adaptive underpinnings is highly questionable. To better appraise the assumptions and implications of such an inference I briefly review the associated evolutionary scenarios.

The Pitfall of the Modern Family Reference

In a scenario that has come to epitomize the idea that pair-bonds were born as parental partnerships, Owen Lovejoy (1981) argued that soon after the *Pan–Homo* split, hominids were already forming nuclear families featuring monogamous pair-bonding, reduced female mobility, paternal provisioning, and continual female sexual receptivity. Lovejoy saw the origin and adaptive aspect of pair-bonds in parental collaboration increasing the reproductive rates of hominids and enabling them to outcompete other primate species. In this scenario paternal provisioning required the transport of food and was the main selective pressure driving the evolution of bipedalism. A closely related scenario attributed the evolution of bipedalism to distinct factors but shared with the previous one the idea that 4 million years ago australopithecines were already practicing monogamous pair-bonding as a strategy of parental collaboration. Helen Fisher (1992, 2006) reasoned that as newly bipedal mothers were moving around in the savanna they were burdened by infants no longer able to cling to them. As a result, mothers needed a male for protection and extra food, and males had to settle for monogamy because they could not defend territories sufficient to sustain more than one female (Fisher 2006, 149). Both scenarios illustrate particularly well the four descriptive features of evolutionary scenarios listed at the beginning of this chapter. They are also remarkable in their anthropomorphic attribution of a modern form of family organization to the earliest hominids.

Earlier versions of the parental cooperation hypothesis ascribed a preponderant role to hunting, the underlying principle being that the evolution of intensive hunting by males coincided, by definition, with the division of labor between males and females, hence with the emergence of pair-bonding and the nuclear family (Washburn and Lancaster 1968; Isaac 1978; Fox 1980; Hill 1982). More recent versions of this hypothesis agree on that basic point but link the development of parental collaboration to the evolution of human life-history traits. For example, Kaplan and coworkers (2000) argued that the dietary shift to hunting was responsible for the evolution of pair-bonding, paternal provisioning, an exceptionally long lifespan, an extensive period of juvenile dependency, brain expansion, and high intelligence. They saw pair-bonding as originating in complementarity-based parental partnerships between females committed to caring for vulnerable young and males providing "calorie-dense, large-package, skill-intensive food resources," that is, large game. "Our hypothesis is that this feeding niche had multiple effects: It increased the premium on learning and intelligence, delaying growth and maturation; increased nutritional status and decreased mortality rates through food-sharing" (179). A closely related view propounds a different causality in which brain expansion and its correlate, increased juvenile dependency, selected for paternal investment. In a recent synopsis of the evolution of the human family, Flinn, Ward, and Noone (2005) argued that the "advantages of intensive parenting, including paternal protection and other care" required "a most unusual pattern of mating relationships: moderately exclusive pair bonding" (553), a pattern that coincided with the evolution of *Homo erectus*.

As noted by Key (2000) and Aiello and Key (2002), most research on the social structure of *H. erectus* has focused on the implications of larger brain size and increased costs of encephalization. But *H. erectus* also differed from earlier hominids in having a remarkably larger body mass. Aiello and Key argued that this factor alone might have favored the evolution of a "cooperative economic division of labor" between males and females, that is, of pair-bonding and the family. They assumed that australopithecine females had a reproductive schedule similar to that of chimpanzees, which reproduce more slowly than human females because they suckle their offspring over a considerably longer period of time. They calculated that the biological costs of reproduction incurred by hominid females as big as *H. erectus* but reproducing like australopithecines would have been 40 percent higher owing to the energetic requirements of a larger body mass. By adopting a humanlike reproductive schedule with its shorter period of lactation, *H. erectus* females would have enjoyed substantially smaller costs of lactation per offspring and would have produced more offspring in their lifetime. But to do so, mothers had to wean their offspring at an earlier age, which meant provisioning them. According to this scenario, cooperative provisioning and pair-bonding were a response to the increase in maternity costs, but in relation to body size rather than to brain expansion.

Whatever the exact causality put forward, all versions of the parental cooperation hypothesis—including others that emphasize the role of concealed ovulation (Alexander and Noonan 1979; Turke 1984)—posit a temporal correspondence between the mother–father association and the parental-collaboration dimension of this association. All see the two aspects as concomitant. In doing so they account for the *origin* of the family in terms of its actual working and *present-day* adaptive function, and here lies a potential pitfall. We encountered that problem with reciprocal exogamy. Lévi-Strauss analyzed contemporary matrimonial exchange and assumed that since all aspects of the phenomenon were functionally related, they had arisen together as a system of exchange right from the outset. In the same manner, all aspects of the human family—sexual, parental, and economic—are functionally related, and it is tempting to assume that they were born as such. But like every complex adaptation, anatomical or behavioral, the human family is most probably the end product of a multistep evolutionary sequence that saw several progressively more elaborated versions succeed one another. On this basis alone the origin of the monogamous family might well have nothing to do with the benefits of parental collaboration (Hawkes, Rogers, and Charnov 1995; Hawkes 2004).

A Two-Step Evolutionary Sequence

What has been lacking in prior discussions about the evolutionary origins of the human family is a consideration of the phylogenetic constraints set by the stem social organization out of which pair-bonding evolved. Stated differently, the focus has been on the adaptive function of pair-bonding, while the question of its progressive and multistage construction over evolutionary time has been somewhat neglected. In keeping with the comparative framework presented in chapter 2, I use principles derived from the study of nonhuman primates as a whole to decompose pair-bonding and the human family into their main building blocks and reconstruct their evolutionary history.

I posited that the earliest hominids had a chimpanzee-like social structure, the promiscuous multimale-multifemale group. At the end point of the hominid evolutionary sequence, the modal composition of contemporary human groups is the cohesive multifamily group in which either all families are monogamous or a majority of families are monogamous and a small fraction polygynous. Up to 80 percent of human societies present this combination. Other possibilities are observed, such as monogamy/polyandry, but they are uncommon. Logically, then, hominids went from the promiscuous multimale-multifemale society to the predominantly monogamous multifamily structure. The primate data suggest that this transition was not direct and that an intermediate stage was the multiharem type of group in which all families are polygynous and a fraction of males are unmated—a structure exemplified by hamadryas and gelada baboons. Accordingly, the evolution of pair-bonding proceeded in two steps. The first step was the transition from sexual promiscuity to the multiharem group; the second, from the multiharem structure to the multimonogamous family structure consisting of a majority of monogamous families and a minority of polygynous ones. Stated otherwise, hominids went from promiscuity to generalized polygyny and from there to generalized monogamy. A direct passage from the chimpanzee pattern of sexual promiscuity to the human pattern of generalized monogamy is unlikely for a number of reasons. First, polygyny is much more common than monogamy in primates and mammals in general, attesting to the intensity of sexual competition among males and the recurrent capacity of males to monopolize more than one female. Second, some primate species exhibit the multiharem structure, but none display the multimonogamous family structure, an observation that further illustrates the importance of male sexual competition. Third, the fact that a majority of human societies are polygynous strongly suggests that the ancestral hominid pattern was generalized polygyny, not generalized monogamy, and that in the course of human evolution some factors operated as strong constraints on the feasibility of polygynous unions, as discussed below.

How then did the first transition occur? A precise answer to this question would simply be guesswork. I prefer to stick to general principles derived from the comparative analysis of primate societies, an approach that provides answers that are generic, if not somewhat abstract, but less conjectural or presumptive. One such principle is that stable breeding bonds in primates and mammals, whether monogamous or polygynous, are primarily mating arrangements rather than parental partnerships. Brotherton and Komers (2003) carried out a phylogenetic analysis of the mating systems and parental care systems of mammals. They found that monogamy had evolved more often in the absence of paternal care than in its presence and, therefore, that it had evolved for reasons other than parental cooperation in childcare. They concluded that in most monogamously breeding species exhibiting parental collaboration, paternal care had evolved *after* monogamy was already established. This would explain why paternal care is absent in many mammals that breed monogamously. Similarly, in nonhuman primates that mate monogamously, direct paternal care (carrying, grooming, protection) is present in a number of species (in callitrichids, siamangs, and titi monkeys), but absent in several others such as gibbons, tarsiers, and many species of lemurs (van Schaik and Kappeler 2003). Paternal care is thus far from being a necessary condition for the evolution of pair-bonds. In all likelihood, the monogamous bond itself operated as a preadaptation favoring the evolution of paternal care, rather than the benefits of paternal care driving the evolution of monogamy (Dunbar 1995; Ross and MacLarnon 2000; van Schaik and Kappeler 2003).

The same principle applies even more to stable polygynous bonds. It is remarkable that in the vast majority of primate species that form polygynous units, whether as autonomous groups or as parts of a multiharem group, direct paternal care is absent. Even though males live with their own offspring and have a high paternity confidence, they do not help the mother raise her young (Smuts and Gubernik 1992). Again, this shows that stable breeding bonds evolved for reasons other than parental care. Moreover, in some multimale-multifemale primate groups that do not have stable breeding bonds, adult males and females sometimes form long-term amicable bonds in the context of which the male takes care of the female's infant. But he appears to do this as a means to obtain the female when she eventually becomes sexually receptive, which means that male care is part of the male's mating effort, not his paternal effort (Smuts 1985; see also van Schaik and Paul 1996).

In sum, the main function of stable breeding bonds in mammals is reproductive. More specifically, enduring breeding bonds exemplify male strategies of mate guarding. They reflect the interplay between a male's attempts to monopolize females for mating purposes and the various social and ecological constraints that affect his capacity to do so (Clutton-Brock 1989b; van Schaik 1996; van Hooff 1999). Prominent among these constraints are predators and the distribution of food. Briefly, if females can forage in small groups owing to the density and spatial dispersion of food, a single male can defend such a group against other males, the outcome being a polygynous unit. But if females forage in larger groups, a single male cannot exclude other males, and the outcome is a multimale-multifemale composition. In multiharem species such as the hamadryas baboon, females forage in small groups defended by a single male, but all such polygynous units coalesce in various circumstances, for example at sleeping sites or when they travel. Such a social structure may be an adaptation to a low food density that cannot support large aggregations. This hypothesis finds support in the observation that savanna baboons, which typically form large multimale-multifemale groups, may sometimes subdivide into polygynous units in harsher ecological conditions (Barton 1999). The exact circumstances that favor the evolution of multiharem structures in extant nonhuman primates, and by extension in extinct hominids, remain to be further investigated by behavioral ecologists. My point here is that if stable breeding bonds in mammals in general reflect various trade-offs between male sexual competition and female feeding constraints, chances are they evolved for similar reasons in our hominid forebears. Thus we do not need a special type of explanation for the human case, and therefore the initial evolution of generalized polygyny in the hominid lineage is best construed within the framework of ecologically constrained mate-guarding strategies.

A possible alternative to the mate-guarding hypothesis is the so-called bodyguard hypothesis. Building on the work of Hrdy (1979), Wrangham (1979), van Schaik and Dunbar (1990), and Smuts (1992), Sarah Mesnick (1997) noted that a female should choose to mate with a particular male based on his ability to defend her or her young from other males. Applying this principle to the human case, she proposed that pair-bonding—whether monogamous, serially monogamous, or polygynous—had evolved in response to the females' need for male protection against other males. This hypothesis assumes that males direct aggression to females and/or their offspring, that paired females or their offspring receive less aggression compared to nonpaired females, and that females are attracted to the males best able to protect them. These three assumptions are verified in gorillas, whose social system may serve to illustrate the hypothesis. Males are known to attempt and succeed in killing infants of other males when the infants are not under the protection of a silverback male (Watts 1989). In this context females look for the most efficient male protector (Wrangham 1979). To explain the evolution of pair-bonding in hominids along those lines, one must assume that in the course of human evolution females increasingly suffered from male violence to the point that male protection became necessary. Conditions responsible for this state of affairs are not specified. But in any case, the bodyguard hypothesis appears to be a variant, if not a correlate, of the mate-guarding hypothesis in two respects. First, the main context for adult males attacking or harassing females is sexual coercion. In the gorilla example, it is the males who create the need for male protectors by using infanticide as a form of sexual coercion. From this perspective, bodyguarding is an integral part of a male's mate-guarding strategy. Second, a male who guards females for his own reproductive benefit is expected not only to prevent other males from copulating with his females but to protect them and their offspring from being attacked by other males. A male defending his females and their offspring is ensuring the physical integrity of his sexual resources and of his own offspring. Again, from this perspective, male protection is an intrinsic feature of mate guarding. In short, the mate-guarding and bodyguard hypotheses look indissociable.

For the two hypotheses to be distinct, bodyguarding must be a *sufficient* condition for the evolution of pair-bonding. This remains to be ascertained. Van Schaik and Dunbar (1990) proposed that monogamous pair-bonds in gibbons had evolved as parental partnerships aimed at preventing infanticide. But as shown by Palombit (1999), various lines of evidence are inconsistent with this hypothesis, and mate guarding per se provides a better explanation for monogamy. One scenario in which bodyguarding is a sufficient condition for the origin of pair-bonding was proposed by Wrangham and his colleagues (1999). They posited that the invention of cooking by *Homo erectus* substantially increased both the diversity of digestible plants and their energy content. In this context, they reasoned, cooking created accumulations of valued food packages at processing areas, where they could be the object of competition. When this happened, hominid females could be exploited by adult males stealing food instead of processing it. To face this threat, females would have turned to stable male bodyguards in exchange for sexual gratification. Although highly speculative, this scenario exemplifies an alternative to mate guarding as the origin of pair-bonds.

All in all, therefore, we are brought back to the conclusion that the origin of polygynous pair-bonds in the human lineage is most parsimoniously explained within the empirically substantiated and widely applicable framework of ecologically constrained mate-guarding strategies. This brings us to the second step in the evolutionary history of pair-bonds, the transition from generalized polygyny to generalized monogamy.

Monogamy as a Special Case of Polygyny

In a multiharem group the distribution of females among males is extremely unequal. By comparison, a pattern of predominantly monogamous pair-bonds amounts to a relatively egalitarian apportionment of females among males. In polygynous species the variance in male reproductive success is high; it is much lower in monogamous species. Why, then, would males settle for lower levels of competition and lower potential reproductive rates? The parental cooperation hypothesis states that monogamy arose as males curbed their mating drive and refrained from monopolizing females to concentrate instead on parental care, presumably because this yielded higher fitness payoffs. But a simpler possibility requiring no specific selective pressures for pair-bonding is that monogamy "replaced" polygyny when the costs of polygyny became too high. This hypothesis, moreover, involves a cultural change as the prime mover.

When the competitive abilities of males are well differentiated, as when they form a clear dominance order, aggression is mostly unidirectional, from higher-ranking to lower-ranking individuals. Rank relations settle conflicts, and the variance in male reproductive success is high as a result. By contrast, in a situation in which males had similar competitive abilities, conflicts would be both more frequent and costlier because they would be indecisive and last much longer—unless, of course, males avoided conflicts. In Barbary macaques, adult males are often observed to avoid fighting with each other over resources, such as females in estrus, and to respect ownership (Sigg and Falett 1985). According to Preuschoft and Paul (1999), this reflects the fact that adult male Barbary macaques have dangerous physical weapons and differ little in their competitive abilities, a situation generating stalemates between them, hence egalitarianism. In that vein consider the following thought experiment. In chimpanzees, male dominance relations are well defined, and higher-ranking males have a higher reproductive success (Constable et al. 2001). Now let us suppose that all males had the same physical strength and fighting abilities. In this situation conflicts would be extremely costly and often indecisive. Males would be better off switching to scramble competition; that is, instead of competing by fighting with each other, they should compete by copulating as often as possible and with as many females as possible. This would translate into high levels of sperm competition and sexual promiscuity—extreme polygynandry. The outcome would be a low variance in reproductive success among males. The costs of aggression would have forced males into a form of egalitarianism.

Let us now carry out the same thought experiment but in a different initial setting: the multiharem structure of the hamadryas baboon.

Hamadryas males differ in their competitive abilities. Accordingly, some have big harems, others have smaller ones, and many others have none. Suppose, again, that all males had the same fighting abilities, so aggressive competition would be extremely costly. Would males switch to scramble competition and become highly promiscuous, like the chimpanzees in the previous experiment? This would be unlikely because hamadryas males are harem builders who defend specific females against other males on a long-term basis. They should, therefore, just keep doing this—forming pair-bonds. However, they would end up forming monogamous bonds, because any male attempting to defend more than one female would be challenged by other, equally powerful males with no more to lose than he had. The outcome would be an egalitarian distribution of females among males—generalized monogamy—because it is the arrangement that minimizes conflicts and hence the costs of aggression. Like extreme sexual promiscuity in the previous situation, generalized monogamy is a form of egalitarianism, but one arrived at from initial polygyny. Monogamy is maximally constrained polygyny.

This reasoning readily applies to the situation of polygynous hominids. A phenomenon whose evolution was progressive, irregular, and largely cultural did equalize the competitive abilities of hominid males, namely, the rise of technology. Any tool, whether it was made of wood, bone, or stone, and whatever its initial function, from digging up roots to killing animals, could be used as a weapon in the context of intraspecific conflicts, provided it could inflict injuries. Armed with a deadly weapon, especially one that could be thrown some distance, any individual, even a physically weaker one, was in a position to seriously hurt stronger individuals. In such a context it should have become extremely costly for a male to monopolize several females. Only males able to monopolize tools or males forming coalitions could do so. But because all males *can* make tools and form coalitions, generalized polygyny was bound to give way eventually to generalized monogamy. According to this reasoning, monogamy was not the end product of selective pressures favoring dyadic pair-bonding per se. It was the mere byproduct of other elements merging together over evolutionary time, namely, prior polygyny and the rise of technology. This explanation is thus considerably more parsimonious than the parental collaboration hypothesis, which requires a whole suite of selective pressures and adaptations, from brain expansion and delayed maturity to hunting and the sexual division of labor. Because the constraints-on-polygyny explanation is so much simpler, it must be disproved before the parental collaboration explanation becomes a solid possibility.

If monogamy did not evolve as a result of specific selective pressures, the drive for polygyny was merely checked, not eliminated. Polygyny could reemerge whenever some males secured more competitive power or were able to attract several females based on attributes other than physical prowess. Human societies amply testify to this reemergence.

Paleoanthropology offers some important clues to the timing of generalized monogamy. The reasoning is based on the assumption that a species' degree of sexual dimorphism provides a reliable anatomical marker of its mating system. Data on living species indicate that the magnitude of sexual dimorphism is minimal in species that mate monogamously, including nonhuman primates. This probably reflects the fact that sexual competition between males, and hence selection for bigger males, is minimal in these species. On the contrary, sexual dimorphism is greatest in groups composed of polygynous units and intermediate in multimale-multifemale groups with a promiscuous mating system (Clutton-Brock, Harvey, and Rudder 1977; Alexander et al. 1979; Plavcan 2001). These empirical findings translate into general principles that apply to fossil species. The degree of sexual dimorphism of australopithecines was substantially higher than that of humans and chimpanzees but somewhat below that of gorillas (McHenry 1992, 1996; Lockwood et al. 1996; but see Reno et al. 2003; and response by Plavcan et al. 2005). This evidence is not compatible with monogamy or with sexual promiscuity; it points to a polygynous mating system. On the other hand, the first hominid species to unequivocally show a modern human pattern of sexual dimorphism was *Homo erectus* (McHenry 1994, 1996). Accordingly, this species has been characterized as having reduced male competition (McHenry 1996), or a monogamous mating system (Wrangham et al. 1999; Flinn, Ward, and Noone 2005). Keeping in mind that such inferences depend on the validity of the assumed correspondence between sexual dimorphism and mating system, generalized monogamy would have coincided with *H. erectus*, and more likely with later stages in the species' evolution.

According to the view presented here, males and females were already forming stable breeding associations when brain expansion and delayed maturity occurred or when the shift to hunting took place. Families, whatever their exact composition, had thrived as mating units for hundreds of thousands, if not millions, of years before they became, in addition, parenting units organized around a sexual division of labor. Implied, then, is that pair-bonding operated as a *preadaptation* for the evolution of parental collaboration; that is, pair-bonding was already present when its parental dimension was called upon.

The Evolutionary History of the Sexual Division of Labor

The evolutionary treatment of the sexual division of labor is scanty, to say the least. Its origin is usually equated with the shift to hunting, and from there it is taken as a given. But the sexual division of labor too has an evolutionary history. The most conspicuous aspect of the phenomenon, sex differences in subsistence activities, is only one of its two basic components. The other component, food sharing, is often taken for granted because sexual specialization always comes along with it. Pair-bonded males and females might have engaged in food sharing through cooperative provisioning well before they specialized in different subsistence activities. Cooperative provisioning does not require sexual specialization; in theory, males and females could bring back similar food types. Sexual specialization is a further dimension of cooperative provisioning and one that requires its own explanation. I shall argue here that this two-step sequence—intrafamilial food sharing followed by sexual specialization—is more parsimonious than the alternative, which sees sexual specialization and food sharing evolving synchronously.

Logically, food sharing between pair-bonded mates required two conditions: the ability to carry food around and the possibility of meeting at a common place where food could be shared. In the absence of food transport, only cofeeding on the spot was possible. This was perhaps the first stage in the evolution of sharing—nonhuman primates commonly practice cofeeding (Bélisle and Chapais 2001)—but transport considerably enriched the basic phenomenon. As soon as hominids' hands were freed, they could "gather"—the term is defined here as collecting food that will be eaten later and at a different place. Whether the benefits associated with food transport selected for bipedalism or bipedalism evolved for reasons unrelated to gathering, the fact is that bipedalism and the ability to gather food were concomitant; the invention of containers constituted a further major improvement. Second, without a common spot to share food, pair-bonded mates would not meet unless they traveled together. Thus, cooperative provisioning did not depend on the existence of a fixed "provisioning locale," or protohousehold. Sharing could take place in constantly changing feeding locales, in trees for example, provided pair-bonded partners kept in view of each other. But then why gather? It might be tempting at this point to invoke food-sharing objectives, again because contemporary humans share gathered food. But a nonanthropomorphic and cognitively much less demanding explanation relates to predator avoidance. Chimpanzees inhabiting mosaic savanna environments in Senegal—a mixture of woodland, grassland, and gallery forest—concentrate their activities in forest patches, moving rapidly and alertly in the more open parts of their environment, which offer fewer shelters (trees) against predators (Hunt and McGrew 2002). Hominids living in such environments could have taken advantage of their ability to gather food as a means to eat in safer locations. Note that this view about the evolution of gathering does not conceive of it as a biological adaptation with its own specific selective pressures. Gathering is simply what bipedal primates would have *learned* to do when they wanted to eat away from predators. The reasoning may be summarized as follows: *bipedalism + predation pressure → gathering.*

This brings us to the origin of intrafamilial food sharing, the first component of the sexual division of labor. Recall that I assume the prior existence of pair-bonds, polygynous or monogamous. A simple explanation for the evolution of intrafamilial food sharing in this context is that it was an ancillary consequence, or emergent property, of the conjunction of two otherwise unrelated phenomena: pair-bonding and gathering. Indeed, the act of gathering implies that a certain amount of food is brought back to a given location. This fact alone created a situation favorable to passive sharing among family members for at least three reasons. First, eating collectively out of the same lump of food is akin to cofeeding at the spot where food was found, the difference being that in the first situation the food source has been brought back "home." In other words, the common practice of cofeeding in nonhuman primates might have operated as a preadaptation for passive sharing of food that had been gathered. Put simply, cofeeding was coopted in a novel context. Second, pair-bonded mates and primary kin were disproportionately more available spatially compared to other individuals; that is, mother, father, and offspring were particularly well positioned to engage in such cofeeding. Third, in nonhuman primates, close kin exibit disproportionate levels of tolerance near defendable food sources (Bélisle and Chapais 2001), and long-term friendships between males and females translate into high levels of proximity and affiliative interactions (Smuts 1985). Putting these factors together, from the time pair-bonded individuals began to gather, they were biased to engage in passive sharing. And passive sharing of food that has been gathered is a de facto form of coprovisioning. This may be summarized as follows: *gathering + pair-bonding → de facto intrafamilal coprovisioning.* According to this view, pair-bonded hominids shared food with one another long before they specialized in different subsistence activities. And they did so simply because they were bipeds who gathered food on a regular basis and maintained close proximity to each other.

We now come to the second component of the sexual division of labor. For sexual specialization to emerge from a situation in which pair-bonded males and females were already practicing passive coprovisioning, one sex merely had to display a bias for gathering certain food types. Interestingly, such biases are clearly manifest in our closest relative, the chimpanzee. Females spend much more time than males fishing for termites and ants and cracking nuts with hammers and anvils (McGrew 1979; Galdikas and Teleki 1981; Boesch and Boesch 1984, 1990; McGrew 1992). As noted by McGrew (1992) these activities are readily compatible with maternal caretaking and infant monitoring; they are offspring-friendly. On the other hand, hunting is essentially a male activity. Males are responsible for 70 percent to more than 90 percent of kills, depending on the population (Teleki 1973; Goodall 1986; Boesch and Boesch 1989; Boesch 1994; Stanford et al. 1994; Stanford 1996, 1999; Mitani and Watts 1999, 2001; Watts and Mitani 2002). Males hunt prey that are more difficult to catch, such as colobus monkeys, more often than females, whereas females hunt smaller prey (small ungulates) more often than males (McGrew 1992). A recent study by Pruetz and Bertolani (2007) confirmed that sexual bias. Female chimpanzees were observed to fashion tools from a branch that they trimmed and sometimes sharpened with their incisors and used to extract lesser bush babies (*Galago senegalensis*), small nocturnal prosimians that weigh about 200 grams, from cavities in hollow branches or tree trunks. Interestingly, most such instances of hunting were performed by females and immature individuals.

The male hunting bias in chimpanzees is attributable to a number of differences between males and females. First, males are larger, stronger, and have bigger canines. As a result, they can more easily overpower large prey and run smaller risks of being injured by them (McGrew 1992). Second, adult females are usually pregnant or lactating and hence less mobile than males (McGrew 1992). Third, males cover longer distances daily than females, 5 kilometers versus 3 kilometers at Gombe (Stanford 1996) and stand a higher chance of encountering prey. To these factors one may add that males have higher levels of social coordination than females and that social coordination is an asset in hunting. Males commonly form political alliances and make boundary patrols, and they often hunt in groups, in which their success rates are higher than when they hunt solitarily (Goodall 1986; Boesch and Boesch 1989; Boesch 1994; Stanford et al. 1994; and Stanford 1996). In sum, hunting in chimpanzees is a predominantly male activity because males sight prey more often than females, are unconstrained by caretaking activities, are better able to overpower large prey, and are used to cooperating with each other. These differences boil down to two biological factors: sexual dimorphism and asymmetries in parental investment. But if these factors account for the existence of a male bias in hunting, they do not explain why chimpanzees hunt in the first place. One obvious possibility is that they do so because they like meat and/or because meat provides important nutrient gains. But another, nonexclusive, possibility is that hunting brings significant social benefits. Males that possess meat have been observed to share it preferentially with their male allies and with estrous females. On this basis it has been suggested that hunting is a mating strategy and/or a political tool (Teleki 1973; McGrew 1992; Stanford 1996, 2001; Boesch and Boesch-Achermann 2000; Mitani and Watts 2001).

Be that as it may, the chimpanzee data suggest that as soon as hominids began to hunt large and/or highly mobile prey, this activity was largely male biased. Notwithstanding that bias, sexual specialization at that stage was merely fragmentary, both sexes gathering the same types of food essentially, though males brought back large and/or difficult-to-catch game more often than females. However incipient, the male hunting bias was already producing a rudimentary form of sexual specialization, which may be expressed as follows: *de facto intrafamilial coprovisioning + male hunting bias → proto–sexual specialization.* It follows from this reasoning that the male hunting bias among hominids had nothing to do, initially, with parental care. It did not evolve as an integral part of a system of cooperative provisioning—for the same reason it does not serve this function in chimpanzees. However, male hunting was providing mothers and offspring with a new, albeit irregular, source of nutrients. Thus it constituted a preadaptation for the next step in the evolution of the sexual division of labor: the transition from fragmentary to complete sexual specialization, or from partial to maximal complementarity between the sexes.

Borrowing from economic models of marriage markets and resource allocation, Gurven and Hill (n.d.) argued that sexual specialization is more productive than the summed production of individuals doing the same thing, and they identified factors favoring sexual specialization in hunter-gatherer societies. Briefly, essential macronutrients—lipids, proteins, and carbohydrates—are found in different types of food. Meat is rich in protein and lipids, whereas plant products are rich in carbohydrates and sugars but low in fat and proteins. This alone means that several categories of food must be obtained to complete an omnivorous diet. Second, different kinds of foods are acquired by different techniques, some of which require extensive learning periods during which individuals become progressively more competent (Kaplan et al. 2000). In this context, Gurven and Hill argued, "specialization is a likely, if not inevitable, efficient outcome that maximizes household utility among cooperating individuals that divide their labor to target different kinds of food." At this point in human evolution, sexual specialization had gone from fragmentary to complete.

To summarize, I have argued that the sexual division of labor was the outcome of a specific concatenation of unrelated events, namely, (1) bipedalism, which rendered gathering possible, (2) pair-bonding, which created food-sharing biases among primary kin, and (3) a chimpanzee-like male hunting bias, which operated as both a template and a springboard for complete sexual specialization. This implies that the sexual division of labor existed in various forms well before it underwent selective pressures relating to its parental function.

Human families as we know them are variants of the same model, the latest in a series of models that thrived in the hominid evolutionary tree. One may envision polygynous or monogamous families that were essentially mating units in which the father's contribution was basically protective, others in which family members enjoyed some level of mutual tolerance around food gathered on a sporadic or regular basis, and still others in which mutual coprovisioning was characterized by various degrees of sexual specialization. But however different, all versions of the hominid family shared the same fundamental feature: the enduring breeding bond. All had in common the father's presence and recognition. In the rest of the book I analyze the far-reaching consequences of fatherhood on hominid social life.

12

Pair-Bonding and the Reinvention of Kinship

There could be no system of kinship through males if paternity was usually, or in a great proportion of cases, uncertain. The requisite degree of certainty can be had only when the mother is appropriated to a particular man as his wife, or to men of one blood as wife, and when women thus appropriated are usually found faithful to their lords.

John McLennan (1865, 65)

Pair-bonding has consistently occupied a central position in models of human evolution among social anthropologists, paleoanthropologists, and primatologists alike. But remarkably, the focus has been on its evolutionary origins, not on its multifarious *consequences* for social structure. As we shall see, the evolution of pair-bonding marked the transition from the deep structure of *Pan*-like societies to that of human society.

The Fundamental Equation of the Exogamy Configuration

Some thirty years ago Robin Fox suggested that the uniqueness of human society lay not in inventing something new but in combining already existing elements, namely, "descent and alliance"—or kinship and pair-bonding. Notwithstanding a few exceptions, such as the hamadryas baboon, Fox was basically right in emphasizing that much of the originality of human society stemmed from the combination of primatelike kin groups and pair-bonding. Equally important, Fox argued that exogamy ("the use of kinship to define the boundaries of marriage") was born when the symbolic capacity superimposed itself on the original kinship/pair-bonding combination. Essentially, then, although he did not put it this way, Fox was implicitly proposing the following "equation": *kin group + pair-bonding + language → female exchange.* It is a basic tenet of this book that Fox was right to a large extent. But, somewhat paradoxically at first sight, it is precisely on this point that he ceased to be convincing. In retrospect, Fox came up with an insightful equation but stopped short of demonstrating its validity. He argued that upon evolving the capacity to symbolically label their kin, hominids were naturally led to treat the members of their own kin group as nonmarriageable and to exchange females with other kin groups. As pointed out earlier, there is a considerable gap in this reasoning, to the extent that the birth of reciprocal exogamy from this perspective looks like some sort of spontaneous generation. But this is not because the causal chain is intrinsically wrong; it is because it is not explicit enough. Fox did not really analyze the consequences of the evolution of pair-bonding for hominid society. He did not go into the details of how pair-bonds transformed the existing kin group, shaping in the process residence patterns, genealogical kinship, affinal kinship, descent, intergroup relations, marriage rules, and so forth. It is not an exaggeration to say that pair-bonding ultimately brought about a true metamorphosis of hominid society. Fox might have perceived this intuitively, but he did not substantiate his intuition.

I shall proceed with the needed demonstration, but with some important differences in relation to the above equation. First, I remove the symbolic capacity. To Fox it was language and the ability to label one's kin symbolically that generated exogamy. Notwithstanding the importance of the symbolic capacity, my aim is to show that prior to the evolution of language human society had already attained levels of structural complexity unequaled in the primate order and was nearing its modern deep structure at a strictly behavioral, nonnormative level. Thus my removal of language from the equation is not to say that it played a minor role in the evolution of female exchange. I do so instead for analytical purposes as a means to characterize the presymbolic state of the exogamy configuration. Second, the equation I am concerned with differs from Fox's in that in keeping with the ancestral male kin group hypothesis the term "kin group" has been replaced by the term "male kin group," that is, by a chimpanzee/bonobo–like kin group. Third, in analyzing the impact of pair-bonding on hominid society I deal with the evolution of several aspects that have not yet been considered, notably the brother–sister kinship complex, unilineal descent groups, between-group pacification, and the content of exogamy rules. The equation I am developing is thus *male kin group + pair-bonding → exogamy configuration (12 components)*

I begin the reconstruction by focusing on how pair-bonding transformed kinship. We saw that kinship is of paramount importance in many primate groups, to the point that it shapes the group's entire social structure. But in other primate species, notably in our two closest relatives, its effect is much less extensive. Kinship relations are relatively silent in chimpanzees and bonobos, a fact that is somewhat paradoxical at first sight, because in line with the ancestral male kin group model this means that our hominid forebears evolved from a somewhat "kinshipless" society into one at the other extreme of the continuum, a society in which the impact of kinship on social relationships is so important as to be unparalleled in the primate order. I argue here that the evolution of pair-bonds was the prime mover behind this drastic transformation. Accordingly, I begin by considering the genealogical environment out of which pair-bonding evolved, using chimpanzees as a model.

Kinship in the Ancestral Male Kin Group

As far as genealogical structure and kin composition are concerned, male philopatry in a chimpanzee-like society is the mirror image of female philopatry in a macaque-like society. Males stay in their natal group while females leave it to breed in other groups. As a result, male localization generates kin groups comprising extensive, multigenerational patrilines (individuals related to a common male ancestor through males), but limited and fragmentary matrilines, as shown in Figure 12.1 (compare with Figure 3.1).

The Issue of Paternity Recognition

In chimpanzees and bonobos kin recognition between parents and offspring is marked by a fundamental asymmetry: mother–offspring recognition is well established, but father–offspring recognition is not. Preferential relationships between mothers and daughters and between

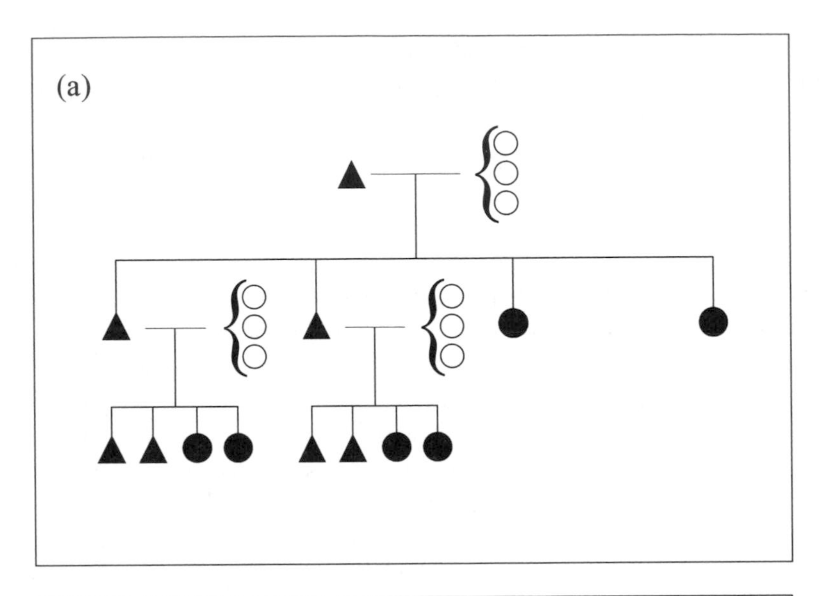

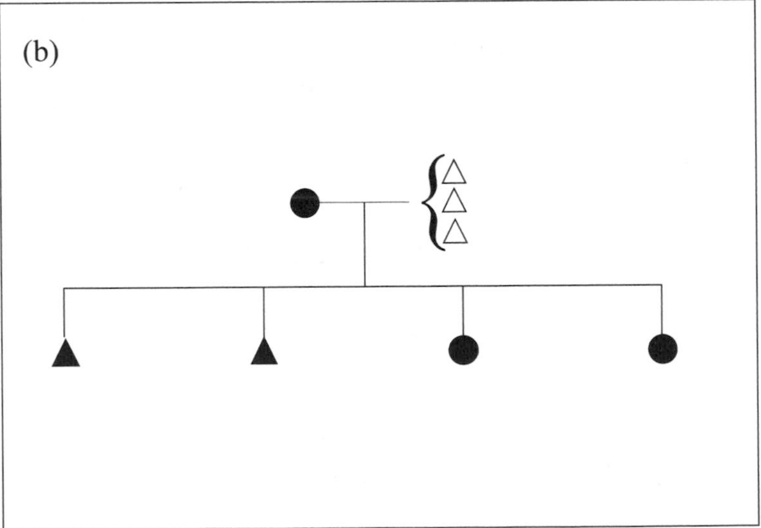

Figure 12.1. Comparison of a single patriline (a) and a single matriline (b) in a male-philopatric primate group. Circles: females; triangles: males. Black symbols: members of the same matriline or patriline; empty symbols: nonmembers. Patrilines are much more extensive than matrilines because males breed in their natal groups whereas females breed elsewhere as a rule.

mothers and sons have been observed in both species (Goodall 1986; Furuichi 1997), manifest in relatively high rates of proximity, grooming, and aiding in conflicts. Mother–son recognition is also apparent in incest avoidance (Goodall 1986; Pusey 1990a; Kano 1992; Moore 1993), although some exceptions have been reported (Constable et al. 2001). Paternity recognition, in contrast, is not documented in chimpanzees and bonobos, probably because familiarity-based mechanisms of paternity recognition are hardly at work in these species.

Mother–offspring recognition in animals stems directly from the uniquely intimate and enduring character of maternal care. For father–offspring recognition to be similarly based on a preferential bond, the father and mother should maintain an enduring preferential relationship—a pair-bond. If the mother maintained a preferential bond with the father during the conception period, if that bond persisted well after the offspring's birth, and, furthermore, if the father–mother bond was exclusive, the father would be in a position to recognize his own offspring as his mate's close infant associate. Reciprocally, the offspring would recognize his father as his mother's closest adult male associate. But we are concerned here with primate species in which a female typically mates with several males and does not normally maintain long-term exclusive relationships with particular ones. Another possibility would be for the father to maintain a preferential bond with the mother only over the days surrounding her ovulation and to identify his future offspring on this basis alone. This is not impossible, but it is more demanding, cognitively speaking.

In any case, such familiarity-based mechanisms of paternity recognition are unlikely in chimpanzees and bonobos for two reasons. First, levels of promiscuity are extremely high in these two species. Females copulate with a large number of males, if not all group males, and at very high frequencies. It has been estimated that female chimpanzees copulate between 400 and 3,000 times per conception, and female bonobos between 1,800 and 12,100 times per conception (Wrangham 1993, 2002). Moreover, consistent, long-term preferential relationships between males and females—hence between fathers and mothers—have not been observed in either species.

Other mechanisms of paternity recognition might be at work, however, namely phenotype matching (defined in chapter 3). In theory, a father might recognize his offspring by comparing its physical characteristics with his own. A number of genetic studies on paternity recognition have been carried out in *female*-philopatric species, principally macaques and baboons. The majority reported no evidence of preferential relationships between fathers and offspring (Barbary macaques, Paul, Kuester, and Arnemann 1992, 1996; Ménard et al. 2001; baboons, Erhart, Coelho, and Bramblett 1997; sooty mangabeys, Gust et al. 1998; for reviews, see Rendall 2004; Strier 2004; Widdig 2007) nor of incest avoidance between fathers and daughters (Kuester, Paul, and Arnemann 1994). However, paternity recognition was reported in one study of baboons in which fathers were observed to intervene preferentially on behalf of their offspring in conflicts with third parties (Buchan et al. 2003). In this study the possibility of phenotype matching could not be excluded, nor could familiarity-based mechanisms. Returning to chimpanzees, one experimental study reported phenotype matching based on visual cues. Females were able to match photographs of mothers with those of their sons (but, intriguingly, not those of mothers and daughters), suggesting that they recognized physical traits shared by close relatives (Parr and de Waal 1999). But this experiment is hardly relevant for understanding paternity recognition, because a father would need to recognize visual similarities between his own body—of which he has an incomplete image at best—and his offspring. This would seem unlikely. Phenotype matching might operate through other modalities, olfaction, for example, but at present this is unknown.

To sum up, paternity recognition in chimpanzees is either absent altogether or incomplete and inconsistent. Even positing some degree of paternity recognition based on phenotype matching, the rates of father–offspring interactions would not be consistent and high enough to reveal the genealogical structure linking males (the agnatic kinship structure), as maternity recognition reveals the uterine kinship structure in female kin groups.

Two Types of Siblings

In primate groups in which mating is promiscuous, most siblings are half-siblings. Siblings born from the same mother are likely to have been fathered by different males and thus to be maternal half-sibs. Reciprocally, siblings sharing the same father are likely to have different mothers and be paternal half-sibs. There are thus few full siblings in such groups. For example, in one group of free-ranging rhesus monkeys, only 5 percent of 324 sibling dyads were related through both parents (Widdig 2002; see also Widdig 2007). The situation must be the same in chimpanzees and bonobos, where mating is highly promiscuous.

Again there appears to be a basic asymmetry between the two categories of siblings as far as kin recognition is concerned. Preferential relationships and incest avoidance between *maternal* siblings are well documented in chimpanzees (Goodall 1986; Pusey 1990a, 2001), as discussed in detail in chapter 13. By contrast, kin recognition between *paternal* siblings is unlikely on theoretical grounds and based on the available empirical evidence. The existence of paternal sibships in primate groups arises from the ability of certain males, often the most dominant ones, to monopolize more than one sexually receptive female. A male who fertilizes a number of females over a relatively short period of time generates a paternal sibship whose members are close in age (Altmann and Altmann 1979; van Hooff and van Schaik 1994; Strier 2004). Paternal siblings may also belong to different age classes, because the reproductive career of a male in a given group often extends over several years.

It has been suggested that preferential cooperation between chimpanzee males of similar age might reflect the effect of paternal kinship. Indeed, if a large proportion of age mates are paternal brothers, males that cooperate with their age mates may do so because they are kin (Mitani et al. 2002; Silk 2002; Strier 2004). Therefore, what primatologists describe as preferential bonds between age mates and ascribe to the effect of age similarity might in fact represent instances of nepotism between paternal brothers. This hypothesis is highly conjectural. One version implies that males would discriminate their paternal brothers among age mates, using phenotype matching. In theory, a male might use his father as a referent and recognize similarities between his father and his paternal brother. But this process assumes paternity recognition, which is at best inconsistent in chimpanzees and bonobos. Another possibility is for a male to use himself as a referent and perceive physical similarities between himself and his brother. This does not appear to be the case, however. A recent study combining long-term observations of wild chimpanzees and molecular genetic techniques found that paternal brothers ($n = 25$) did not affiliate or cooperate preferentially (Langergraber, Mitani, and Vigilant 2007). Another study reported no relation between the participation of male chimpanzees in joint hunting and the males' degree of genetic relatedness, including paternal relatedness (Boesch, Boesch-Achermann, and Vigilant 2006). In sum, it appears that nepotism among siblings in chimpanzees is limited to maternal siblings.

Ego's Other Kin

There is every reason to believe that chimpanzees do not recognize their relatives on either their mother's side or their father's side, and for distinct reasons. Let us first consider ego's *matrilateral* kin. In male-philopatric species males live with their mothers, but as a rule they do not live with their mothers' kin because the mothers are immigrants whose relatives stayed behind. Thus males do not normally grow up with their maternal grandfather, maternal grandmother, and maternal uncles, for they live in another community and are strangers. Essentially, then, chimpanzees do not recognize their matrilateral relatives because mothers do not normally raise their children in the group in which the mothers were born. This is illustrated in Figure 12.2.

Interestingly, however, mothers sometimes do breed in the group

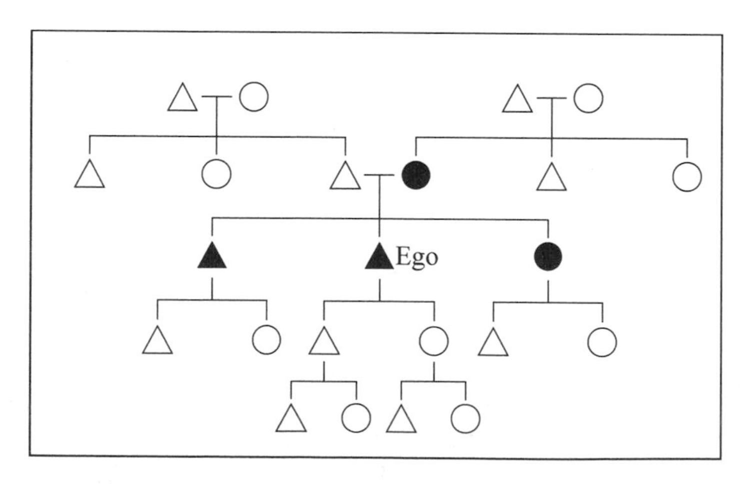

Figure 12.2. Ego's domain of kindred (recognized kin) in a male-philopatric, chimpanzee-like group in which males and females mate promiscuously, assuming that all females breed outside their natal group. Five generations are pictured, but normally only three coexist. Black symbols: ego's kindred; empty symbols: ego's other kin. Only primary and secondary kin are shown.

in which they were born, a situation that provides us with an opportunity to better appraise the *potential* for chimpanzees to recognize their matrilateral kin. In the Gombe colony of wild chimpanzees, 50 percent of the females stayed and bred in their natal group. A male born from a mother who bred in her natal group could thus be living with his maternal grandmother and his maternal uncles. Jane Goodall has provided anecdotal but revealing accounts of preferential bonds between grandmothers and grandsons and between maternal uncles and their sororal nephews. For example, she described the relationships of a young male with his three maternal uncles. The younger uncle was "fascinated by his young nephew," carrying him and playing with him on a regular basis. When the young uncle died at eight years of age, the nephew "lost not only his principal playmate but also his adolescent male role model," wrote Goodall (1990, 112, 169–170). The nephew also "developed friendly bonds" with his two other uncles, both high-ranking adults. Preferential bonds between maternal uncles and their nephews are particularly interesting in view of the importance of the brother–sister kinship complex and avuncular relationships in human kinship. Goodall also outlined the development of a strong bond between a grandmother and her grandson from the time of his birth. The grandmother frequently approached, groomed, and played with her grandson. Clearly, the kin-recognition potential of chimpanzees encompasses not only primary maternal kin (mothers, daughters, and maternal siblings) but secondary maternal kin as well. However, female dispersal does not normally allow the recognition of one's secondary uterine kin.

The situation is different with ego's *patrilateral* kin, but the outcome is apparently the same. The males being localized, ego lives with its male kin, which include its father, paternal grandfather, paternal uncles, and the uncles' sons (patrilateral parallel cousins)—but not its paternal aunts and their offspring (patrilateral cross-cousins). But given that chimpanzees do not maintain long-term bonds with their fathers, ego can hardly recognize its father's kin on this basis. Thus, although male chimpanzees and bonobos are surrounded by several categories of agnatic kin, they probably do not discriminate them.

Overall, then, the domain of kindred in our two closest relatives is normally quite limited. It generally includes primary uterine kin, specifically the maternal fraction, namely, mother–daughter dyads, mother–son dyads, and maternal siblings. Secondary and tertiary kin can hardly engage in kinship-based favoritism either because they do not grow up in the same group or, if they do, they have no means to recognize each other. As we saw, however, the recognition of secondary kin lies well within the cognitive abilities of chimpanzees. Assuming that early hominids had a chimpanzee-like social structure and formed male kin groups, one may infer that male relationships in these groups were not differentiated on the basis of generation and degree of kinship. Nor were there any group-wide kinship structures such as those observed in female kin groups. It is precisely in this sense that we may speak of a "kinshipless" society. However, such a social structure was just one (big) step away from a society with full-fledged kinship networks, and that step was the recognition of paternity.

Fatherhood

The evolution of long-term breeding bonds (pair-bonds) in the course of hominid evolution transformed a promiscuously mating group into a group composed of biparental families (whether monogamous or polygynous). From that time on, children were in a position to recognize their fathers, and the fathers their children. In theory, the developmental processes involved in father–child recognition are similar to those underlying the recognition of uterine sisters, in which the mother acts as the mediator between the two. A child and his father are bound to become disproportionately familiar with each other by virtue of their common preferential bond with the female who is a mother to one and a "wife" to the other. If only on that basis, father and child are biased to interact with each other and develop a preferential bond. But simultaneously and independently, the child is also in a position to learn his father's identity by recognizing the relationship between his mother and his father and the specific characteristics of that bond. Specifically, his father may be that particular adult male who spends a disproportionate amount of time with his mother, protects her against other males, has sexual interactions with her, and so on.

Importantly, father–child recognition does not require or imply any form of paternal care, contrary to mother–child recognition, which follows directly from maternal care. Father–child recognition is similar to sibling recognition, not to mother–child recognition. It is not a direct, dyadic process, but one based on the mediation of a third party. Thus, as long as the mother–father bond lasts, father–child recognition ensues from it. This has implications regarding the evolution of pair-bonding itself. It means that father–child recognition was not dependent upon pair-bonds evolving as parental partnerships. As discussed in the previous chapter, pair-bonding probably evolved for reasons that had nothing to do with paternal care. But pair-bonds created favorable grounds for the subsequent evolution of preferential interactions between fathers and children, including protection, food sharing, and a host of cooperative partnerships, from coalition formation to hunting.

The idea that father–child recognition unfolds not so much from paternal care but from the father's long-term association with the mother bears upon one well-recognized anthropological fact. In the vast number of human societies, legal fatherhood is determined on the basis of the mother–husband association. "Almost all societies," wrote Sahlins, "adhere, implicitly or explicitly, to the dictum of the Napoleonic code in this respect: the father of the child is the husband of the mother" (1960, 81). And as noted by Holy, "The principle of the Roman law according to which the child's pater is a man who can prove that he married the child's mother, or the principle of proverbs such as 'children belong to the man to whom the bed belongs,' is the basis for assigning legal fatherhood in many societies" (1996, 17). This is precisely what one would expect if fatherhood had evolved from a nonhuman primate legacy. Indeed, what better criterion of paternity than a male's long-term association with the known mother? Any male of reproductive age may claim paternity for a child born to a given mother, but the mother's primary male associate is the most probable sire among them.

Primate studies validate the claim that long-term breeding bonds per se may occasion father–offspring recognition and favor the evolution of some forms of paternal care. Consider, for example, our closest primate relative exhibiting stable breeding bonds, the gorilla. Most gorilla reproductive units include a single reproductive male living permanently with a number of females and their offspring. Infanticide by adult males is common in mountain gorillas, accounting for nearly 37 percent of infant mortality (Watts 1989). The most straightforward proof that fathers distinguish their own infants from unrelated ones is that only extragroup adult males, that is, nonfathers, kill infants. To avoid infanticide, females form a lasting association with a powerful adult male, with whom they reproduce, and the ensuing familiarity between the father and his offspring presumably inhibits infanticide. There is also some evidence that daughters are inhibited from mating with the silverback, who is likely to be their father. Females often emigrate before they start breeding, but in situations where they were observed to start mating in their natal group, they most often did so with a younger male (Stewart and Harcourt 1987; Pusey 1990a). Finally, father–offspring recognition in gorillas is also manifest in various forms of paternal nepotism: defense against predators, protection against extragroup males during interunit encounters, marked levels of tolerance toward infants (who are highly attracted to the silverback), and interventions in the infants' disputes (buffering them against possible injuries). Interestingly, the silverback can also serve as an "attachment figure with maternal orphans" (Stewart 2001). Immatures who lose their mothers (either through emigration or death) greatly increase their association with the silverback (Watts and Pusey 1993; Stewart 2001).

The Institutionalized Denial of Paternity

Implicit in the above reasoning is the idea that fatherhood has evolutionary roots and is biologically grounded, in the same manner that motherhood has phylogenetic origins. But if motherhood is commonly construed as a cultural extension of the mammalian mother–daughter bond, fatherhood is often thought of as a cultural construct. "Institutionalized fatherhood, unlike motherhood," stated Fortes, "comes into being not by virtue of a biological . . . event, but by ultimately juridical, societal, provision, that is, by rule. Fatherhood is a creation of society" (1983, 20). Among the main arguments in support of such a position is the observation that not all cultures recognize the procreative and genealogical dimension of paternity. For example, in societies organized around matrilineal descent groups, the father's role as genitor of his own children may be negated, as first described in detail by Malinowski for the Trobrianders (Malinowski 1929). The father is said to contribute no physical substance to his children, and the mother to be impregnated by spirits. In these societies the mother's brother has several of the father's attributions and is called by anthropologists the (avuncular) pater. This type of evidence, it is argued, proves that fatherhood is not a universal feature of human societies.

Two points must be made in relation to this claim. First, the people themselves are not necessarily ignorant about biological paternity. Rather, they choose to deny it for ideological reasons, as anthropologists have argued. Referring to the Trobrianders, Holy (1996) summarized some of the evidence for the recognition of physical paternity. Although biological paternity is denied, the child is always said to physically resemble its father, never its mother or maternal relatives. That resemblance is said to derive both from the father nursing his child in his arms and from the copulatory act that shapes the fetus. After the father's death his kinsmen visit his children in order to see him in them. Thus, concluded Holy, "it would be wrong to assume that the Trobrianders have no notion whatsoever about the genealogical connection between the child and its father" (19–20). Second, even positing that human beings in general are unaware of the consanguineal link between fathers and their biological children, this is no argument to conclude that fatherhood is a "creation of society" with no biological basis. For that matter, nonhuman primates are wholly unaware of the consanguineal links connecting them to their kin, and they nonetheless discriminate them. The important point is whether biological paternity in humans commonly translates into social situations that allow fathers and offspring to form preferential bonds. And this is apparently the case. Not only is the recognition of physical paternity manifest in what people say about the father and in how they treat him, but, more importantly from an evolutionary perspective, it is equally manifest in the existence of special types of bonds between children and their biological fathers. Here I can do no better than quote Fortes himself about the dual nature of fatherhood in societies where paternity is most explicitly denied, those with matrilineal descent:

> It has long been known . . . since Malinowski's famous description of Trobriand matrilineal fatherhood . . . , that paternal care, unlike the ideal undivided maternal care, is compounded of two elements. There is the jural element of authority and responsibility that has the sanction of society at large . . . And there is, side by side with this, the effective element of altruistic devotion based on attachment not jural right. And as Malinowski showed and has been frequently confirmed since then, this two-sided fatherhood that is vested wholly in the genitor . . . in patrilineal systems . . . *is split between genitor and avuncular pater in matrilineal systems. Malinowski described the matrilineal father* [the

genitor] *as making free gifts to his son quite in the spirit of what I am here attributing to the ideal or paradigmatic mother, that is without expectation of reward or return.* (1983, 24, my emphasis)

Thus matrilineal societies do not eliminate fatherhood and the associated bonds; rather, they split fatherhood between the biological father and the maternal uncle, and both types of relationships translate into distinct types of long-term bonds with children. The reason physical paternity is still recognized in societies that negate the procreative role of the biological father is that the ideological denial of paternity does not eliminate the special bond linking the mother to the biological father. As argued previously, this bond alone is all that is needed for a father to recognize his wife's children among all others, and for a child to recognize his father among other adult males. To prevent paternity recognition altogether, one would need to eliminate the conjugal bond in the first place. Institutionalized fatherhood may be a creation of society, as Fortes argued, but institutionalized fatherhood is itself derived from biological fatherhood.

The Development of Agnatic Kinship Structures

As soon as hominids could recognize their father, they were in a position to recognize their father's relatives, including their paternal grandfather and grandmothers, their paternal uncles and aunts, and their patrilateral cousins. The processes involved in the recognition of patrilateral kin may be readily inferred from those described in relation to uterine kinship. In that model, the mother was the central reference point from ego's perspective. Accordingly, ego's father should be the central reference figure here. To recognize, say, his paternal grandfather, a male must witness a preferential bond between his father and his father's father. In this way the male is disproportionately exposed to his grandfather—kin recognition based on ego's own experience with the grandfather—and is simultaneously in a position to recognize his grandfather as that individual who interacts in some specific ways with his father—kin recognition based on ego's recognition of others' relationships. Crucially, grandfather–grandson recognition is possible only if fathers and sons engage in enduring, or lifetime, bonds with each other. This raises the question of how such bonds might have evolved.

One simple process is suggested by relationships between adult and adolescent male chimpanzees. As they start to travel independently of their mothers upon reaching adolescence, male chimpanzees often attempt to form bonds with particular adult males. Anne Pusey has described such bonds in the Gombe colony (1990b). One adolescent male associated with his older brother, an association that persisted into adulthood. Another adolescent male associated with a young adult male who was probably not a maternal relative. When the community split into two distinct groups, the two were still together. The third case is particularly significant. An adolescent male formed an association with the group's alpha male, who had been a close associate of his mother at a time when the adolescent was still traveling with his mother. Again, the association between the two males persisted into adulthood. This case shows that young males may be biased to establish a bond with their mothers' male friends. Considering that from an evolutionary angle the father is no more than the mother's primary adult male associate, the last example illustrates how pair-bonding may have led to long-term bonds between fathers and sons. Jane Goodall used the term "follower" to describe such a relationship between a youngster and a particular adult male, stressing that the bond "is almost entirely initiated and maintained by the follower" and that "the advantages to the adolescent are obvious; fascination for the older male assists the younger in loosening childhood dependency on his mother, and his exposure to male behaviors (such as hunting and patrolling) provides opportunity for observing and learning these patterns" (1986, 202). Goodall also noted that as he matured, the adolescent male received an increasing amount of aggression from the adult male and that such experience taught him about male temperament in general and rendered him "more adept at finding ways of avoiding trouble and raising his status in the adult male dominance hierarchy."

Returning to the issue of pair-bonding, the chimpanzee evidence suggests that enduring father–son bonds might have been initiated and maintained by the sons themselves, hence independently of and prior to the evolution of active forms of paternal care. It also shows that the development of preferential bonds between fathers and sons merely required fathers to be selectively *tolerant* toward their sons, a condition that would be facilitated by their high degree of relatedness and associated familiarity.

With the evolution of pair-bonding and lasting father–son bonds, the domain of kindred in male kin groups expanded to a considerable extent. This can be seen by comparing Figures 12.2 and 12.3. Besides his mother and maternal siblings, a male could now recognize his children, father, father's kin, and brothers' children. But he still could not recognize his mother's kin and his sisters' children because they lived in other groups. *Bilateral kinship,* the recognition of kin on both the father's and mother's side, would have to await the evolution of the first social entities encompassing distinct local groups, the primitive tribe (chapter 14).

I began this chapter with a paradox. Kinship is a central organizing factor in small-scale human societies, but humans apparently evolved from primates with a social structure in which kinship's influence was minimal. The solution to this paradox is simple. In a chimpanzee-like male kin group, the agnatic relatedness structure is present but is not socially discernible because father–offspring links are not recognized; the genealogical structure thus lies dormant. To reveal it, paternity recognition is needed, and this is precisely what pair-bonds accomplished. With paternity recognition, the role of agnatic kinship in structuring social

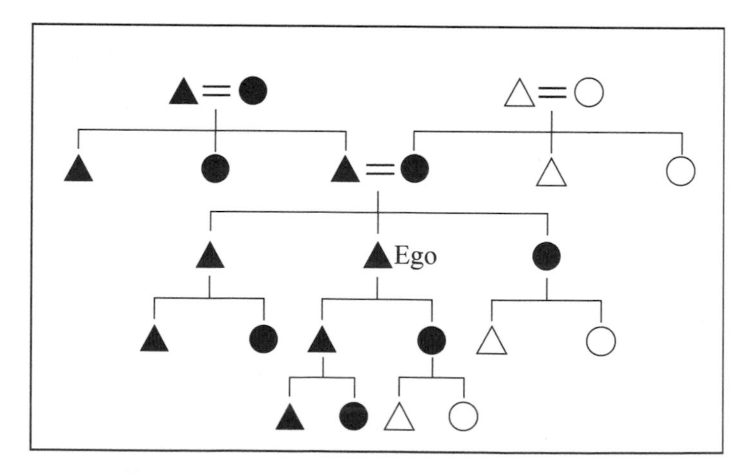

Figure 12.3. Ego's domain of kindred in a male-philopatric group in which males and females form pair-bonds. Five generations are pictured, but normally only three coexist. Black symbols: ego's kindred, assuming that ego recognizes its uncles, aunts, nieces, and nephews; empty symbols: ego's other kin. Only primary and secondary kin are shown.

life in male kin groups became comparable in extent to the role of uterine kinship in structuring female kin groups.

Agnatic kinship structures are similar to uterine kinship structures as described in chapter 3. They are a correlate and byproduct of the enduring nature of father–son bonds, in the same manner that uterine structures are a byproduct of the lasting mother–daughter bond. All males have fathers, and all males are related to each other through males. As a result, the father–son bond alone connects all males and reveals the group's genealogical structure. Earlier I argued that matrilineal dominance systems arose in female kin groups when mothers began to help their immature daughters against the females to whom the mothers were dominant (Figure 3.3). The application of this principle to male kin groups helps explain the origin of agnatic kinship structures. Suppose that fathers similarly began to consistently help their immature sons against the males they were dominant to and that sons, as a result, socially inherited their father's dominance rank. One would automatically obtain a "patrilineal dominance system," a group-wide social pattern entirely predictable from knowledge of the group's agnatic kinship structure. Such speculation finds support in the observation that mothers have a marked influence on their sons' status in the male dominance order in bonobos (Furuichi 1997). If males benefit from the interventions of female protectors against other males, one may infer that they would benefit even more from their fathers' help. This principle has a general applicability. Inasmuch as fathers and sons maintain lasting bonds, any recurrent social pattern characterizing the father–son relationship—for example, grooming, food sharing, coalition forming, or cooperative hunting—is bound to produce a group-wide social pattern paralleling the group's entire agnatic structure. In short, the evolution of cooperative partnerships between fathers and sons produced the first kinship structures in male kin groups.

13

Biparentality and the Transformation of Siblingships

> This structure [the avunculate] rests upon four terms (brother, sister, father, and son) . . . [It] is the most elementary form of kinship that can exist. It is, properly speaking, the unit of kinship.
>
> *Claude Lévi-Strauss (1963, 46)*

Chimpanzees and bonobos consistently recognize a single parent, their mother. They have monoparental families. Human beings, as a rule, recognize both their mother and their father. This raises a basic question: How did biparentality affect relationships between siblings? This is not a trivial issue. From an evolutionary perspective, the nature of bonds between siblings in our species appears to be the single most original characteristic of human kinship. A particularly puzzling phenomenon is the importance of the brother–sister bond in so many human societies. The brother–sister kinship complex integrates phenomena like the avunculate—the special relationship between a brother, his sister, and the sister's son—cross-cousin marriage prescriptions, and various kinship terminologies. The brother–sister kinship complex has no equivalent in nonhuman primates. Not only is it uniquely human, it is a cornerstone of human society's deep structure. Referring to the avunculate, Robin Fox wrote: "It seemed to me worthwhile to review this relationship, which has been as important to anthropological theorizing as the incest taboo, and to ask whether or not it is indeed the first true cultural incursion into nature" (1993, 193). Fox's own response was that the avunculate was very close to "the defining principle of humanity and culture." The epigraph further attests to the cardinal importance of brother–sister bonds in Lévi-Strauss's scheme.

As we shall see in this chapter, the brother–sister kinship complex stems from several distinct phylogenetic components that eventually found their way to our species and merged. Prominent among these, I shall argue, was pair-bonding and its correlate, biparentality. The evolutionary impact of biparentality on sibling relations is a topic more complex than it appears at first sight, and to my knowledge one that has not yet been investigated. On theoretical grounds, pair-bonding should have altered sibling bonds both because it created a whole new structure of relationships within the family and because it brought about changes in life-history traits, notably a reduction in the interbirth interval. The two routes—structural and life-historical—are independent of each other in that each affects sibling bonds whether or not the other route is operating. Together they transformed relationships between siblings by strengthening them considerably. In so doing they set the stage for the development of the brother–sister kinship complex and the evolution of units of primary agnates (fathers, sons, and brothers) whose cohesiveness is unparalleled in nonhuman primates. To understand how these changes came about, one needs a comparative base line; one must know about the nature of sibling relations in the ancestral male kin group prior to the evolution of pair-bonds. Accordingly, I start by describing sibling relations in our two closest relatives, beginning with brothers.

Chimpanzee Siblingships

Jane Goodall described bonds between maternal brothers in chimpanzees that began in infancy and endured well into adulthood. Such fraternal friendships involved cotraveling, cofeeding, play, mutual grooming, reciprocal support, and conesting. "A friendship of this kind," she wrote, "can have far-reaching consequences for a male eager to rise in the dominance hierarchy, because brothers typically support each other during social conflicts and almost never take sides against their sibling" (1986, 177). In contrast with her observations, two studies carried out in different populations of wild chimpanzees reported no preferential relationships between uterine kin in general (Goldberg and Wrangham 1997; Mitani, Merriwether, and Zhang 2000; Mitani et al. 2002). In these studies maternal kinship was assessed on the basis of mitochondrial DNA haplotype sharing by males. This technique identifies all categories of uterine kin in a given group but indiscriminately so, that is, without differentiating between maternal brothers and other categories of maternal kin, such as cousins. The two studies reported that none of the dyads identified as uterine kin affiliated or cooperated preferentially with each other, whereas several other pairs of males did. If one assumes that uterine kin in these populations did include some pairs of maternal brothers, which appears most likely, one must conclude that fraternal nepotism was absent in these populations. But in a recent study carried out in the same population studied by Mitani et al. (2000, 2002), Langergraber, Mitani, and Vigilant (2007) used molecular genetic techniques that discriminated between maternal siblings and other uterine kin among adolescent and adult males. They found that maternal brothers did affiliate and cooperate more often than unrelated males. The discrepancies between the two sets of studies remain to be clarified, but all in all the available evidence points to the existence of fraternal nepotism in chimpanzees, whether the latter occurs consistently or inconsistently across pairs of brothers and across populations (see also Boesch, Boesch-Achermann, and Vigilant 2006; Hashimoto, Furuichi, and Vigilant 1996).

In nonhuman primates in general, maternal siblings learn to recognize each other through their common bond with their mother. By virtue of growing up in close proximity to the same maternal figure, siblings undergo proximity and familiarity biases for developing enduring preferential bonds with one another. They experience what will be referred to here as a period of shared developmental familiarity near a parental mediator. In chimpanzees two factors determine the duration of that period. The first is that males become sexually mature at about nine years of age and maintain a remarkably close association with their mothers until that age. Pusey (1990b) reported that juvenile males spent a median of 88.5 percent (range: 84–95 percent) of their time with their mothers (see also Goodall 1986, 166.) The attainment of puberty is followed by a rapid decline in mother–son association, which correlates with testicular growth. Upon reaching adolescence, males spent most of their time away from their mother and close to adult males and estrous females (Pusey 1983, 1990b; Goodall 1986, 81). The second factor is the duration of the interval between consecutive births, about five to six years in this species. Combining the two factors, it follows that two brothers born consecutively will experience at least four to five years of shared developmental familiarity near their mother. Nonconsecutively born brothers, in contrast, experience a much shorter period of developmental familiarity because they differ in age by at least ten to twelve years; the older brother will already be spending most of his time away from his mother when his younger brother is born. On this basis alone, one expects consecutively born brothers to form different types of bonds than other pairs of brothers do, for example, stronger cooperative partnerships. Owing to the paucity of information on sibling bonds in chimpanzees and bonobos, this remains to be ascertained.

Also, if only because they are closer in age consecutively born brothers have more in common in terms of social status, skills, competitive abilities, and mutual value as social partners compared to brothers more separated by age. This reasoning finds support in the observation that male chimpanzees cooperate preferentially with males of similar age. John Mitani and his colleagues (2002) analyzed cooperative partnerships involving alliances, meat sharing, and boundary patrols among male chimpanzees. They found that males selected their partners on the basis of age similarity and dominance-rank similarity. For that matter, one does not expect uterine kin to invariably constitute the best of all available partners regardless of the cooperative activity; relative social *competence* should take precedence over kinship whenever cooperating with a more competent nonkin is likely to be more beneficial than cooperating with a less competent kin (Chapais 2006).

The two factors—developmental familiarity and closeness in age—have convergent effects that lead to the following prediction. The nature and/or the strength of relationships between chimpanzee brothers should vary across pairs of brothers in relation to the magnitude of the age difference between them, and brother nepotism should be somewhat inconsistent across pairs. Although the available data do not make it possible to test this hypothesis satisfactorily, this discussion sets the base line against which one may better appreciate the consequences of pair-bonding for sibling bonds. Pair-bonding, I shall argue, transformed siblingships both by enriching the period of shared developmental familiarity and by reducing the average age discrepancy between siblings. I examine each effect separately.

Fatherhood and the Evolution of Strong Brotherhoods

The father's enduring association with the mother meant the arrival of a new and powerful parental mediator who would bring brothers closer to each other. As soon as fathers and sons could recognize each other, an immature male was in a position to develop a long-term cooperative bond with a closely related adult male, his father. The importance for a male of being able to cooperate with a parent of the *same sex* is clear. Excluding reproductive activity, most instances of cooperation in primates take place between same-sex individuals. Cooperative activities among females are widespread because they share several objectives relating to resource acquisition, social strategies, and breeding, which they can achieve through collaboration. In female kin groups, females cooperate in grooming, allomothering (taking care of another female's infant), aiding in conflicts, acquiring dominance status, and cofeeding (Chapais 2006; Silk 2006). Mothers and daughters are primary lifetime cooperative partners in these species (Fairbanks 2000; Silk, Altmann, and Alberts 2006b), basically because, in addition to sharing a high degree of kinship, they are of the same sex. This principle applies even to chimpanzees, a species in which females normally disperse away from their mothers. In the Gombe colony a fraction of females stay in their natal group and establish long-term friendships with their mothers. Goodall described such relationships between mothers and adult daughters as "the strongest of all bonds among adults." (1986, 174). Similarly, same-sex cooperation between males is relatively frequent in nonhuman primates because males, too, share various goals that they can achieve cooperatively. In chimpanzees, for example, males join forces in territorial defense, alliances in conflict, hunting, and meat sharing (Mitani et al. 2002). One may thus reason that if fathers and sons recognized each other in male-philopatric groups—as mothers and daughters do in female-philopatric groups—they would undergo selective pressures for forming lifetime cooperative bonds.

The role of parental mediators in shaping sibling bonds is decisive, and I take a brief detour to describe one particularly relevant manifestation of this, namely, the greater strength of bonds between maternal half-sisters compared to paternal half-sisters in macaques and baboons. While preferential bonds between maternal sisters have long been documented in these species, studies that looked for preferential bonds between paternal sisters had yielded negative results until lately (Fredrickson and Sackett 1984; Erhart, Coelho, and Bramblett 1997; Kuester, Paul, and Arnemann 1994). But in a more recent study that controlled for various confounding factors, Anja Widdig and her colleagues (2001) found that paternal sisters did affiliate at higher rates than unrelated females in rhesus macaques. This is all the more significant because paternal sisters do not maintain preferential bonds with their fathers, or if they do, the bonds are in no way comparable to those of maternal sisters with their mothers. Accordingly, Widdig and colleagues resorted to phenotype matching, not to familiarity-based processes, to explain their findings. The important point here is that while maternal sisters undergo a long period of developmental familiarity near a particularly effective parental mediator, their mother, paternal sisters do not. On this basis alone, one would expect maternal sisters to affiliate and cooperate at much higher rates than paternal sisters do. This was precisely the case. The affiliation index of maternal sisters (a composite measure based on spatial proximity, grooming, and approaches) was 171 times higher than that of paternal sisters (Widdig et al. 2001, Table 1). This discrepancy supports the idea that the strength of bonds between siblings is markedly affected by their history of developmental familiarity near a parental mediator. Similar results are apparent in data reported for savanna baboons (Silk, Altmann, and Alberts 2006a and b).

The centripetal force exerted by parental mediators on siblings helps us appraise the effect of pair-bonding on the evolution of hominid siblingships. In theory, pair-bonding should have made it possible for males to do what females do in female-philopatric groups: establish a lifetime bond with a same-sex parent. This, in turn, should have had important repercussions for sibling bonds. I use the term "brotherhood" to refer to brothers who maintain preferential bonds with each other as adults. Based on the previous discussion, I assume that *strong* brotherhoods in chimpanzees are concentrated among pairs of consecutively born brothers and that weaker ones characterize pairs of brothers with a larger age discrepancy, an assumption that remains to be empirically substantiated. It follows that if a chimpanzee mother gives birth to only two sons, born nonconsecutively, no strong brotherhood would result. This situation is illustrated in Figure 13.1a, in which two mothers each give birth to a daughter and two sons born nonconsecutively. After the evolution of pair-bonding, however, the same mothers would each produce one strong brotherhood (13.1b). Indeed, from the time fathers and sons first engaged in lifetime bonds, all brothers, whether they were born consecutively or nonconsecutively, would experience a period of shared developmental familiarity near their father. Older brothers would still be close associates of their father when their younger brothers were

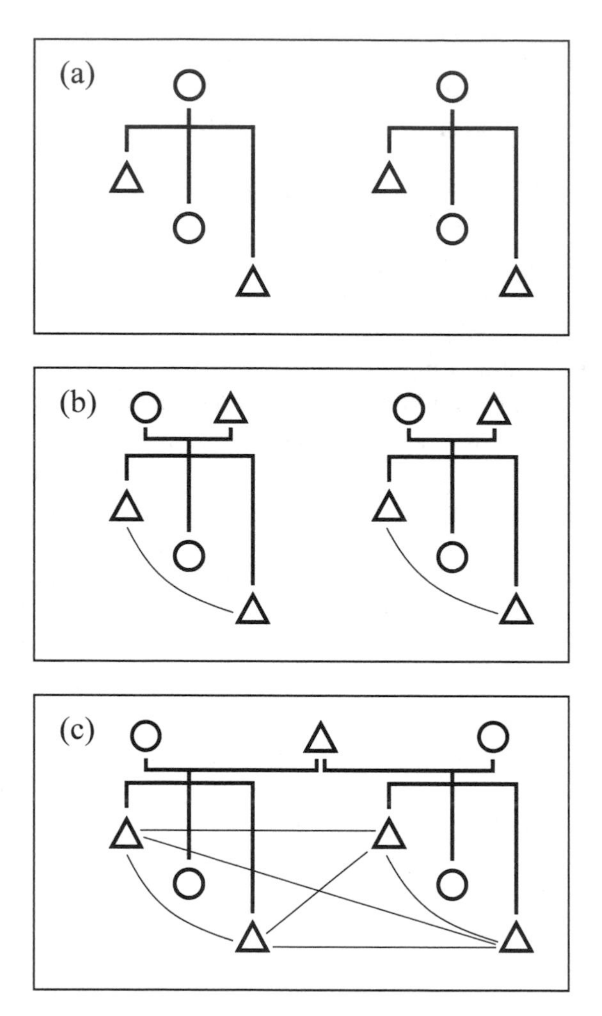

Figure 13.1. The composition of strong brotherhoods before and after the evolution of pair-bonding. In the three situations shown here, two mothers each give birth to one daughter and two sons born nonconsecutively. In a chimpanzee-like group with promiscuous mating (a) there are no strong brotherhoods. Monogamous pair-bonding (b) produces two strong brotherhoods (thin lines), each consisting of two full brothers. Polygynous pair-bonding (c) produces a single larger brotherhood consisting of four brothers (two pairs of full brothers and four pairs of paternal half-brothers).

born, and subsequently thereafter. Also, because father and mother were primary associates, lifetime bonds between fathers and sons should favor the prolongation of *mother*–son bonds beyond adolescence, so brothers would be pulled toward each other by two distinct parental mediators. Finally, polygynous pair-bonding created a new category of brotherhoods: those composed of paternal half-brothers who shared a period of developmental familiarity near their common father only (Figure 13.1c).

To summarize, pair-bonding transformed brother bonds in various ways. First, it *strengthened* brotherhoods because brothers were now brought closer together through two parental mediators, one of which was a same-sex parent with whom they could engage in various types of cooperative activities. Second, pair-bonding increased the *number* of strong brotherhoods because all brothers, independent of their birth order and whether they were related through the father only or through both parents, could experience extensive developmental familiarity biases. Fraternal nepotism became more consistent, if not generalized. Third, pair-bonding enlarged the average *size* of strong brotherhoods. From mostly dyadic in chimpanzees, strong brotherhoods could now include more than two brothers. In short, pair-bonding generated larger and more cohesive units of primary male kin, or *primary agnates* (a father and his sons), so male kin groups now comprised a number of units of primary agnates. These units were like thick kinship knots in a group-wide agnatic web, just as matrifocal units are in female-philopatric societies.

Fatherhood and the Brother–Sister Bond

Very little data are available on brother–sister bonds in chimpanzees and bonobos, which may reflect the fact that preferential relationships between brothers and sisters are less frequent than between brothers. Among the factors responsible for this is female dispersal. Female chimpanzees emigrate at eleven years of age on average (Boesch and Boesch-Achermann 2000, 44; Nishida et al. 2003, 108). As a result, a majority of brother–sister dyads in any social group are composed of an older brother and his preadolescent sister, more rarely of a younger brother and his older sister. For a male to know his older sister before she leaves the group, the two siblings must be born consecutively. If they are not, the older sister may already have left the group before her younger brother is born. Thus male chimpanzees come to know only a fraction of their older sisters but all of their younger sisters.

Maternal brothers and sisters may associate at high rates prior to the sister's first estrus, but such dyadic associations drop abruptly thereafter (Goodall 1986; Pusey 1990b, 216). In the absence of sufficient hard data on brother–sister relationships, a reasonable assumption is that the strength of any such bond correlates with the duration of the corresponding period of developmental familiarity. Accordingly, preferential bonds between brothers and sisters should be concentrated among consecutively born siblings, those with an age difference of five to six years. For example, if we consider a mother who had two sons born consecutively and a daughter born afterward, the sister had an extensive period of developmental familiarity with the younger of her two brothers, but not with the adolescent one, who is ten to twelve years older. The sister, therefore, should have a stronger affiliative bond with the younger brother (Figure 13.2a). Based on this assumption, the evolutionary consequences of pair-bonding for brother–sister bonds should have been no less important than its consequences for brotherhoods. Pair-bonding should have strengthened brother–sister bonds through at least three different processes involving three categories of individuals: fathers, grandmothers, and sisters-in-law (Figure 13.2b). First, assuming that fathers maintained lifetime preferential bonds with their sons, and provided they maintained preferential bonds with their daughters until they left their natal group, the father could act as a parental mediator between his daughters and his sons, his presence ensuring a period of shared developmental familiarity between younger sisters and their older brothers. The father would have been especially well positioned to do so because sons maintained lifetime bonds primarily with their father.

Second, pair-bonding set the stage for consistent grandmothering in male kin groups, an evolutionarily novel pattern in itself. Grandmothering is defined here as any form of care of her grandchildren by a grandmother, whether or not the grandmother is herself nursing an infant. In female kin groups such as macaques, daughters breed in their mothers' group and grandmothering is common. For example, in experiments that we carried out in Japanese macaques, grandmother–granddaughter dyads had the highest scores for aiding and food sharing next to mother–daughter dyads (Chapais et al. 1997; Chapais, Savard, and Gauthier 2001; Bélisle and Chapais 2001). But in male kin groups, such

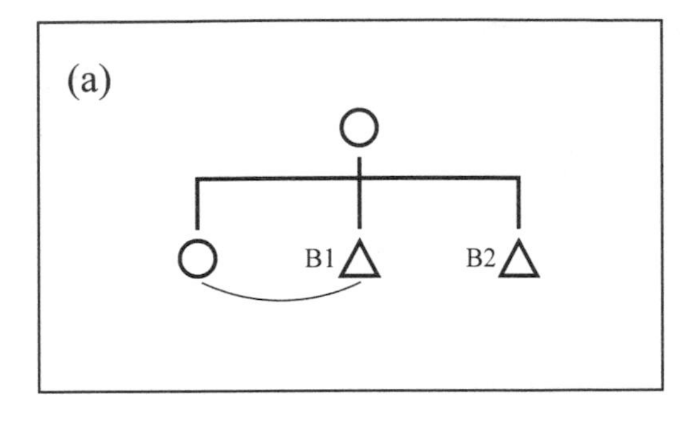

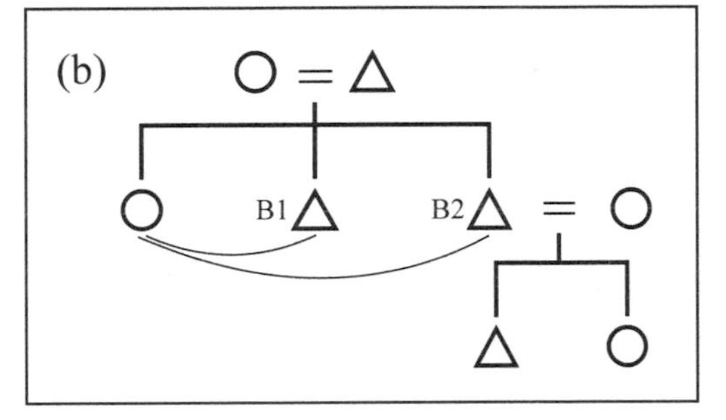

Figure 13.2. Comparison of brother–sister bonds before and after the evolution of pair-bonding. In each of the two situations, a mother has two sons born consecutively and a daughter born afterward. In a chimpanzee-like group (a), the sister shares an extensive period of developmental familiarity with her younger brother, B1 (thin line) but not with her older brother (B2). After the evolution of pair-bonding (b), three factors bring about a period of developmental familiarity between the sister and her older brother, B2: the presence of a new mediator (the father), grandmothering activities between B2's mother and offspring, and preferential bonds between the sister and her brother's wife and children.

as chimpanzees and bonobos, preferential relations between grandmothers and grandoffspring are uncommon, although there is some anecdotal evidence pointing to the chimpanzee's potential for such interactions, as noted in chapter 12. Grandmothering directed to a *daughter's* offspring is normally not possible because daughters breed in groups other than their mothers'. And grandmothering directed to a *son's* offspring is not possible because the fathers themselves do not recognize their offspring.

The evolution of pair-bonds in male kin groups rendered consistent grandmothering possible. Third parties could now recognize the special bonds linking fathers to their offspring. A father's mother, in particular, could recognize her sons' children, and from then on she had ample opportunities to take care of them. Grandmothering is particularly well developed in humans, in which a female who has reached the age of her last birth still has a twenty-year life expectancy ahead of her (Hawkes, O'Connell, and Blurton-Jones 2003, 211). The important point here is that grandmothering activities directed to a son's offspring favored the extension of bonds between mothers and their sons beyond adolescence and well into adulthood. In doing so, grandmothering was extending the mother's role as mediator of developmental familiarity between her sons and daughters. To put it simply, younger sisters would keep meeting their adult brothers because their mother was taking care of the brothers' children. Grandmothering had the effect of a centripetal force within the family.

Finally, pair-bonding should have brought brothers and sisters closer together through a third, unrelated process. Younger sisters growing up with pair-bonded brothers could now recognize two new categories of individuals: their maternal brothers' "wives" (their "sisters-in-law") and their brothers' offspring (their nieces and nephews). Younger sisters should have developed affiliative bonds with both categories of individuals. From the younger sister's viewpoint, a sister-in-law is a close associate of her brother and a same-sex social partner with whom cooperation is possible. As to the younger sister developing friendly bonds with her brother's offspring, this idea finds support in the observation that male chimpanzees may develop friendly bonds with their sisters' offspring, as described in chapter 12. Both categories of relationships would increase the rates of friendly interactions between younger sisters and older brothers.

To summarize, pair-bonding brought sisters and brothers closer together (1) by introducing a new mediator of developmental familiarity between them, the father; (2) by lengthening the mother–son bond through grandmothering, hence increasing the mother's impact as a mediator of familiarity between her sons and daughters; and (3) by creating opportunities for friendly interactions between sisters and their brothers' offspring and wives. In short, pair-bonding created a better-integrated, more closely knit three-generational family. This is apparent when comparing Figure 13.2a and b.

The Added Effect of Shorter Interbirth Intervals

I have so far been concerned with strictly structural consequences of the father's presence on relationships within the family. In closing I consider some life-history consequences of pair-bonding on the dynamics of sibling bonds. The mean duration of the interval between successive births is considerably shorter in humans than in chimpanzees. Chimpanzees have a birth interval 1.6 times longer than humans (66.7 months/41.3 months) according to Kaplan et al. (2000), or 1.9 times longer (48.8/25.5) according to Key (2000) and Aiello and Key (2002). Whatever the exact figure, the discrepancy is substantial. The length of gestation being similar in humans and chimpanzees, the difference comes mainly from the period of lactation. Human mothers have much shorter birth intervals because they begin weaning their offspring at an earlier age. And, as we saw, human mothers are in a position to curtail the period of lactation because they feed their offspring with solid food. Given that provisioning activities may be performed by third parties and that food sharing is a common practice in our species, it is likely that the father's association with the mother, among other factors, contributed to reducing the interbirth interval.

A shorter birth interval meant that successively born siblings were closer in age: two or three years in humans versus five or six in chimpanzees. This must have had significant repercussions on sibling relationships. As pointed out earlier, closeness in age favors the formation of cooperative partnerships among male chimpanzees, presumably because similar-age individuals resemble each other in terms of social status, skills, competitive abilities, and value as social partners. Moreover, a shorter birth interval meant that any two siblings born successively spent more time near their parents before the older sibling reached adolescence. Human brothers thus experienced longer periods of developmental familiarity near parental mediators compared to chimpanzees. The two factors converged in favoring cooperation between consecutively born brothers. Importantly, even siblings not born successively could now undergo periods of developmental familiarity near their parents. In chimpanzees, such siblings are ten to twelve years apart minimally, but in humans the difference reduces to four to six years, which means that they could grow up in close proximity to each other. All in all, the reduction of the interbirth interval in the course of human evolution should have resulted in more hominid sibling pairs experiencing developmental familiarity biases and in more cooperative partnerships.

I have focused here on the impact of pair-bonding on sibling relationships. But this is not to say that pair-bonding was the sole factor involved. Other factors affecting sibling bonds may have contributed to strengthening them. One of these was the extension of the juvenile period. The prereproductive phase is about 1.4 times longer in humans than in chimpanzees (Kaplan et al. 2000). This change alone—that is, regardless of the interbirth interval—implied that any two siblings spent more years in close intimacy with their parents, resulting in a substantial extension of the period of shared developmental familiarity, hence in stronger biases for cooperative partnerships.

Building upon a chimpanzee-like male kin group structure, pair-bonding created a whole new type of family. From a bigenerational and monoparental (mother–offspring) unit, the hominid family evolved into a biparental unit integrating three generations of individuals and some affines as well, that is, some sort of extended family. Within this novel family, primary kin were tightly knit together compared to the situation in chimpanzees. Any individual had preferential bonds with its two parents. Owing to this structural factor and to the added effect of shorter birth intervals, any individual was now in a position to develop preferential bonds with each of its siblings, not only with those close in age.

Based on the assumption that fathers and sons developed lifetime cooperative partnerships, pair-bonded families included a well-defined core of primary agnates whose cohesiveness stemmed, fundamentally, from the benefits of cooperating with same-sex close kin. Importantly, daughters (or sisters) were an integral part of such units. In chimpanzees, females have loose bonds with their brothers, and only a fraction of them, and, as far as we know, no special bonds with their fathers. Pair-bonds changed that situation drastically. Henceforth, among a young female's most basic bonds were those with her primary kin: her mother, father, and brothers. This simple fact may be seen as the single most important factor enabling the evolution of the brother–sister kinship complex. Strong brother–sister bonds could not emerge out of a chimpanzee type of family with its loose siblingships. Only pair-bonding and biparentality could produce the required levels of cohesiveness among family members. Pair-bonding is certainly not a sufficient condition for the evolution of the brother–sister kinship complex, but it was a necessary one. In the following chapters I consider other factors that further increased the importance of the brother–sister bond and set the foundation of the brother–sister kinship complex.

14

Beyond the Local Group: The Rise of the Tribe

By binding together a whole community with ties of kinship and affinity, and especially by the peacemaking of the women who hold to one clan as sisters and to another as wives, it [exogamy] tends to keep down feuds and to heal them when they arise, so as at critical moments to hold together a tribe which under endogamous conditions would have split up.

Edward Tylor (1889b, 267)

It does not seem unwarranted to assert that the human capacity to extend kinship [beyond the local hunting-gathering band] was a necessary social condition for the deployment of early man over the great expanses of the planet.

Marshall Sahlins (1960, 81)

Until now I have been concerned with the consequences of pair-bonding on relationships *within* the local group. From here on I move to the level of between-group relations. Primate studies indicate that an individual's social horizon ends at the boundaries of its local group. The extension of social structure beyond the local group and the existence of organized social entities encompassing several distinct groups are uniquely human phenomena. For the sake of simplicity I use the term "tribe" and the expression "tribal level of organization" in a generic sense to refer to such supragroup entities. Lévi-Strauss argued that human society was born with the matrimonial exchange of women between distinct male kin groups. He thus believed that intermarriage—exogamy—was an intrinsic aspect of supragroup social structures. Without marriage and intermarriage there could be no alliance between groups, and hence no tribe.

My aim here is to demonstrate that pair-bonding, the evolutionary precursor of marriage, was indeed a *prerequisite* to the formation of solid and enduring alliances between hominid local groups. In arguing that pair-bonding predated the tribe, I shall also justify the implicit idea that the tribe could not evolve directly from the ancestral male kin group exemplified by chimpanzee society.

If these propositions proved true, Lévi-Strauss's much-debated views about marriage generating intergroup alliances would receive strong support from evolutionary theory and comparative primatology. But at the same time his conception of how the whole system arose—as a cultural construct born ex nihilo—would be untenable.

Male Pacification as a Prerequisite for the Tribe

Why is a direct evolution of the tribe from a chimpanzee-like (or bonobo-like) society unlikely? Briefly stated, because intergroup relations in our two closest relatives are dominated by avoidance and hostility. In chimpanzees hostility between groups is the norm in all five populations that have been studied on a long-term basis, whether these were artificially provisioned ($n = 2$) or not ($n = 3$) (Newton-Fisher 1999; Wrangham 1999, Boesch and Boesch-Achermann 2000; Watts and Mitani 2001; Watts et al. 2006). Aggressive episodes are initiated and conducted by adult males and may include remarkably violent episodes, such as the intentional killing of outsiders (in four of the five populations). Importantly, the targets of intergroup aggression include infants and adult males and sometimes mothers (Goodall 1986; Wrangham 1999; Wilson and Wrangham 2003; Watts et al. 2006). Bonobos also are territorial, with intergroup encounters characterized by high rates of aggression (Fruth and Hohmann 2002), but lethal attacks have not been observed in this species. Critiques to the effect that chimpanzee violence is an artifact of human provisioning or habitat disturbance (Power 1991; Sussman 1999) are thus insupportable.

The implications of chimpanzee and bonobo territoriality are clear: for the tribe to have evolved directly from a *Pan*-like ancestor, the nature of relationships between males would have fundamentally changed, from hostility to tolerance, minimally. Given that conflicts between primate males stem from their competing for feeding territories and females, the evolution of the tribe directly from a chimpanzee-like group could occur only through a reduction of male competition for these resources. Current socioecological theory indicates that animals compete aggressively with one another when resources that are limited in relation to the number of individuals are distributed spatially in such a way that they may be aggressively defended economically, that is, at reasonable costs to the aggressor (for a general discussion, see Pusey and Packer 1997). For example, resources occurring in small packages widely scattered over large areas are not economically defensible, either by a single individual or by a group. But fruit trees, or territories packed with enough food to satisfy the dietary requirements of the group, are defensible. It follows that any significant decline in levels of feeding competition between male hominids—any move in the direction of intergroup pacification—should correlate with dietary adaptations and changes in patterns of resource exploitation.

One such possibility that has received ample empirical support from paleoanthropology is the adaptation of hominids to more open environments, such as savannas or woodlands, in which resources are spread out over large territories and less economically defensible. For that matter, a dependence on widely dispersed food sources is precisely the key feature of the diet of modern hunter-gatherers. And the forager diet is consistently invoked to explain the egalitarian nature of their society, the pacific character of relations between bands, and the flexibility of postmarital patterns. The following quote summarizes well the basic idea: "Foragers acquire foods that are spatially dispersed, such as tubers, honey, game, fruit, and berries. Foraging for these foods requires large day ranges. Depletion and seasonal fluctuation in food require that camps move every so often, and therefore home ranges are much larger than those of our closest relatives [chimpanzees and bonobos]. In short the forager diet favors mobility" (Marlowe 2004, 283). Modern foragers testify to the capacity of hominids to evolve peaceful intergroup relations at the intratribal level (Knauft, 1991). But they do not testify to their ability to achieve this independently of the evolution of pair-bonding. Pair-bonds characterize all hunter-gatherer societies and constitute an essential element of their subsistence pattern, with its clear sexual division of labor. As mentioned previously, if males are in a position to specialize in hunting, it is because females specialize in gathering. Therefore, the fact that pair-bonds are a correlate of the hunting-gathering way of life only reinforces the hypothesis that they had something to do with the evolution of the tribe.

Even if one posits that prior to the evolution of pair-bonding some changes in subsistence patterns did take place, bringing about a significant reduction in levels of *feeding* competition among hominids, the fact is that male *sexual* competition remained. Mountain gorillas are male philopatric and enjoy relatively low levels of feeding competition, but they nevertheless exhibit high levels of sexual competition, between-group intolerance, and aggressive encounters (Sicotte 1993, Watts 1994). The permanent state of sexual competition between primate males in general is certainly not extraneous to the fact that even though primate groups may tolerate each other in some circumstances, they have not evolved a tribal level of social organization. In nonterritorial primate species, groups mingle peacefully around the same resources—water holes, fruit trees, or seeping sites (Cheney 1987)—but they display no tribal organization. Nor does our close relative the bonobo, despite exhibiting sporadic peaceful intergroup meetings.

To pacify territorial males that were naturally hostile, some novel centripetal force operating between groups and inhibiting male violence was needed. In what follows I argue that intergroup pacification required a *preexisting* structure of affiliative bonds linking members of distinct groups and that such a pacifying structure was a correlate of pair-bonding.

Females as Peacemakers: The Consanguinity Route

The basic principle underlying the role of females in pacifying intergroup relations is that as the dispersing sex, females had the opportunity to act as connectors between their birth group and their group of adoption. After pair-bonds evolved, they could do so through two distinct routes: directly, via preexisting bonds with their natal kin, and indirectly, as intermediates between their natal kin and their "husband"—in other words, as connectors between "affines." I examine each route to pacification separately.

Hereafter I use the expression "bonded kin" for consanguines who develop a preferential bond with each other. In chimpanzees and bonobos, as we saw, bonded kin are most often limited to mother–daughter dyads, mother–son dyads, and maternal siblings. Crucially, one consequence of female dispersal is the allocation of bonded kin in different male kin groups. Let us posit that some females of group A transferred to group B and that some females of group B transferred to group A. Let us further assume that groups A and B, or various fractions of the two groups, come into contact near their common border from time to time, mingling to some extent, and that such meetings create opportunities for bonded kin to get in touch again and pick up their relationship where they had left it. This assumption—distinct local groups initially meeting peacefully on a sporadic basis—must itself be substantiated, and I will come back to this important issue shortly. Taking it for granted for now, the point is that since bonded kin have a past history of amicability, they constitute potential *appeasing bridges* between interbreeding communities. That is to say, assuming that interbreeding groups come into contact sporadically, bonded kin should, minimally, direct no aggression to each other, and they might even engage in affiliative interactions. The idea that kinship bonds should breed peace between groups follows directly from what we know about the nature of relationships between close relatives in nonhuman primates. It is also derived from ethnographic data on interband relations in hunter-gatherer societies in which kinship is "often a synonym for peace," as stated by Sahlins (1960, 82).

Prior to the evolution of pair-bonds, meetings between distinct male-philopatric groups provided opportunities for only two categories of bonded kin to renew contact consistently: mother–daughter dyads and brother–sister dyads. We saw earlier that older brothers were in a position to develop preferential bonds with their younger sisters prior to the sisters' emigrating, but that only a fraction of these dyads (those born consecutively) would be expected to do so. Hence only a fraction of all brother–sister pairs could act as appeasing bridges between interbreeding groups. I also exclude sister–sister dyads on the basis of life-history considerations. Female chimpanzees give birth every five to six years on average, and they emigrate at around eleven years of age. Thus only consecutively born sisters will have had an opportunity to develop strong preferential bonds prior to the older sister leaving the group; hence only a few sister dyads could act as appeasing bridges between groups.

The evolution of pair-bonding altered that situation substantially. The number of bonded kin who could act as appeasing bridges between interbreeding groups increased markedly. They included mother–daughter dyads, as before, but also a larger proportion of brother–sister dyads, because biparentality ensured that sisters experienced a period of shared developmental familiarity with most of their older brothers before emigrating. Moreover, a transferred female's bonded kin now comprised her *father* and her *patrilateral* kin, including her paternal grandfather and grandmother, her paternal uncles, and possibly her patrilateral cousins, as illustrated in Figure 14.1. I refer to all such dyads as "primary kinship bridges." Crucially, a large number of primary kinship bridges would be composed of a transferred female and a *male* relative. This is important because intergroup conflicts in chimpanzees are initiated and conducted by males. The fact that males would be an integral part of appeasing bridges between groups is thus all the more significant. Minimally, a male should have been somewhat inhibited from attacking his female kin—his daughter, granddaughter, sister, niece, and so on; that is to say, the transferred female should have benefited from some kind of immunity from her male relative. If more than one female had moved to group B, any group-A male would be likely to recognize more than one female

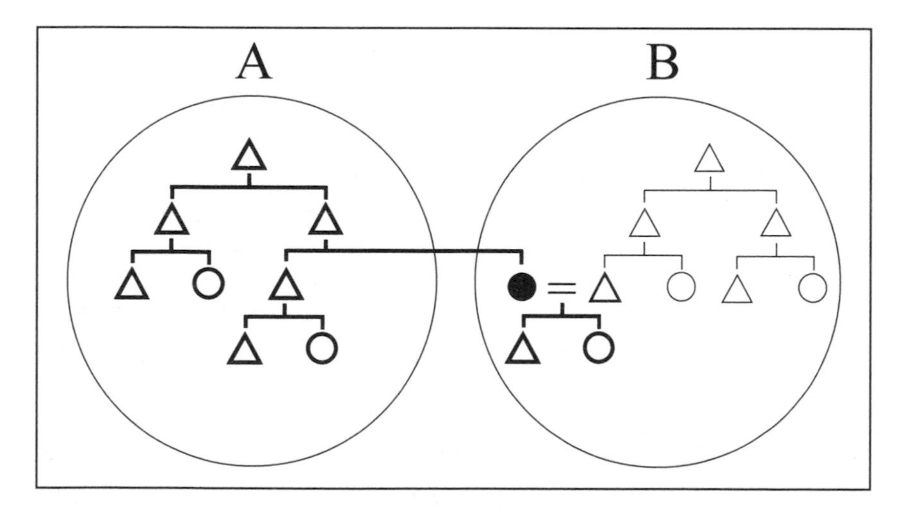

Figure 14.1. "Appeasing bridges" activated during meetings between two interbreeding male-philopatric groups. The structure illustrated here is generated by a single female from group A (black circle) after she has transferred to and pair-bonded in group B. Primary kinship bridges connect the transferred female to her consanguineal kin living in A. Secondary kinship bridges are established between the transferred female's offspring and their matrilateral kin living in group A. Affinity bridges link the transferred female's "husband" to her relatives living in A. For the sake of clarity, the genealogical structure here includes four generations instead of the usual three.

kin in group B—for example, a daughter, a sister, and a niece. Thus, group-A males would be collectively inhibited from attacking several group-B females. Of course, the reciprocal also applied: group-B males would be collectively inhibited from attacking their female relatives living in group A.

In short, one consequence of pair-bonding and paternity recognition was that a fraction of males in each of two interbreeding groups should have refrained from attacking a fraction of females living in the other group. One thus obtains a reciprocal, though fragmentary, state of tolerance between two interbreeding groups. That such a state of mutual tolerance might also favor positive interactions between groups, such as grooming interactions between fathers and daughters, cannot be excluded.

Another major consequence followed directly from the establishment of primary kinship bridges between dispersed females and their kin. A female's immunity against aggression should have extended to her offspring. A male who refrained from attacking his daughter or sister should also refrain from attacking the individual that his daughter or sister carried on her back or belly, his own grandoffspring or nephew. Or, if he did not refrain initially, the mother would remind him to do so by defending her offspring. Male chimpanzees have been observed to attack and kill the infants of isolated mothers when they come across them at their common border. From the time males could recognize such females as their close kin, the overall frequency of male infanticidal attacks should have dropped. Moreover, given that mother–offspring bonds extend over several years, the offspring themselves would learn not to fear certain outside males who were friendly with their mother. Eventually, the offspring would be in a position to meet these males with relative impunity, even if their mother was absent. The developmental process envisioned here is particularly significant because it involves the formation of appeasing bridges between males, the aggressive sex. Pacifying the males would have been especially effective in preventing intergroup conflicts.

An interesting correlate of the latter process is that from the offspring's standpoint, the pacified males living in the other group were their *matrilateral* kin: their maternal grandfather, uncles, and cousins. I refer to these kin dyads as "secondary kinship bridges." In male-philopatric groups, female dispersal normally precludes the recognition of one's matrilateral kin. The only way for a male in such a group to recognize his matrilateral kin is if he has opportunities to witness interactions between his mother and her kin in the context of peaceful meetings between his group and his mother's natal group, which is precisely what the developmental process just described is all about. The first situations in which males were exposed to their matrilateral kin occurred when, as immatures and still in their mother's arms, they witnessed the immunity she enjoyed on the part of some specific outside males they came across at their groups' common border.

Already at this stage in the evolution of hominid society, a transferred female's kinship network extended beyond her local group. Through procreation the female was progressively building up a kinship network of her own in her group of adoption, while keeping alive on a sporadic basis the kinship network in her group of origin. The first and most primitive of all *supragroup* kinship structures was born. Remarkably, according to this reasoning, the primitive tribe emerged from the combination of three otherwise ordinary primate phenomena: (1) nepotism, the driving force behind the establishment of enduring social bonds; (2) pair-bonding, which considerably extended the domain of kin recognition and the number of bonded-kin dyads prior to females moving out of their birth group; and (3) female dispersal, which apportioned bonded kin to different groups, generating appeasing bridges in the process.

Females as Peacemakers: The Affinity Route

In parallel with the aforementioned processes, pair-bonding facilitated intergroup pacification through a no less fundamental route, affinity. The basic idea was stated long ago by Edward Tylor (quoted in the chapter epigraph). Let us reexamine this old principle in the light of evolutionary theory. Affines are the consanguines of one's spouse or, reciprocally, the spouses of one's consanguines. The evolution of affinal kinship in the course of human evolution required no more than the merging of consanguineal kinship with pair-bonding. Cognitively speaking, the recognition of in-laws stems merely from the ability to recognize preferential bonds between others. As discussed earlier, many nonhuman primate species possess this ability, so there is every reason to believe that it was part of mankind's primate legacy. Thus, for affinal kinship to emerge in the course of hominid evolution, no cognitive abilities beyond those already displayed by nonhuman primates were needed. In this sense, early hominids were cognitively preadapted to recognize their affines.

Following the evolution of pair-bonding, meetings between interbreeding groups provided opportunities for affines to come into contact. When a female born in group A and pair-bonded in group B came into contact with her natal kin living in group A, the "husband" was likely to meet with his wife's relatives, for example, his "father-in-law" and "brothers-in-law" (Figure 14.1). But why should relationships between affines be nonaggressive, peaceful, and even amicable to begin with? From an evolutionary perspective, the answer is straightforward: such relationships were bound to be, fundamentally, relationships between potential *allies*. Affines, such as brothers-in-law, share a vested interest in the same female, one as a husband, the other as a brother. Both in-laws derive benefits from the female's well-being, the husband through his own reproductive interests and long-term cooperative bond with his wife, the brother by virtue of his genetic relatedness with her—through inclusive fitness benefits. Crucially, this shared interest is not impeded, in fact not nullified, by sexual competition between the two males. Owing to incest avoidance, a brother does not compete with his sister's "husband" for sexual access to her.

Now two individuals who share a common concern for a third party's well-being and are not competitors for that third party are somewhat biased toward interacting positively with each other. To take an analogy, parents are allies in parental care because they share a vested interest in the same child, an interest unhampered by sexual competition, for that matter. They are bound to cooperate with each other even though their contribution to their child's well-being and the nature of their relationship with it may differ considerably. Minimally, therefore, brothers-in-law shared a common interest in protecting the female who stood as wife to one and sister to the other. They were de facto partners. Moreover, brothers-in-law should have refrained from attacking each other if only because an attack on the brother by the husband would prompt the sister to protect her brother, and an attack on the husband by the brother would prompt the wife to defend her husband. When defending her brother, a sister is in effect increasing her inclusive fitness benefits, and when defending her husband a wife is attending to her reproductive interests. It is precisely here that the expression "females as peacemakers" takes its fullest meaning. Females were more than mere links through which affines could become disproportionately familiar to each other; they were intermediaries and intercessors through which they could become allies and partners.

The importance of the affinity route in the pacification of intergroup relations can hardly be overstated. The processes described under the consanguinity route to pacification involved adult males refraining from attacking their *female* kin and the latter's immature offspring. But in the processes just described, adult males refrain from attacking other adult *males*. While the consanguinity route to intergroup pacification is about the formation of appeasing bridges between adult males and adult females (and between adult males and immature males), the affinity route is about appeasing bridges between adult males, the individuals directly responsible for intergroup conflicts.

This brings me to an important point. Gorillas display female transfer between independent polygynous units and stable breeding bonds between the silverback male and his females. Gorillas thus meet the two basic prerequisites for intergroup pacification. On this basis one might expect a female who transferred into another polygynous unit to act as a peacemaker between her new silverback leader—her "husband"—and her natal silverback, who was presumably her father or brother; one might expect the female to pacify the two "in-laws." Yet there is no tribal level of organization in gorillas. Why? I see two reasons. First, the high degree of sexual dimorphism that characterizes gorillas is hardly compatible with a female interposing herself between silverbacks that are twice her size. The risks of injury would simply be too high. One general principle follows. For females to be in a position to act as efficient peacemakers between adult males, levels of sexual dimorphism must be relatively low. Interestingly, the foregoing reasoning satisfies this condition implicitly, because pair-bonding comes along with low levels of sexual dimorphism (see chapter 11). The sexual dimorphism issue thus provides a supplementary argument in support of the view that pair-bonding was a prerequisite for pacification. Pair-bonding favored the advent of the tribe not only because it created the potential for consanguinity and affinity bridges between groups, but also because it coincided with a smaller power differential between males and females. One may infer from this that female peacemaking was unlikely in highly dimorphic hominid species such as australopithecines but became a possibility with the evolution of *Homo erectus*.

The second reason why pacification is unlikely in a gorilla-like species is that for a female to act as an effective peacemaker between one of her male relatives and her "husband," she must have a vested interest in the latter. This would especially be the case if a pair-bond was a cooperative partnership that provided the female with some important benefits rather than an asymmetric relationship primarily serving the male's reproductive interests. Cooperative partnerships involving proximity, male protection, food sharing, and cooperation in childcare are absent in gorillas, but they are the norm in humans. This suggests that peacemaking by females and the evolution of the tribe required not only pair-bonding, but pair-bonds that were mutually beneficial.

The Initial Impetus

Combining the consanguinity route and the affinity route to pacification, and considering that the two sets of processes worked reciprocally between interbreeding groups, the evolution of pair-bonds meant that the basic ingredients of peace were now in place. But the structure of appeasing bridges just described was nonetheless *latent*. Bonded kin and affines were distributed in adjacent interbreeding local groups, ready to engage in pacification, so to speak. But if two interbreeding groups never met at their common border, if bonded kin—fathers and daughters, brothers-in-law, and so on—never came into contact, the bridging structure could not be activated. In other words, pair-bonds were a necessary condition to pacification but not a sufficient one; they were a catalyst, not the sole determining factor of pacification. For pacification to get going, interbreeding groups had to meet in a nonhostile manner in the first place, if only on a sporadic basis. What then, could have been the initial impetus to pacification between territorial groups?

The bonobo evidence is particularly revealing at this point. Genichi Idani (1990) described a few peaceful encounters between wild groups of bonobos, which took place most often when two groups met at the same artificial feeding site. Early in the encounter tension was evident, but as time passed the two groups would eventually cofeed peacefully. The meetings involved various types of affiliative interactions, mostly between females. It is notable that the males were the most aggressive participants. The bonobo evidence thus indicates that territorial male kin groups may meet peacefully on a sporadic basis in special circumstances, perhaps because bonobos are less aggressive than chimpanzees, owing possibly to lower levels of competition for food (White 1996; Wrangham et al. 1996; Wrangham 2000). It is noteworthy that although bonobo groups may sometimes meet peacefully, they have no tribal level of organization. This makes sense in the context of the present model: bonobos do not form pair-bonds. If they did, they would have a multifamily group structure, and one might predict that they would meet peacefully on a regular basis.

Another possible initial impetus to pacification relates to the processes involved in the fission of a primate group into two independent subgroups (group fissions are described in chapter 18). Following a group fission in a female-philopatric species such as macaques, the two nascent subgroups are initially mutually tolerant, even mingling together. Meetings may involve grooming between previously familiar females and high rates of male transfer (Cheney 1987). But as time goes by, relations revert to avoidance and hostility, and the two subgroups eventually become fully autonomous social units. Applying this principle to the present situation, recent common filiation might provide the initial impetus needed for pacification to start between recently fissioned male kin groups (A and B). If meetings between groups A and B were initially peaceful owing to their common descent, and if the two groups began to interbreed—group-A females pair-bonding in group B, and group-B females in group A—kinship and affinity bridges would be activated right from the onset. For example, fathers and brothers would be in a position to keep in touch with their daughters and sisters right after the fission. As a result, a father would recognize his daughter's "husband" and, reciprocally, the husband would recognize his "father-in-law." This should prevent the two males from becoming progressively estranged from each other, a situation that would normally prevail in the absence of pair-bonds.

The process just envisioned is logical, but in examining the few documented cases of group fissions in chimpanzees, its reality is not immediately obvious. In one case no initial postfission harmony between the nascent subgroups was reported, contrary to the macaque pattern. Eventually the smaller group was systematically and brutally attacked by the larger one and finally exterminated (Goodall 1986; Wilson and Wrangham 2003). Thus recent common descent from the same group does not by itself necessarily promote peace between highly territorial male kin groups such as those formed by chimpanzees. And whether postfission harmony would prevail after the evolution of pair-bonds cannot be ascertained.

Future studies on chimpanzees and bonobos might establish that recent group fissions promote peace between local communities or that peaceful meetings between groups occur independently of common ancestry and reflect, instead, lower levels of between-group competition. In either situation, one obtains the initial impetus needed for pacification.

The Prelinguistic Tribe

Implicit in the foregoing discussion is that the basic parameters of the tribe—its size and the number of local groups involved—are defined by the exact pattern of female transfer between groups. It is female circulation that delineates the boundaries of the tribe. The main factors include whether females circulate bilaterally or unilaterally between local groups, whether they all transfer to the same group or are distributed in two or more groups, and the extent to which local groups vary in size. Let us first consider a simplified "tribal system" characterized by the bilateral circulation of females between two same-size groups (Model 1, Figure 14.2a). In this model all group-A females (two in the figure) transfer and form pair-bonds in group B, while all group-B females (two as well) transfer and pair-bond in group A. This system maximizes the number of appeasing bridges between groups because all females of each group end up in the other group (for a total of four transfers in the figure). Moreover, the system produces a *symmetric* pattern of pacification. Group-A males have both female kin and male affines in group B, and group-B males have both female kin and male affines in group A. Consequently, adult males in both groups are expected to exhibit similar levels of tolerance toward the other group.

Based on what we know about the patterns of transfer in chimpanzees and bonobos, the bilateral circulation of females within a closed two-group system is not realistic. Bilateral circulation per se has been documented in chimpanzees; for example, one study on the Mahale chimpanzee population reported that 76 percent of the 29 females who moved out of group A had emigrated into group B, and 55 percent of the females who had immigrated into group A came from group B

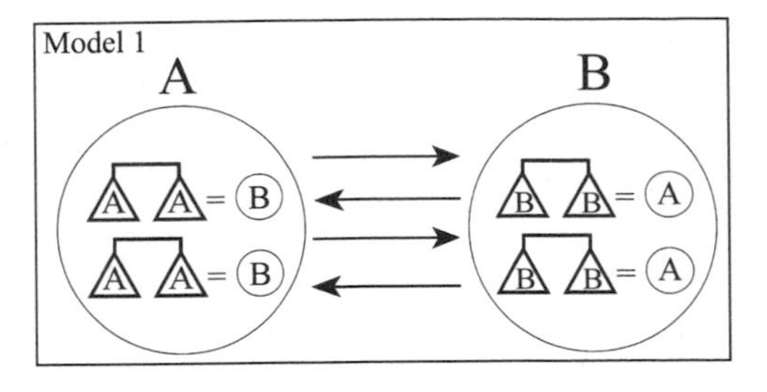

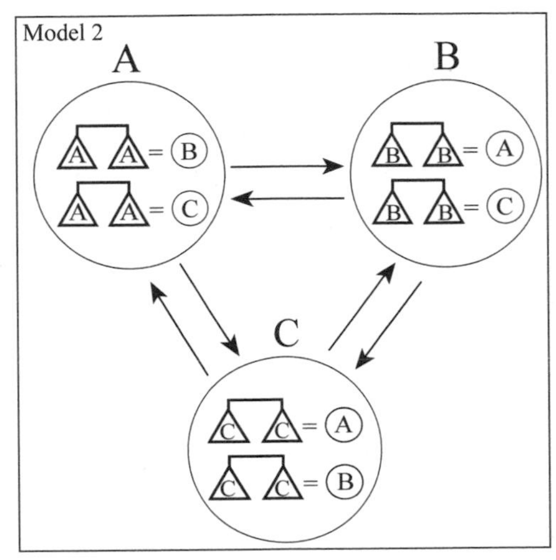

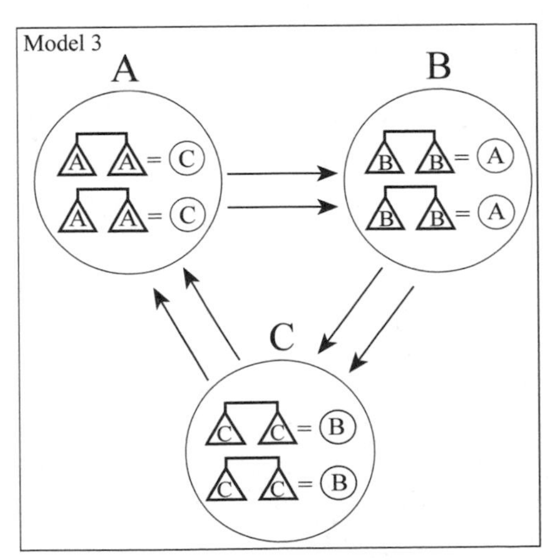

Figure 14.2. Three models of female circulation between male-philopatric groups after the evolution of pair-bonding. In all three models, individual families are illustrated by a single dyad of brothers, two per group. Each group produces two emigrating females. Arrows indicate the direction of transfer of a single female. Model 1: bidirectional transfer between two groups. Model 2: bidirectional transfer among three groups. Model 3: unidirectional transfer among three groups.

(Kawanaka and Nishida 1974). But bilateral circulation may take place among more than two communities. Although quantitative data on the exact destination of all females from a given group are not available, some authors have mentioned that females transfer into two or three communities (Goodall 1986, 89, 513; Nishida et al. 2003). Thus a more plausible pattern of transfer involves more than two local groups. Model 2 (Figure 14.2b) illustrates a system of bidirectional circulation between three same-size groups. Group-A females emigrate equally often into groups B and C, group-B females equally often into groups A and C, and group-C females equally often into groups A and B. As in the previous model, pacification is symmetrical between any two groups because males in each of them have both female kin and male affines in the other group. However, compared to the previous situation, pacification is expected to proceed at a slower pace because the females are distributed in two different groups instead of just one. As a result, the number of appeasing bridges between groups A and B, for example, is half the number in Model 1 (two transfers instead of four). The same principles apply in a four-group system, with the number of appeasing bridges divided by three, and so on as the number of groups increases.

Models 1 and 2 both posit that female circulation is bidirectional between groups. But in theory females might circulate unidirectionally, at least in relation to certain groups. To assess the effect of unidirectional transfer on pacification while controlling for other factors, Model 3 (Figure 12.2c) pictures an idealized pattern of unidirectional transfer between three same-size groups. The A females transfer and pair-bond in group B, the B females move to group C, and the C females to group A. The number of appeasing bridges between any two groups is still two, as in Model 2, but the pattern of pacification is now asymmetrical. Group-A males have both female kin and male affines in group B. In contrast, group-B males have no female kin in group A, but they have male affines in it (the consanguines of their spouses). One thus expects group-A males to be familiar with and tolerate a larger number of B individuals compared to group-B males in relation to the A individuals. The same asymmetry characterizes the other two pairs of groups (B–C and C–A). The important point here is that even if female circulation is strictly unilateral between two groups, pacification is nonetheless expected.

In all three models, group size was a constant. But primate groups vary substantially in size. Let us consider a pattern of bidirectional female circulation between a large group and a small one, a situation matching the reality of chimpanzees in one study, in which transfer was bilateral between two groups numbering thirty and seventy individuals respectively (Kawanaka and Nishida 1974; Nishida 1979). If transfers are frequent between the two groups, most of the adult females living in the small group would originate from the large group, but only a fraction of the adult females living in the large group would come from the small group. This introduces a different kind of asymmetry. While most males in the smaller group would have female kin and male affines in the large group, only a fraction of males in the larger group would have female kin and male affines in the small group. As a result, most males in the small group would be pacified in relation to the larger group, but only a fraction of males in the large group would be pacified in relation to the small group. Those with no female kin in the small group would not be.

All of the models just described are somewhat idealized, and a large number of variants are obviously possible. For example, female circulation might take place bilaterally between more than two groups, as in Model 2, but with females distributing themselves *unequally* among different groups. Or transfer might be bilateral between certain groups in a given system but close to unilateral between other groups in the same system. But however idealized, the models are sufficient to establish a few basic principles. First, pacification between any two groups takes place whether female circulation is unidirectional or birectional, but bilateral circulation promotes further congeniality because tolerance between males is symmetrical. Second, the pace at which pacification proceeds between any two groups crucially depends on the proportion of females who transfer into the other group. The more groups that females transfer into, the fewer appeasing bridges there are between any two groups, and the lower the pace of pacification between them. Therefore, pacification should be less intense, at least initially, in tribal systems composed of several local groups than in systems consisting of two or three. Third, bilateral circulation between local groups of substantially different size works to the disadvantage of the smaller group, as far as pacification is concerned; in extreme cases, pacification might even be prevented. Other things being equal, therefore, intergroup harmony is facilitated between groups of similar size. Combining the three principles, one may infer that intergroup pacification would have been favored in situations where (1) female circulation was bidirectional, (2) the number of local groups was small, and (3) they were not too dissimilar in size.

These principles readily point to the essential nature of the primitive tribe. At this stage in its evolution, the tribe is merely a set of interbreeding local groups whose members interact peacefully on a regular though infrequent basis and whose social boundaries are somewhat blurred owing to between-group familiarity. The main factors affecting the size and composition of a tribe are those that determine the spatial distribution of the local groups, namely, the spatial and temporal distribution of food resources and predators and the presence or absence of geographical barriers between groups. In accordance with this reasoning, it has been reported that the home ranges of two chimpanzee communities between which female transfer was frequent did overlap extensively, whereas the home ranges of communities between which female transfer was infrequent did not overlap (Kawanaka and Nishida 1974; Nishida 1979). The spatial distribution of communities is thus an important factor affecting the circulation of females. On this basis alone one may envision the emergence of geographical clusters of local groups, or *regional tribes* whose groups "exchanged" females and enjoyed varying levels of peaceful relations with one another, whereas groups belonging to different tribes were unfamiliar and hostile to each other: in short, a system in which peace prevailed at the *intra*tribal level but not at the *inter*tribal level.

Clearly, then, the evolution of the tribe did not eliminate intergroup hostility. Instead it brought about a major change in the level of social structure at which hostility was taking place. This principle helps resolve the discrepancy between chimpanzees and human foragers with regard to intergroup patterns of violence. Compared to chimpanzees, human hunter-gatherers are much more egalitarian and display substantially lower levels of intergroup competition and violence. So striking is the difference that some authors have spoken of a phylogenetic discontinuity between chimpanzees and human foragers in patterns of violence (Knauft 1991; Kelly 2000). But, as argued by Lars Rodseth and Richard Wrangham, the local band of hunter-gatherers is not the right social unit for a meaningful comparison with chimpanzees; the tribe is.

> In chimpanzees, the community is the largest social unit that can be mobilized for purposes of aggression. In humans there is no inherent

> limit on the size of the unit that can be mobilized . . . In this light, the question is not so much: "Why did human foragers evolve such peaceful societies?" but "Why did humans in general evolve the capacity for segmental organization?"—a capacity that tends to create peace at lower levels of social organization even as it mobilizes people for potential conflict at higher levels. (2004, 398)

The present discussion provides an answer to this question. The primitive tribe is a likely candidate for the most basic level of "segmental organization" to evolve among hominids. The first tribe would have been a *prelinguistic* entity discernible, essentially, in a pattern of interactional regularities between members of distinct local groups. In the same manner, *within*-group patterns of interactions studied by primatologists—mating systems, dominance hierarchies, play groups, and so forth—are manifest solely as patterns of interactional recurrences. Admittedly, such embryonic versions of the tribe must have been comparatively loose compared to their later symbolic elaborations. But as we shall see in subsequent chapters, even at this stage in the evolution of supragroup social structures, some profound changes would have taken place that made the tribe a meaningful social entity. According to the present view, the primitive tribe thrived for a long time before language evolved, providing the basic framework from which the linguistically organized tribe would emerge.

Alternatively, as suggested by Rodseth and colleagues, language might have been a prerequisite to the tribe (Rodseth et al. 1991; Rodseth and Wrangham 2004). According to these authors, the key to the evolution of intergroup social networks was the extent to which hominids could maintain social relationships in the absence of physical proximity—the "release from proximity"—which they attributed to the ability to refer to events that took place elsewhere and at other times. "In humans," they wrote, "language makes it possible for relationships to be decoupled from interaction for months or years at a time." This possibility "depends on uniquely human capacities to (1) monitor the behavior of absent others through gossip and other verbal reports, (2) reach precise (i.e. verbalized) understandings of rules and agreements, and (3) enforce such rules and agreements through criticism and the mobilization of public opinion" (Rodseth and Wrangham 2004, 393). In other words, after they have moved into other groups, women may still hear from their relatives and talk about them with others, including their husbands and in-laws, and in this manner maintain lifelong bonds with absent kin. True, the effect of language in releasing social relationships from physical proximity must have been tremendous. But for the reasons discussed above, symbolic communication would have been superimposed upon the preexisting tribe; it would not have created the tribe.

Lévi-Strauss argued that intergroup alliances were born with marriage and the exchange of women between male kin groups. The present analysis shows that pair-bonding and female circulation between male kin groups—behavioral exogamy—were both a prerequisite and the driving force leading to the most primitive of intergroup structures. The convergence is clear, but the processes are drastically different: a cultural creation in one case, the outcome of a specific concatenation of evolutionary events in the other. The primary function of pair-bonds also differs. In the present scheme, their importance hinges on the role they played in pacifying intergroup relations. To Lévi-Strauss marital unions were the basic material of intergroup *alliances*. The two functions are of course compatible and intimately related, but it appears that pair-bonds were a means to pacify males well before they were a means to unite them.

15

From Male Philopatry to Residential Diversity

> The incest taboos which regularly prevail within the nuclear family exert [an] extremely important effect on social organization. In conjunction with the universal requirement of residential cohabitation in marriage, they result inevitably in a dislocation of residence whenever a marriage occurs. Husband and wife cannot both remain with their own families of orientation in founding a new family of procreation. One or the other, or both, must move.
>
> *George Peter Murdock (1949, 16)*

Murdock provides us here with the classical anthropological explanation for the very existence of postmarital residence patterns. Men and women are born and raised in their so-called family of orientation. But upon marriage, owing to the incest taboo which prevents them from marrying their primary kin, the husband or the wife or both leave their parental families and spend the rest of their life, or a substantial part of it, elsewhere, where they establish a family of procreation. It is thus the incest taboo that accounts for the universal phenomenon of dual-phase residence or, as Murdock put it, the "dislocation of residence." This reasoning has much in common with Lévi-Strauss's in that both attributed outmarriage to the incest taboo, and the incest taboo to culture, although they had different explanations for the incest taboo itself. The problem with this type of explanation, as we saw, is that hominids avoided breeding with their primary kin and reproduced out of their family well before marriage and the incest taboo were institutionalized. Postmarital residence patterns have a long evolutionary history, which is contained in part in the ancestral male kin group hypothesis discussed previously. I argued that the residence pattern of hominids, immediately after the *Pan–Homo* split, was a chimpanzee-like pattern of male philopatry and that this pattern had homologous components with human patrilocality. This implies that primate philopatry patterns and human postmarital residence patterns are evolutionarily connected, a proposition that is not immediately obvious. There are some profound differences between the two sets of phenomena in terms of the social processes involved and their consequences for the social networks of individuals. Here I spell out these differences before showing how philopatry patterns and residence patterns are related.

Some Serious Discrepancies

Moving to another social group, as female chimpanzees do, and relocating into one's husband's group upon marriage, as women do, have little in common. In humans, not only does female "dispersal" coincide with marriage, but it is the marriage that determines the wife's movement to her new residence group. Marriages are between-group agreements that are often part of enduring mutual obligations between families. This is best exemplified by bridewealth, a common currency in matrimonial exchange. When a man and his relatives acquire a wife, they transfer bridewealth, for example cattle, to the wife's family, and in so doing they obtain rights to the wife's future children. Marital unions are an integral part of contractual relationships between families. Other clear examples of mutual obligations between intermarrying families are the levirate, the obligation for a man to marry the wife of his deceased brother, and the sororate, the obligation for a woman to marry the husband of her deceased sister. Thus changes in residence upon marriage involve much more than the spouses' individual interests. By contrast, in chimpanzees and bonobos, female dispersal is an individual affair, a social process involving the female herself and her networks of social relationships, in particular the ones she is developing in her new group. It is a process that unfolds in the absence of long-term social relationships between members of the two groups, least of all in the context of between-group agreements. Clearly, the proximate and developmental processes leading individuals to stay in their natal group or move out of it differ in some fundamental ways between *Pan* and humans.

A second major discrepancy between primate philopatry patterns and human residence patterns has to do with the effects of changes in residence on the social networks of dispersing individuals. Postmarital residence patterns in humans involve individuals moving between groups embedded in larger social entities. In patrilocality, for example, women may move between clans of the same band or phratry or between moieties of the same tribe. Women moving between two patrilocal moieties upon marriage are still members of the same tribe, and as such they may have frequent opportunities to come into contact with their kin or to hear about them. As emphasized by Rodseth and coworkers (1991), in humans both the dispersing sex and the resident sex are able to maintain lifetime bonds with their families. But in chimpanzees and nonhuman primates in general, the members of the dispersing sex become physically separated from and unable to interact with their kin after they have transferred. Thus dispersal in nonhuman primates and changes in residence in humans have drastically different consequences for the social networks of the dispersing sex.

A third disparity is that while postmarital residence patterns are remarkably diverse and flexible, primate dispersal patterns vary little across populations of the same species. Besides patrilocality, human groups exhibit the following broad residence patterns: matrilocality (spouses live with or near the wife's parents), bilocality (spouses live near either the husband's parents or the wife's parents), neolocality (both spouses leave their natal home to live elsewhere), or avunculocality (males live with their maternal uncles, wives move to their husbands' location—virilocal marriage—and their sons return to live with the mothers' brothers). No less significant is that each broad category constitutes only an ideal or modal type, allowing a fair degree of residential flexibility depending on the society. For example, several hunter-gatherer societies were classified by Ember (1975, 1978) as "patrilocal with a matrilocal alternative." This referred to societies in which most couples conformed to the rule of patrilocal residence, but a certain proportion of couples lived matrilocally either permanently or temporarily before residing with the husband's family. Patrilocality, therefore, is not necessarily equivalent to complete or permanent male localization. Moreover, patrilocality includes a variety of situations in terms of physical distance between married women and their kin. Women may travel long distances to settle in their husband's group, in which case they have relatively few opportunities to interact with their relatives, or they may move just a few meters away and keep interacting with them on a daily basis. Similar variation ranges around a modal type characterize other human residence patterns.

The philopatry patterns of nonhuman primates, in contrast, are relatively rigid. I have already discussed the situation in chimpanzees and bonobos, and I shall consider just one other example. Sufficient behavioral data have now accumulated on species of macaques *(Macaca)* to allow genus-wide comparative analyses (Fa and Lindburg 1996; Thierry 2000; Thierry, Iwaniuk, and Pellis 2000). There are nineteen extant species of macaques, all descended from a common ancestor that lived some 5 million years ago (Delson 1980; Fa 1989; Fooden 1980). *Macaca* has the widest geographical distribution of all nonhuman primate genera, ranging throughout Asia, with relic populations in North Africa. Correlatively, macaques exhibit the widest ecological diversity of all nonhuman primates. Not surprisingly, their behavioral variation is also extremely pronounced. For example, some species exemplify a "despotic" type of social organization, with strict dominance hierarchies and low levels of reconciliation after conflicts, while other species exhibit a much more relaxed type of social organization. The phylogenetic tree of macaque species has been reconstructed from molecular data. By mapping the behavioral traits of macaques onto that phylogeny, Bernard Thierry and his colleagues found that several aspects of social organization correlated with phylogeny. This indicates that the traits in question are under substantial genetic influence or that they covary with elements that are themselves under genetic influence (Thierry 2000; Thierry et al. 2000). The main point here is that despite considerable interspecific variation in geography, habitat, behavioral ecology, demography, and phyletic subgrouping, all species of macaques exhibit the same sex-biased pattern of female philopatry. This findings points to the phylogenetically conservative aspect of philopatry patterns in nonhuman primates.

Clearly the discrepancies between primate dispersal patterns and human postmarital residence patterns can hardly be exaggerated. Nevertheless, the phylogenetic connection between the two sets of phenomena is no less evident.

The Emergence of Residential Diversity

Chimpanzees and bonobos, like primates in general, have a *dual-phase* residence pattern: females spend a prebreeding phase in their natal group, followed by a breeding phase elsewhere. Dual-phase residence is thus a phylogenetically primitive pattern. At some point in their evolution, hominids went from chimpanzee-like promiscuity to stable breeding bonds. The evolution of pair-bonding and its integration to a chimpanzee-like male philopatry pattern transformed that system into one having a pre–pair-bonding phase (or "premarital" phase) spent in the natal group, followed by a pair-bonding (or "marital") phase spent in the new group. Structurally speaking, therefore, "postmarital" residence emerged from the mere integration of pair-bonding and male philopatry, a fusion that generated an embryonic and strictly behavioral (nonnormative) form of patrilocality. To anthropologists such as Murdock, prior to the institutionalization of the incest taboo, residence was supposedly a single phase spent in one's natal group. The primate data tell us that residence has always been a dual-phase process.

Immediately after the evolution of pair-bonding, female hominids were still moving between socially independent communities as chimpanzees and bonobos do. Dispersing females had no means to keep interacting with the relatives they had left behind. After the evolution of the tribe, however, dispersing females were in a position to maintain bonds with their family even after relocating. The state of mutual tolerance prevailing between "intermarrying" groups meant that females came into contact with their kin on a more or less regular basis. Thus both the resident sex and the dispersing sex could maintain lifetime bonds with their families, as in the human situation. The idea that the transition from male philopatry to humanlike patrilocality coincided with the evolution of the tribe helps solve the problems raised by the first two discrepancies mentioned earlier. Between-group agreements are an evolutionarily recent dimension of human residence patterns. The primate evidence suggests that well before female relocalization was the object of such agreements, females were moving freely between groups and establishing stable breeding bonds on their own in their new group. Following the advent of the tribe, they could maintain lifetime bonds with their kinsmen, but they were not under their control nor exchanged by them. That would come later.

This brings me to the solution to the third disparity. With the rise of the tribe, the conditions were set for the emergence of evolutionarily novel residence patterns, including matrilocality and bilocality. The underlying principle is simple. In chimpanzees, females move between territorial groups aggressively defended by males. In such a context, the evolution of any new dispersal pattern, say bilocality, requires some fundamental changes in relationships between males living in distinct groups. A change from patrilocality to bilocality in humans involves no major structural difficulty because the relocalization of adult males upon marriage takes places between groups whose members are already familiar with each other, either as consanguines or affines. This suggests that some level of peaceful relations between local groups in hominids was a prerequisite for the diversification of postmarital residence patterns. Upon the rise of the tribe, residential flexibility could flourish for the simple reason that males were now moving between nonhostile groups. Henceforward, hominids could adjust residence patterns to local conditions, for example to resource distribution, without such adjustments involving evolutionary—biological—changes in the nature of male relationships and in levels of feeding or sexual competition. Stated bluntly, males could remain as competitive and xenophobic as before at the between-tribe level and nonetheless move between distinct local groups that belonged to the same tribe.

The case of matrilocality illustrates particularly well the idea that peaceful relations between local groups were a prerequisite for residential diversity. Matrilocality, a relatively rare postmarital residence pattern, is characterized by the localization of women and a change of residence by men upon marriage. Superficially, it is the reverse pattern of patrilocality. But an essential feature of matrilocal societies is the lifetime bonds between females and their kinsmen. This may be best illustrated by an important correlate of matrilocality, matrilineal descent. As has long been recognized, and articulated most clearly by David Schneider, matrilineal descent in humans is not the mirror opposite of patrilineal descent (Schneider 1961; Schneider and Gough 1961). In matrilineal descent systems, the line of authority runs through kinsmen, as in patrilineal descent groups. The difference is that the kinsmen are matrilineally related. For example, inheritance of property is typically avuncular, from men to their sister's sons, rather than from women to their daughters. The important point in relation to the present argument is that for any form of political control by men over their kinswomen to be associated with human matrilocality, localized women must maintain lifetime bonds with their kinsmen, notably their brothers, even if they live elsewhere. Only through such bonds can "absent" men be in a position to exercise control over their kinswomen. This statement of the obvious implies that peaceful and regular contacts between intermarrying local groups were a necessary condition for the advent of matrilocality. Accordingly, humanlike matrilocality is not possible or observed in nonhuman primates, as discussed in chapter 19.

The same reasoning applies to another postmarital residence pattern intimately linked to matrilineal descent. Avunculocality involves lifetime relationships between brothers and sisters and between maternal uncles and their sororal nephews. It implies that brothers and sisters maintain lifetime bonds even though brothers live with their maternal uncles, and sisters with their husbands' own maternal uncles. Like matrilocality, therefore, avunculocality presupposes the existence of some sort of supragroup level of social structure. In sum, if one posits that early hominids formed territorial male kin groups, humanlike matrilocality, bilocality, and avunculocality could hardly evolve prior to the advent of the tribe, even less before the evolution of pair-bonding, itself a prerequisite to the tribe.

In contrast with this view, Lars Rodseth and Shannon Novak recently argued that residential flexibility—a mixture of patrilocality, matrilocality, and bilocality—had evolved directly from ancestral male philopatry. Implicit in their reasoning was that the advent of residential flexibility was independent of the evolution of pair-bonding, male pacification, and the primitive tribe. Rodseth and Novak's argument was based on differences between chimpanzees and humans in the nature of female relationships. They contrasted the relatively solitary existence of female chimpanzees with the comparatively higher gregariousness of women: "Even when they transfer to unrelated groups," they wrote, "women everywhere tend to be far more gregarious than female chimpanzees," and "even unrelated females in the most extreme patriarchal societies regularly form close and enduring friendships" (2006, 203). They rightly noted that in this respect women resemble female bonobos more than female chimpanzees. Now if one assumes that female bonobos are more gregarious because they experience lower levels of feeding competition, women might be more gregarious because they underwent a decrease in levels of feeding competition in the course of human evolution. "Whatever the precise mechanism," Rodseth and Novak argued, "if scramble competition were reduced, an important change in hominid social behavior would tend to follow: a shift away from the ancestral, chimpanzee-like pattern of consistent male philopatry and toward a mixed pattern of patrilocal, matrilocal, and bilocal residence" (208). Stated otherwise, following changes that reduced levels of feeding competition between them, hominid females became more gregarious so they could either transfer between groups, as before, or remain in their natal group. For some reason males as well would have become facultatively philopatric: they could either stay in their natal group or transfer between local groups.

This scenario is unlikely because it does not address the central obstacle to residential flexibility: intergroup male hostility. First, Rodseth and Novak mistakenly assumed that the main reason female chimpanzees move between groups is to reduce feeding competition. But as discussed previously (chapter 10), female transfer in chimpanzees has more to do with the fact that males are localized and that in this context females must avoid inbreeding. True, lower levels of feeding competition might well increase female gregariousness and sociality *within* groups, but that would not be expected to give rise to female localization, even less to male transfer. On the contrary, female localization in nonhuman primates is associated with high levels of feeding competition, both within groups and between groups (Wrangham 1980; van Schaik 1989; Sterck, Watts, and van Schaik 1997; Isbell and Young 2002). Second, granting that bonobos experience lower levels of feeding competition compared to chimpanzees, the fact is that female bonobos are not philopatric; they still disperse out of their natal group. Third, positing that females become philopatric, or that a fraction of them are, does not mean that males start moving between rival groups; males are still philopatric. That is, one does not obtain a humanlike pattern of bilocality in which both sexes move between groups. For this to take place, male pacification is required. Thus the process envisioned by Rodseth and Novak is unlikely to produce a mixed pattern of matrilocality, patrilocality, and bilocality, and hence is unlikely to explain the evolution of residential flexibility in hominids. Prior changes in the nature of intergroup relationships appear to be of paramount importance.

I have argued that the rise of the tribe brought about the diversification of postmarital residence patterns. But why should residence patterns diversify in the first place? Why are some societies matrilocal and others bilocal or patrilocal? If primate studies can shed light on the origin of stem social patterns such as incest avoidance, pair-bonding, and sex-biased residence, they can hardly enlighten issues relating to the ensuing cultural diversification of these stem patterns. Answers to such questions belong to the fields of sociocultural anthropology and human behavioral ecology. However, primatology has its own hypotheses about and models of sex-biased philopatry in primates, and these provide a general principle that helps explain part of the variation in human residence patterns, notably the direction of causality between residence patterns and sex-biased patterns of cooperation (chapter 19).

Ancestral Patrilocality and Grandmothering

I have already discussed several critiques of the idea that male philopatry was the stem residence pattern of the *Pan* and *Homo* lines. In closing this chapter, I consider one last critique addressed to the ancestral male kin group hypothesis, which is best responded to by considering the evolution of residential diversity. In a recent discussion of the ancestral patrilocality model, Helen Alvarez stated that this hypothesis "places strong constraints on development and evaluation of alternative hypotheses about ape and human evolution" (2004, 438). She was concerned here with the "grandmother hypothesis," which aims to explain the evolution of postmenopausal longevity in humans. The grandmother hypothesis posits that women of childbearing age benefited from residing near their mother and other female kin who could help them provision their weaned offspring (Hawkes et al. 1998; Hawkes 2003; Hawkes, O'Connell, and Blurton-Jones 2003). Grandmothers who helped their daughters reproduce by taking care of their grandchildren would have enjoyed a selective advantage over grandmothers who did not. One adaptive consequence of this aiding pattern would have been an increase in longevity well past menopausal age among hominids. If one posits that male philopatry was the hominid ancestral pattern, Alvarez continued, "hypotheses assigning a central role in human evolution to cooperation between aging mothers and maturing daughters cannot be entertained" (439). That is to say, the grandmother hypothesis is not compatible with male philopatry; it requires matrilocality or bilocality or a flexible pattern of residence.

The grandmother hypothesis is concerned with grandmothering activities directed specifically at a *daughter's* offspring and thus requires the coresidence of female kin. According to my argument above, the coresidence of mothers with their reproductive daughters became possible with the emergence of the tribal level of organization, under either matrilocality or bilocality. Thus, if the rationale underlying the grandmother hypothesis is correct, grandmothering and postmenopausal longevity would have developed after the tribe had evolved. Based on various lines of evidence, Hawkes and colleagues argued that postmenopausal longevity coincided with the appearance of *Homo erectus* 1.9–1.7 million years ago (O'Connell, Hawkes, and Blurton-Jones 1999; Hawkes, O'Connell, and Blurton-Jones 2003). The question thus becomes: Had the tribe, and hence pair-bonding, already evolved at that stage in human evolution? If they had, the grandmother hypothesis and the present reasoning are compatible. Thus, contrary to Alvarez's assertion, the ancestral patrilocality hypothesis does not mean that "cooperation between aging mothers and maturing daughters cannot be entertained." Rather it implies that such cooperation evolved subsequent to the onset of the tribe. Now some authors have argued that pair-bonding—and presumably primitive tribal organization as well—had already evolved at the *H. erectus* stage (Wrangham et al. 1999), but Hawkes and colleagues argued that pair-bonding had not yet evolved. This is an empirical issue that may be settled only with more data on the anatomical and behavioral correlates of pair-bonding, grandmothering, and life-history traits.

Be that as it may, it is possible to envision a major variant of the grandmother hypothesis that can thrive under male philopatry, regardless of the pattern of proximity between grandmothers and their adult daughters. Earlier I argued that from the time pair-bonding and biparentality merged with male philopatry, grandmothering activities directed to a *son's* offspring became a possibility. Paternity recognition afforded aging females with opportunities to take care of their sons' children by helping their daughters-in-law. Grandmothering activities directed to a son's offspring are common in humans. In terms of selective advantages, whether a grandmother takes care of her sons' children or her daughters' children, she obtains similar fitness benefits: grandmothers are equally related to both sons and daughters. One might argue, however, that grandmothers are *on average* more related to their daughters' children than to their sons' children, given that the latter may have been fathered by males other than their sons. A son's paternity can never be taken for granted, contrary to a daughter's maternity. Hence, grandmothers would benefit more (genetically speaking) by helping their daughters than their sons. This objection is weak, however. Under male philopatry, grandmothers would be better off taking care of their sons' offspring, even if this meant taking care of a nonbiological grandoffspring from time to time, than ignoring them.

Thus there is no reason that grandmothering and postmenoposal longevity could not have evolved under male philopatry and primitive patrilocality, and no reason that these traits should be restricted to a matrilocal or bilocal situation. Of course grandmothering could shift to or encompass the daughters' children under bilocality or matrilocality.

16

Brothers, Sisters, and the Founding Principle of Exogamy

> It is this general structure [that which underlies cross-cousin marriage and avuncular relationships], of all the rules of kinship, which, second only to the incest prohibition, most nearly approaches universality.
> *Claude Lévi-Strauss (1969, 124)*

> As I predicted, it would be the avunculate and not the incest taboo that would become the defining principle of humanity and culture.
> *Robin Fox (1993, 227)*

These citations by Fox and Lévi-Strauss have much in common. Fox saw in the special relations linking an uncle to his sister's children (the avunculate) the cornerstone of mankind's originality in the social sphere. He thus opposed the avunculate to Lévi-Strauss's incest taboo as the "defining principle of humanity." Notwithstanding that opposition, Fox's position is much closer to Lévi-Strauss's than it appears. Although Lévi-Strauss consistently asserted that the defining principle of humanity lay in the incest taboo—"the transition from nature to culture"—he also identified another structure, which he described as "second only to the incest prohibition" in its universality and importance. In chapter 7, for lack of a more suitable expression, I subsumed the various dimensions of that universal structure under the phrase "brother–sister kinship complex." Included in it are the following practices: preferential marriages between cross-cousins (matrilateral or patrilateral); proscriptive marriages between parallel cousins; matrimonial privileges of maternal uncles over their sisters' daughters, and of nephews over their paternal aunts; special relationships between maternal uncles and their sisters' sons (avunculate); and various associated kinship terminologies. Structurally, all aspects are derived from the brother–sister bond. Given that the avunculate is an integral part of the brother–sister kinship complex, Fox's defining principle of humanity is incorporated in what Lévi-Strauss described as a universal structure second only to the incest prohibition. In this sense, Lévi-Strauss and Fox basically agree about the brother–sister bond lying at the very heart of mankind's uniqueness as a social species.

The present phylogenetic analysis supports the view that the brother–sister kinship bond may be the single most original feature of human society. But as we shall see, the defining principle of the exogamy configuration lies not so much in the brother–sister kinship complex itself as in a closely related kinship structure that results from the conjunction over evolutionary time of a number of otherwise primatelike phenomena.

The First Step: Outmarriage

Etymologically, "exo-gamy" means marrying out of one's group, or outmarriage. Lévi-Strauss's conception of exogamy went far beyond its etymological sense. Outmarriage as he saw it was an integral part of the substantially more inclusive phenomenon of matrimonial exchange, which incorporated the conjugal bond itself, the circulation of females between groups, incest prohibitions between primary kin, and reciprocity-based agreements between intermarrying families. Lévi-Strauss repeatedly emphasized the indissociability of marriage and female exchange: "No matter what form it takes," he wrote, ". . . it is exchange, always exchange, that emerges as the fundamental and common basis of the institution of marriage" (1969, 478–479). To him outmarriage was born with matrimonial exchange.

The primate evidence belies this view: it indicates that outmarriage emerged independently of and prior to female exchange. And, moreover, that it did so somewhat "spontaneously" upon the evolution of pair-bonding, in the absence of any other specific condition or prerequisite. From an evolutionary perspective, primeval outmarriage is simply the merging of two otherwise typical primate patterns: kin-group outbreeding and pair-bonding. Outbreeding through female dispersal is the rule in chimpanzees and bonobos, and it was presumably the rule in the ancestral male kin group as well. Females were thus moving between local groups from time immemorial before pair-bonds evolved. When they did, the new mating system superimposed itself on male residence and female dispersal. Henceforward females kept emigrating from their birth group into a new group, as before, but instead of mating promiscuously in it, they formed a pair-bond in their new group and began to breed. Females were by then practicing a strictly behavioral form of exogamy, one manifest solely as behavioral regularities in dispersal and mating; they were practicing the phylogenetic precursor of outmarriage. Accordingly, female exogamy at that stage did not include any exchange dimension. Females were transferring between groups on their own initiative, not as part of transactions between males. Hominids were thus involved in pre-exchange forms of exogamy long before they built upon that phylogenetically old substrate to eventually exchange females and forge intergroup alliances on this basis. Clearly, female circulation coupled with pair-bonding creates a natural intermediate term that smoothes the transition to cognitively sophisticated phenomena such as matrimonial exchange and reciprocal exogamy.

Also of interest at this point is the primate evidence indicating that it was female exchange, not male exchange, that was primeval. Here lies a further and somewhat unexpected element of convergence with marriage alliance theory. Lévi-Strauss held that female exchange was the primary and most binding form of reciprocity between men, an argument he based on a strictly structural analysis. The primate evidence vindicates this conclusion but by providing different types of arguments, including a temporal one. Female circulation was primitive, chronologically speaking, and this may explain in part why a female bias in dispersal is widespread cross-culturally. But other causal factors are also involved. The reasons invoked by Lévi-Strauss to explain the primacy of female exchange were (1) that throughout the world women are the "most precious possession," and (2) that in humans, reciprocal transactions commonly transcend the object of exchange, giving rise to social partnerships (the "synthetic nature of the gift"). Lévi-Strauss's reasoning on these points makes perfect sense from an evolutionary perspective. In particular, it fits nicely with sexual selection theory. Throughout the animal kingdom, females are certainly the most "precious possession" males compete for, as has been overwhelmingly documented ever since Darwin ([1871] 1981) first explained why this was so. Male competition for females is one of the most consistent behavioral regularities characterizing sexually reproducing species everywhere in the animal world;

moreover, it is one of a very small number of such universally valid regularities, and one of the best-documented ones. The potency of male sexual competition in structuring the social life of animals, whether within or between groups, is matched only by that of female competition for physical resources.

The reason males commonly attempt to monopolize several females, whereas females rarely attempt to monopolize several males, is that in most species it is the females who do the bulk of parental care. Females are thereby physiologically limited in the rate at which they produce offspring. They can hardly increase their reproductive success by fighting with each other to maximize the number of males they mate with. For their part, males who attempt to maximize the number of females they inseminate do increase their reproductive success (Trivers 1972; Clutton-Brock and Packer 1992; for a recent synthesis, see Clutton-Brock 2004). Thus, whenever females have lower reproductive rates than males owing to biological discrepancies in parental investment, male sexual competition predominates over female sexual competition. Nonhuman primates are no exception to that rule (see contributions in Kappeler and van Schaik 2004), and therefore so were our hominid forebears. True, the evolution of pair-bonding and paternal care significantly reduced the discrepancy in parental investment between the two sexes, bringing about some profound changes in sexual strategies. But male sexual competition is still very much alive. Thus Lévi-Strauss's argument about the prominent value of women to men transculturally can hardly be more compatible with evolutionary theory, in particular with sexual selection theory.

It is also the case that reciprocity in nonhuman primates has some sort of "synthetic nature" that transcends the specific currency of exchange. For example, reciprocal grooming between any two individuals is commonly part of long-term, multidimensional cooperative relationships that involve not only acts of grooming but contact, proximity, reconciliation, tolerance at food sites, aiding in conflicts, and so forth. Social partnerships in nonhuman primates commonly transcend their constituent building blocks; they are more than the sum of their parts (Hinde 1979, 1987; Kummer 1979; van Hooff 2001). Given that primates in general build cooperative partnerships by exchanging services (Kappeler and van Schaik 2006), one may infer that our hominid forebears, assuming that they were eventually able to exercise some control over their kinswomen, were *preadapted* to form strong partnerships on this basis. Stated otherwise, the old primate abilities involved in the emergence of cooperative partnerships from simple acts of exchange were coopted for a new use, and a particularly potent one, when females, the "most precious possession" became an object of exchange.

Thus both reasons invoked by Lévi-Strauss to explain female exchange apply to primates in general, and in this sense his arguments are perfectly compatible with evolutionary theory and the primate data. But, then, why are humans the only primate in which males exchange females? Some elements of the answer have already been discussed. For humanlike female exchange to evolve, three basic prerequisites were needed. As it happens, the three factors are not found together in any known nonhuman primate species, but they co-occurred in hominids at some point in their evolution. The first condition was the circulation of females between male kin groups. In relation to this point, Elman Service came close to a similar argument. In adopting Lévi-Strauss's exogamy theory in its broad outline, Service proposed that women were exchanged by men because of the primeval character of patrilocality. In short, females were traded because exchange was a necessity and because males were already localized for reasons of their own. The second prerequisite to female exchange in the context of marital unions was the existence of long-term breeding bonds. As soon as these two conditions were met, a behavioral form of outmarriage was born de facto. But a third factor brought hominids one big step closer to female exchange. Antagonists can hardly engage in exchange. Thus, for males to be in a position to trade their kinswomen, a state of mutual tolerance and peaceful contact between intermarrying male kin groups was needed. The advent of the tribal level of organization thus removed a prohibitive obstacle to the very possibility of transactions between males belonging to distinct groups.

At the dawn of the primitive tribe, female exchange was a structural possibility, but it was still no more than a theoretical one. Significantly, however, in the very processes that led to the pacification of between-group relations lay a major catalyst for matrimonial exchange. Among all the males belonging to distinct intermarrying groups, some were better positioned than others to develop preferential bonds. It was within these specific male clusters, or nascent male partnerships, that female exchange would eventually emerge, as I now argue.

Affinal Brotherhoods and the Origin of Exogamy Rules

Lévi-Strauss has consistently emphasized the cardinal importance of the overall kinship structure in governing mate assignment among exogamous units, and, as pointed out earlier, he was the object of much criticism on this point. Marriage often does not take place between relatives, and even when it does, the marriageability of individuals is not assessed strictly on the basis of genealogical relations. In societies prescribing cross-cousin marriage, spouses are often not real cross-cousins but classificatory ones. The evolutionary perspective sheds new light on this old and important debate. After the primitive tribe had evolved, females were emigrating into local groups in which they had kin and affines with whom they were already familiar to some extent. This original conjunction set the stage for the emergence of substantial kinship constraints on marital unions.

One consequence of pair-bonding was the strengthening of bonds within families and its correlate, the emergence of cohesive units of primary male kin—or primary agnates. By comparison to the situation in *Pan,* in which the father is not recognized, hominid females were growing up in close contact with their fathers and brothers. After leaving her birth group and forming a pair-bond with a male in another community, a female would start developing bonds with her "husband's" relatives. In so doing she would become a member of yet another unit of primary agnates, this time as a wife and affine. Prior to the evolution of intergroup pacification, a female's dual membership in distinct units of primary agnates was a sequential process in her life. But from the time she could maintain bonds with her family after emigrating, she *simultaneously* belonged to two distinct units of primary agnates, either as a consanguine or as an affine, as illustrated in Figure 16.1. Therefore, once the tribe had evolved, any pair-bond could act as a link between the two corresponding units of primary agnates. As argued previously, male affines were biased to become allies for two reasons: they had a vested interest in the same female, either as kin or as affine, and they were not sexual competitors for that female. Moreover, relations of affinity provided the single source of ties linking adult males from different groups. I use the expression *affinal brotherhoods* to refer to the network of social relationships comprising two units of primary agnates united by one or more pair-bonds. Affinal brotherhoods may be seen as the primeval

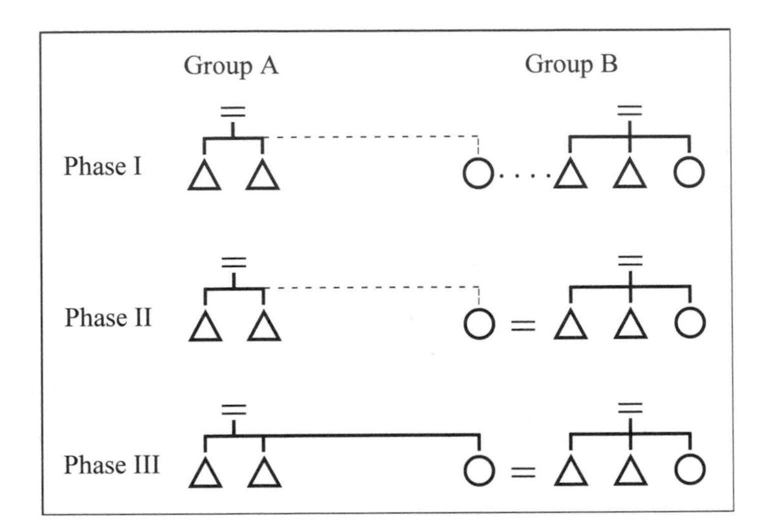

Figure 16.1. The evolution of affinal brotherhoods between two male-philopatric groups. A single pair of brothers is illustrated in each group. The sister of the A brothers emigrates and breeds in group B. In a chimpanzee-like group (Phase I), brother–sister bonds are inconsistent. After the female emigrates, she is cut off from her male kin (dashed line) and has short-term sexual relationships with males (dotted line). After the evolution of pair-bonding (Phase II), the female has stronger bonds with her brothers but is still cut off from them after emigrating. However, she has affines (here, siblings-in-law). Once the tribe has evolved (Phase III), the female maintains contact with her male kin (heavy line) and connects the two units of primary agnates she belongs to.

knots in the emerging web of social relationships embracing distinct interbreeding local groups.

The core of the present argument is that precisely because affinal brotherhoods constituted primary connections between exogamous patrilocal groups, they set up basic constraints on the formation of pair-bonds between them. Put simply, upon transferring into another group, females would have been more or less biased to form pair-bonds with their affines, for example their brothers-in-law. Before describing the type of mating arrangements this led to, I briefly examine the proximate processes through which affinal brotherhoods could influence the formation of pair-bonds. Affinal brotherhoods meant not only that brothers-in-law were developing preferential bonds with each other but that all affines were. The model envisioned here is pictured in Figure 16.2. For the sake of simplicity, the model posits interbreeding between two male-philopatric ("patrilocal") groups and focuses on a single affinal brotherhood: two families initially linked through a single pair-bond, that of M1 and F1. The reasoning starts with the idea that as a young female (Ego in the figure) grows up in daily contact with her older brother (M1) and his wife (F1), she becomes disproportionately familiar with her sister-in-law. As this bias goes on over several years, Ego and her sister-in-law have several opportunities to develop a preferential relationship, much as consanguineal kin would. Interbreeding groups mingle on a regular basis. Every time they do so, the sister-in-law (F1) gets in touch with her own kin, in particular with her father and her

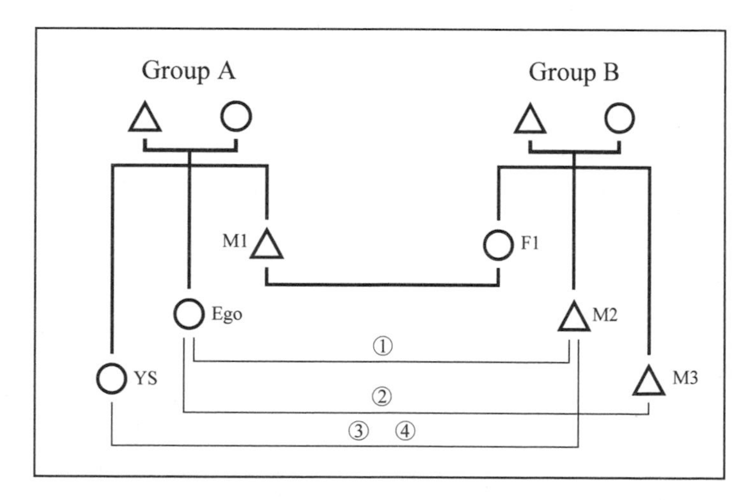

Figure 16.2. The processes involved in the development of various types of mating biases between two male-philopatric groups at the dawn of the primitive tribe. The model posits an initial pair-bond between male M1 living in group A and female F1 born in group B but now living in group A. This pair-bond produces familiarity biases between Ego and her brother's affines, M2 and M3. (1) Pair-bond between Ego and M2. In conjunction with the M1–F1 pair-bond, this produces the behavioral equivalent of *sister exchange* between M1 and M2. (2) Pair-bond between Ego and her brother-in-law M3 following M2's death: the behavioral form of the *levirate*. (3 and 4) Pair-bond between M2 and his sister-in-law YS, producing the behavioral equivalent of *sororal polygyny*, if Ego is alive, or the *sororate*, if Ego is dead.

brothers (M2 and M3). In this context, two mediators are expected to predispose young female Ego to interact preferentially with her sister-in-law's brothers. The sister-in-law stands as a natural intermediate and intercessor between her own brothers and her friend Ego. And Ego's older brother himself, because he maintains preferential bonds with his brothers-in-law, also stands as a natural intermediate and intercessor between his younger sister and his brothers-in-law. Given that Ego undergoes these kinds of familiarity biases on an ad hoc basis, and repeatedly over several years before emigrating, she is in a position to develop *predispersal friendships* with her brothers' affines. Such friendships, in turn, are expected to markedly orient the direction of dispersal, inducing the female to transfer preferentially into her predispersal friends' group. They are also expected to translate into mating biases. Indeed, compared to all other male–female pairs, predispersal friends stand out as disproportionately familiar with each other. But because they live apart in their respective natal groups, their levels of familiarity are sufficiently low so as not to deter sexual attraction between them—unlike siblings living together from childhood. Stated otherwise, affines are optimally familiar: they are familiar enough that they experience no fear inherent to unfamiliarity but not so familiar as to undergo the Westermarck effect.

From a structural angle, four types of mating biases are possible in this context. First, if Ego forms a pair-bond with her brother's affine M2, the resulting arrangement is the behavioral equivalent, structurally speaking, of *sister "exchange"* between brothers-in-law (between M1 and M2). Second, if male M2 died, Ego might form a pair-bond with his brother M3, her own brother-in-law and a predispersal friend with whom she was also disproportionately familiar. This is the behavioral equivalent, in strictly structural terms, of the widespread practice of levirate (a man marrying the wife of his deceased brother). Third, given that Ego's younger sister (YS) also underwent familiarity biases with males M2 and M3, she might also pair-bond with her brother-in-law M2, in which case one obtains a form of *sororal polygyny,* M2 being "married" both to Ego and YS. Alternatively, if Ego died, YS might form a pair-bond with Ego's husband, her brother-in-law, thus producing the structural equivalent of the widespread practice of sororate, (a woman marrying the husband of her deceased sister). Sororal polygyny and the sororate are structurally similar, a man marrying sisters in both cases, but simultaneously in the first situation and sequentially in the latter.

In sum, affinal brotherhoods might have provided the primeval constraints orienting mate selection in the context of pair-bonding, hence affecting the organization of exogamy. But clearly, if such processes were at work among hominids, they could hardly have generated more than mere *biases* in mate selection; that is, they would have produced tendencies rather than consistent regularities. Compared to explicit marriage prescriptions and institutionalized rules, primatelike familiarity differentials between potential mates impose relatively loose constraints on mate selection. Exceptions to the patterns described previously must have been numerous. Nonetheless, it is difficult not to establish a connection between such biases, however imperfect, and their more systematic counterparts in human societies. Indeed, sister "exchange," the levirate, sororal polygyny, and the sororate are widespread types of unions in humans, as noted long ago by George Murdock in the first statistically oriented cross-cultural survey of human societies. Citing figures based on 250 societies, Murdock concluded that "both the levirate and the sororate are exceedingly widespread phenomena" (1949, 29, 31). In reviewing sex regulations relating to kinship, he distinguished between *preferential mating,* "a cultural preference for marriage between persons who stand in particular kin relationships to one another, such as cross-cousins or siblings-in-law of opposite sex," and *privileged relationships,* a permissive regulation according to which "sexual intercourse is permitted before marriage [to another person] and frequently afterwards as well." With respect to the latter, he added that "the most illuminating of privileged relationships are those between siblings-in-law of opposite sex, who are, of course, frequently potential spouses under the sororate and levirate" (268). After dismissing a number of explanations for the generalization of preferred marriages and permissive adulterous sex between siblings-in-law—for example, group marriage—he argued that they reflected "extensions of the marital relationships" and the attraction of people to persons who most closely resemble one's spouse. "The persons who universally reveal the most numerous and detailed resemblances to a spouse," he wrote, "are the latter's siblings of the same sex . . . who are likely to have similar physical characteristics . . . [and] have almost identical statuses since they necessarily belong to the same kin groups—family of orientation, kindred, sib, etc." (269).

Murdock's explanation thus conceived of mate selection as significantly affected by simple processes such as familiarity differentials, physical similarities, and social compatibility. His explanation has much in common with the foregoing model. This being said, the approximate character of the processes described by Murdock makes his explanation somewhat unsatisfying and incomplete. Lévi-Strauss was to provide a much more comprehensive interpretation in his marriage alliance theory. To him, bilateral marriages between brothers-in-law embodied the reciprocity dimension of marriage: brothers trade their sisters with other males to obtain wives in return, and they build alliances in the process. In such a system sister exchange, the sororate, and the levirate are facets of reciprocal obligations between exchanging units. The sororate ensures that a widower gets another wife, namely his deceased wife's sister; the levirate, that a widow obtains another husband, her deceased husband's brother. The widespread character of these marital practices would thus reflect the universal structure of reciprocal exogamy; they would not express mere patterns of familiarity or physical similarities, as envisioned by Murdock. Nonetheless, Lévi-Strauss's theory and the present model fit together particularly well: affinal brotherhoods, with their inherent structural constraints, would have provided the initial blueprint, however imprecise, out of which the first rules governing marriage between affines emerged. Simple familiarity differentials inducing behavioral regularities would have paved the way for the elaboration of normative rules, in the same manner that incest avoidances paved the way for incest prohibitions.

From Siblings-in-Law to Cross-Cousins

But what about cross-cousin marriage, described by Lévi-Strauss as the "elementary formula for marriage by exchange" (1969, 129) and "the simplest conceivable form of reciprocity" (48)? In the present model cross-cousin marriage is intimately related to rules of marriage between affines. In fact, marriage rules affecting cousins are derived from marriage rules involving siblings-in-law. Let us consider two males who marry each other's sister, as illustrated in Figure 16.3. The two males (M1 and M2) become brothers-in-law, and the two females (F1 and F2), sisters-in-law (16.3a). In a situation of sister exchange, sisters-in-law are married to brothers-in-law. Now, if these individuals extend the exchange principle to their own children, that is, if M1 and M2 exchange their respective daughters as wives for their sons, this results in marriage between cross-cousins (16.3b). Indeed, M1's son (M3) marries female

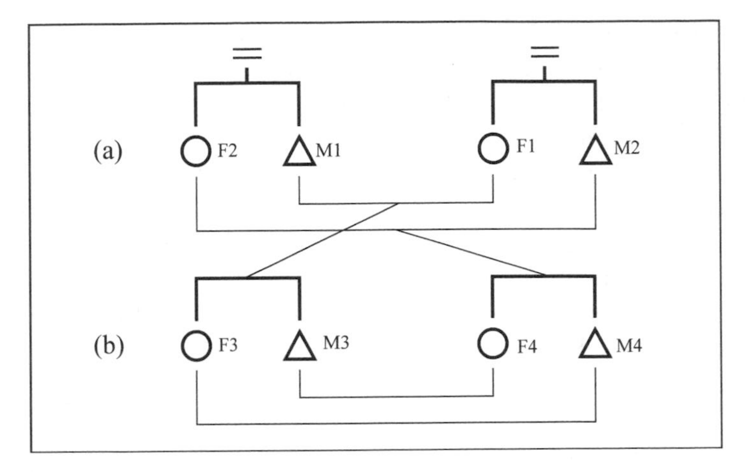

Figure 16.3. Structural relations between pair-bonds involving siblings-in-law and pair-bonds between cross-cousins. (a) Bilateral pair-bond between brothers-in-law M1 and M2, the behavioral equivalent of sister exchange. (b) When extended to children, sister exchange produces pair-bonds between double cross-cousins.

F4, who is simultaneously his patrilateral cross-cousin (the daughter of his father's sister) and his matrilateral cross-cousin (the daughter of his mother's brother). In other words, male M3 marries his double cross-cousin. The same applies to male M4, who marries his double-cousin F3. Structurally speaking, the extension of bilateral marriage between brothers-in-law (sister exchange) to the next generation produces marriages between double cross-cousins.

Correlatively, marriages between *parallel* cousins are not possible. Parallel cousins are the offspring of either two brothers or two sisters. In the present model, brothers are philopatric and breed in the same group. Therefore their sons and daughters (patrilateral parallel cousins) do not constitute potential mates because daughters breed outside their birth group. Similarly, the sons and daughters of two sisters (matrilateral parallel cousins) who transfer and breed in the same group are not potential mates because the daughters move out of that group to breed. Thus marriage asymmetries between cross-cousins and parallel cousins are a correlate of the sex-biased dimension of residence patterns. Sex-biased dispersal eliminates parallel cousins as potential mates while leaving cross-cousins as potential ones. For that matter, such a situation prevails in any male-philopatric primate species. In chimpanzees and bonobos, for example, male localization and female dispersal create the same type of asymmetry in the availability of cross-cousins and parallel cousins. The major difference from the human species is that cross-cousins, although they are potential sexual partners, are not biased to be, because there is no tribal level of social structure in chimpanzees. As a result, females maintain no contact with their brothers after they have emigrated, and cross-cousins undergo no familiarity biases. Cross-cousins are no more than potential mates among others, and potential mates who do not recognize each other as kin.

In Lévi-Strauss's alliance theory, kinship permeates exogamy rules, setting the most fundamental constraints on marital unions in the form of prescriptions, preferential unions, and proscriptions. The ultimate source of kinship's potency in organizing exogamy lies in the brother–sister bond; it stems from brothers renouncing marriage with their sisters in order to exchange them with other males. And from this angle, cross-cousin marriage is the extension of female exchange to the males' daughters. Lévi-Strauss thus placed the brother–sister bond at the very heart of reciprocal exogamy. But he did so by mistakenly assuming that brothers and sisters were spontaneously attracted to each other and therefore had to counter that natural tendency for alliance purposes. All dimensions of reciprocal exogamy were seen as unfolding from this principle. Notwithstanding this important discrepancy, Lévi-Strauss's reasoning dovetails rather nicely with the present model. Phylogenetically speaking, kinship constraints on exogamy do originate importantly from the brother–sister bond. But the significance of that bond follows from the merging of four elements at some point in human evolution: (1) the very existence of preferential bonds between brothers and sisters—of brother–sister nepotism—whose intensity was substantially enhanced owing to the effect of biparentality on primary kinship bonds; (2) the fact that brothers and sisters commonly ended up in distinct kin groups as a result of sex-biased dispersal, a pattern that created as many latent connections between groups; (3) the rise of the tribal level of organization, which activated these connections by allowing brothers and sisters to keep in touch despite physical distance; and (4) pair-bonding, which in its amalgamation with brother–sister ties created new types of bonds, those between affines. I have emphasized here the brother–sister bond, but an equally important bond was that between the female and her father. Before moving to another group, a female was linked both to her brothers and to her father, and the four elements just described apply equally well to the father–daughter bond.

The founding principle of the exogamy configuration would thus reside in the coupling of brother–sister (and father–daughter) bonds with pair-bonds and the concurrent creation of affinal kinship. The elementary core of the exogamy configuration, the irreducible unit of exogamy, would be a *sister (and daughter) linking her brother (and father) to her husband,* that is, a sister linking brothers-in-law, as illustrated in Figure 16.1, Phase III—or a daughter linking a father-in-law to his son-in-law. This triadic type of relationship bridging two interbreeding units would be the true *elementary unit of exogamy.*

The "Atom of Kinship" Revisited

This brings me to a further point of convergence between Lévi-Strauss's structural analysis of human society and the present evolutionary analysis, namely, his concept of the "atom of kinship." It is most significant that the brother-sister-husband triad just described is the core of that atom. In a discussion of the avunculate, Lévi-Strauss stated that the relation between a maternal uncle, his sister, and his nephew must be seen as one type of relationship within a general structure, which he described as "the most elementary form of kinship that can exist" and "properly speaking, *the unit of kinship*" (1963, 46). The structure rests upon four terms: a brother, his sister, the sister's husband, and their son. "This elementary structure," Lévi-Strauss continued, is "the true *atom of kinship* . . . it is the sole building block of more complex systems. For there are more complex systems; or, more accurately speaking, *all kinship systems are constructed on the basis of this elementary structure,* expanded or developed through the integration of new elements" (48, my emphasis). If the atom of kinship was so fundamental in Lévi-Strauss's mind, one would expect it to be closely connected to the incest taboo, which stood as *the* principle par excellence in his alliance theory. And indeed, wrote Lévi-Strauss, "the primitive and irreducible character of the basic unit of kinship is actually a direct result of the universal presence of an incest taboo. This is really saying that in human society a man must obtain a woman from another man who gives him a daughter or a sister" (46). The atom of kinship is thus the immediate outcome of the incest impediment between siblings: brothers are linked to their sisters' husbands because they cannot mate with their sisters. To Lévi-Strauss, the atom of kinship is the closest correlate of the incest taboo. In fact, the two are indissociable.

This interpretation is obviously at odds with the present "atom of exogamy," whose emergence has nothing to do with brothers renouncing their sisters. Moreover, Lévi-Strauss's atom of kinship and the present atom of exogamy display some structural differences. First, Lévi-Strauss places the emphasis on sister exchange between brothers. But from an evolutionary perspective, the sister's bond with her own father is equally significant. Second, his unit of kinship includes four terms, whereas the atom of exogamy includes only three. The son is present in the atom of kinship but not in the present model. In Lévi-Strauss's scheme, the son is a necessary element because a brother, upon renouncing having children with his sister, elects to lend her to another male (her husband). The son "is therefore the product, indirectly, of the brother's renunciation," as Fox put it (1993, 192). Or, more bluntly, a brother reproduces *via his sister's husband,* so to speak. Accordingly, he has rights over the sister's sons, and in these rights rests the special relationship between maternal uncles and sororal nephews. To Lévi-Strauss, then, the avunculate "is nothing but a corollary, now covert, now explicit" of the incest taboo (1963, 51), and as a result, "we do not need to explain how the maternal uncle emerged in the kinship structure: he does not emerge—he is present initially" (46).

In the present model, in contrast, the son is absent from the atom of exogamy. The emergence of the brother-sister-husband triad has nothing to do with brothers handing over their sisters to other males and gaining rights over their sororal nephews in the process. The formation of bonds between brothers-in-law through the mediation of a sister is not dependent upon males developing special relationships, whether indulgent or authoritative, with their sisters' children. To Lévi-Strauss, however, the children were a central objective of female exchange. In his terms, "the child is indispensable in validating the dynamic and teleological character of the initial step, which establishes kinship on the basis of and through marriage" (47). According to the present analysis, affinal brotherhoods did not initially embody avuncular relationships. Special relationships between maternal uncles and sororal nephews are a further development of human society. Accordingly, the behavioral equivalent of the avunculate is nowhere observed in nonhuman primates.

Interestingly, Fox was not of this opinion. He thought that the avunculate existed in "primate nature" in a "shadowy form" (1993, 229). But his conclusion was based on his somewhat misreading the primate data. Fox was referring here to female-philopatric groups such as macaques, which exhibit matrilineal dominance structures (described in chapter 3). In these groups daughters inherit their mother's rank in the dominance order, so whole matrilines are ranked in relation to one another. For their part, sons leave their natal group and are not part of the group's matrilineal hierarchy. Although Fox was aware that males move between groups in these species, he apparently believed that male dispersal was a highly variable pattern and that males often stayed in their birth group. He reasoned that in this situation they formed a hierarchy in which they ranked according to their mother's rank. As a result, he wrote, "the son of a high-ranking female would be depending on his maternal "lineage, which would include his mother's brothers. There is at least then the possibility of some sort of rudimentary 'avunculate' in this type of system," in which a male's "maternal uncles would be part of his route to success" (217, diagram 4.12). Fox might have been implicitly referring here to some exceptional situations in which the sons of top-ranking females have indeed been observed to stay in their natal group and occupy high ranks. I have described such a case in some detail (Chapais 1983). In a large, free-ranging group of rhesus monkeys, the two mature sons of the alpha female occupied the first and second highest ranks in the male dominance order. All lower-ranking males were nonnatal. The two males clearly owed their high rank to their mother's alpha status. Significantly, however, they had no coresident maternal uncles, and the latter played no part in their special status. As a rule, males move out of their natal group around puberty, so adult brothers and sisters do not live together. Nor, as a result, do adult maternal uncles and sororal nephews. The avunculate is thus not present, even in "shadowy form," in macaque societies.

The primate situation that comes closest to an avuncular relationship, and for which the term is wholly inadequate in any case, has been described earlier. In chimpanzees, exceptionally, a young male may coreside with his mother's brothers if the mother breeds in her birth group instead of emigrating. Jane Goodall mentioned such a situation in which a young male had three maternal uncles. The youngest one was "fascinated by his young nephew" and interacted with him on a regular basis (1990, 112–113). The nephew also "developed friendly bonds" with his two other uncles. These observations are interesting in that they suggest that kin recognition between uncles and nephews may lie within the abilities of our closest living relatives, a possibility that finds support in the observation that in macaques, nephews appear to discriminate their maternal aunts (Chapais et al. 2001). One may infer on this basis that at the dawn of the primitive tribe a male's domain of kindred included his maternal uncles in addition to his coresident mother, father, and patrilateral kin.

I have just emphasized the differences between the most elementary *unit of exogamy* revealed by a phylogenetically comparative analysis and Lévi-Strauss's most basic *unit of kinship*. In spite of the discrepancies, the congruence between the two units is most remarkable. They are, in effect, basically the same. The atom of exogamy (or atom of kinship) is uniquely human. It is also the most primitive of all human kinship structures. Its evolution coincided with the birth of human society; it marked the onset of mankind's deep social structure. From this root structure all known human societies and kinship systems would emerge and diversify.

With the core unit of reciprocal exogamy defined, I have nearly completed the evolutionary reconstruction of human society. The end product is, I believe, the most advanced stage in the evolution of reciprocal exogamy that hominids had reached prior to the advent of language, cultural norms, and institutions. Stated slightly differently, a society featuring the elementary unit of exogamy would be the most sophisticated stage of social organization that primates had attained prior to the evolution of the symbolic capacity. The last component of human society to be integrated into the present scheme is unilineal descent, which is the object of Part IV.

IV

Unilineal Descent

17

Filiation, Descent, and Ideology

The germ of government must be sought in the organization into gentes [clans] in the Status of savagery.

Lewis Henry Morgan ([1877] 1974, 5)

Of the twelve building bocks of the exogamy configuration, I have now dealt with eleven: multimale-multifemale group composition, kin group outbreeding, stable breeding bonds, uterine kinship, agnatic kinship, incest avoidance, the brother–sister kinship complex (and proto-exogamy rules), affinal kinship, postmarital residence, matrimonial alliances, and the tribal level of social structure. The remaining component is descent. The reason I left it for the end is that besides being a big topic in itself, descent is a facultative manifestation of kinship in human societies, and therefore its treatment as a component of the exogamy configuration is not evident at first sight.

Ever since its discovery by Lewis Henry Morgan in the late nineteenth century, the *clan* organization, with its unilineal descent structure—or *gens* organization, as Morgan (1974, 61–87) referred to it—has proven to be one of the most prominent and conspicuous exemplifications of kinship's potency in organizing human affairs. This undoubtedly reflects to a large extent the sex-biased character of unilineal descent groups. Human kinship is basically symmetrical in relation to sex; it is bilateral. But actual relationships between relatives are far from being symmetrical in relation to sex. Humans often single out one specific line of filiation, matrilineal or patrilineal, whose members enjoy considerable privileges compared to nonmembers. For example, succession to office (status) and inheritance of property may be transmitted from fathers to sons exclusively. If the unilineal descent of succession and property are clear instances of sex-biased favoritism among kin, unilineal descent *groups* provide an even more striking illustration of the all-inclusive impact of sex-biased kinship structures on social life. Many societies are divided into subgroups called lineages or clans, whose members regard themselves as related to a common ancestor, either through males only or through females only. For example, a patrilineage (or patrilineal descent group) includes all individuals, males and females, descended from a single male ancestor, but through males only. Both sons and daughters are members of their father's patrilineage, but while the sons' children are part of their father's descent group, the daughter's children are not; they belong to the husband's patrilineage. Thus a patrilineal descent group is a bisexual group in its membership, but a unisexual one as far as its perpetuation is concerned.

The question of the origin of unilineal descent groups has intrigued anthropologists from the time they were first described. Among the issues debated by historical evolutionists was whether descent through the female line had arisen prior to descent through the male line, as held by Morgan, McLennan, and others, or whether patrilineal descent predated matrilineal descent, as contended by Henry Maine, among others, a debate that Marvin Harris qualified as "one of the most heated and useless discussions in the history of the social sciences." Useless because, in Harris's terms, "both groups were wrong, constituting one of those rare cases of diametrically opposed positions about which it is impossible to say that either contained a grain of truth" (1968, 187). True, both groups were wrong because, owing to the paucity of ethnographic data by the end of the nineteenth century, the evolutionists' understanding of descent was plagued with several important errors and misconceptions. To take just one example, matrilineal descent was commonly associated and confused with a state of "mother-right" and political domination by women (matriarchy). The evolutionists were also wrong in that the reasons they gave in support of the chronological priority of either matrilineal or patrilineal descent were entirely or partially flawed. This being said, the question they were debating was itself pertinent. Descent groups do have an origin, and from an evolutionary perspective the issue of the chronological priority of matrilineal versus patrilineal descent is a sound one.

My aim in the next three chapters is to show that unilineal descent groups are sophisticated cultural offshoots of primate kinship structures. And, if the "germ of government must be sought in the organization into gentes" (the clan organization), as Morgan thought, that germ was already present in embryonic form in the group-wide kinship structures of nonhuman primates. Such a claim is far from self-evident, if only because unilineal descent groups are not a universal feature of human societies. Descent groups are nearly absent or faintly manifest in the simplest societies, those of hunter-gatherers, and they are absent altogether in the most complex ones. In fact, descent groups are most in evidence, as Fortes put it, "in the middle range of relatively homogeneous, pre-capitalistic economies in which there is some degree of technological sophistication and value is attached to rights in durable property," such as pastoral and agricultural economies (1953, 24). If descent groups are not a universal phenomenon, if they are not part of mankind's unity, like motherhood, for instance, on what basis should one expect primate studies to have any bearing on their existence? The answer I shall provide here is that descent groups are a facultative manifestation of underlying organizational principles that do have a universal character and are biologically grounded. While these unitary principles lie more or less dormant in many social contexts and societies, they are activated in other contexts, wherein they produce, depending on the circumstances, a large array of social patterns, including descent groups in segmentary societies, lineage-based structures in nonsegmentary societies, unilineal patterns of succession to office or inheritance of property in societies without descent groups, and systems of double descent with a partitioning of rights and duties along the two lines.

Thus it is these principles, and not descent groups per se, whose evolutionary origins I seek to establish by showing that some primate societies exhibit similar principles and rudimentary forms of "descent groups." I begin by describing the main properties of human descent groups as these were spelled out more than fifty years ago by British social anthropologists, these properties forming the basis of so-called lineage theory.

The African Model of Unilineal Descent Groups

The properties of unilineal descent groups were described in the middle of the last century by a number of anthropologists, Radcliffe-Brown, Fortes, and Evans-Pritchard being among the most prominent. In a classical article published in the early 1950s, Fortes set out to summarize the main characteristics of African unilineal descent groups. "What I wish to convey by the example of current studies of unilineal descent group structure," he wrote, "is that we have, in my belief, got to a point where a number of connected generalizations of wide validity can be made about this type of social group," generalizations, he continued, that "seem to hold for both the patrilineal and matrilineal groups" (1953, 24). The results of this endeavor came to be known as the *African model of unilineal descent groups*. There is obviously no point in providing here a detailed account of that model, about which so much has been written. My aim is to focus on the main properties of unilineal descent groups inasmuch as these are pertinent for stating both the similarities and the differences between human descent groups and their primitive primate counterparts.

The most basic property of unilineal descent groups is that genealogical kinship alone defines an individual's status and affiliation to a given subgroup in its society, so "membership of a clan is normally determined by birth," as stated by Radcliffe-Brown (1950, 50). In a patrilineal descent group, an individual belongs to his father's lineage; in a matrilineal descent group, to his mother's. As pointed out earlier, both sons and daughters are members of their father's patrilineage, but only the sons' children are members of their father's group: membership is unisexually transmitted. But there is more. As argued by Scheffler (1985), a patrilineal descent rule means that an individual's paternal filiation is both a *necessary* and a *sufficient* condition for that individual to be recognized as a member of the patrilineage. It is a necessary condition in that other types of links, for example those between the individual and his mother's kin, or those between the individual and his spouse, do not confer membership in the mother's group or the spouse's group, respectively. Only patrifiliation confers group membership. The descent rule is also a sufficient condition for membership in that no other conditions are needed. In this sense, patrilineal filiation is nearly omnipotent.

Intimately related to the question of descent as a criterion of membership is the idea that "descent is fundamentally a jural concept" and that a unilineal descent group has a jural identity that transcends the spatial proximity of its members, even though lineage and coresidence often coincide. "A lineage cannot easily act as a corporate group if its members can never get together for the conduct of their affairs," specified Fortes.

"It is not surprising therefore to find that the lineage in African societies is generally locally anchored; but it is not necessarily territorially compact or exclusive . . . Spatial dispersion does not immediately put an end to lineage ties . . . For legal status, property, office and cult act centripetally to hold dispersed lineages together and to bind scattered kindred" (1953, 36). In sum, the members of a lineage might live in different villages, but they retain their membership in their descent group and enjoy its associated privileges.

Another aspect of unilineal descent groups relates to the way they evolve over time and give rise to higher levels of social structure. With each new generation, lineages grow in size, eventually splitting into two branches or segments, a process called lineage *segmentation*. Each segment is a sublineage that will grow and eventually split up into smaller sublineages. Lineage fissions thus take place along natural kinship lines. In a patrilineage, every man is the potential founder of a new sublineage. For example, one brother and his descendants, or a group of brothers and their descendants, could leave the main lineage and set up their own. The split does not end relationships between sublineages; they remain bonded as part of some higher-level structure. The result of this repeated process of fission is a hierarchical segmentary structure that resembles a tree, with the end products of segmentation—the smallest lineages—all being part of the same highest-order segmentary level, that of the clan, or the tribe. For example, in Evans-Pritchard's (1940) classical description of the segmentary structure of the Nuer, two or more so-called *minimal* lineages issuing from a common ancestor are part of the same *minor* lineage. Two or more minor lineages derived from the same ancestor belong to the same *major* lineage, and so on with segments of greater inclusiveness, up to the most inclusive level, that of the clan.

A related but distinct property, "the most important feature of unilineal descent groups," insisted Fortes, ". . . is their *corporate* organization" (1953, 25). Radcliffe-Brown referred to a corporate group as one "whose members come together occasionally to carry out some collective action . . . [and which] possesses or controls property which is collective" (1950, 41), such as land. Fortes further emphasized that the corporate aspect of a lineage was best defined from the viewpoint of other lineages or groups: "Where the lineage is found as a corporate group," he wrote, "all the members of a lineage are to outsiders jurally equal and represent the lineage when they exercise legal and political rights and duties in relation to society at large. This is what underlies so-called collective responsibility in blood vengeance" (26). In other words, the members of a lineage are undifferentiated when viewed from outside. Juridically, they are a single person.

When the two preceding aspects of the African model of unilineal descent groups (corporateness and segmentary structure) are considered together, they give rise to another fundamental dimension of descent groups, which I refer to here as their *nested* (or multilevel) structure of allegiance. In a conflict between two individuals belonging to different minimal lineages (A and B) of the same minor segment (borrowing from Evans-Pritchard's terminology), all members of minimal lineage A would ally against all members of lineage B. But in a fight opposing individuals from two different minor segments, lineages A and B would sink their differences and ally against the other minor segment. The same reasoning applies to higher levels of organization up to that of the clan, such that a conflict between individuals belonging to different clans would involve each of the two clans in its entirety. Thus, according to the African model of descent, the groups of lineages that oppose each other in any conflict are in *balanced opposition,* and alliances are formed at the same structural level, or order, of segmentation. Implied in the model is that the corporate dimension of descent applies at each structural level of segmentation; each segment is a corporation whose members are jurally equal.

Corporateness is also seen as multidimensional. For example, the structure of ancestor worship may parallel that of allegiance in conflicts, with the minimal lineages of the same minor segment combining to worship their common ancestor, and the minor lineages of the same major segment uniting to worship their own common ancestor, and so on.

A distinct aspect of unilineal descent groups is their primary role in marriage regulations and incest prohibitions. The segmentary organization of the society determines the marriageability of individuals, and hence the boundaries of the exogamous unit. But although these boundaries are always defined in reference to kinship and genealogical structure, they are not immutable; they change as the segmentary structure itself changes. "The lineage range within which the rule of exogamy holds," Fortes remarked, "is variable and can be changed by a ceremony that makes formally prohibited marriages legitimate and so brings marriage prohibitions into line with changes in the segmentary structure of the lineage" (1953, 38). Finally, a unilineal descent group is characterized by a structure of authority—"not only the lineage but also every segment of it has a head, by succession or election" (32)—and by its perpetuity in time. By that is meant "not merely perpetual physical existence ensured by the replacement of departed members" but a "perpetual structural existence" (27); that is, the lineage's rights and duties transcend the death of its individual members.

To summarize, the African model of unilineal descent groups holds that descent is the single most important criterion defining an individual's status in his society. It sees descent groups as undergoing repeated fissions along kinship lines, a process that gives rise to a treelike structure whose segments, from the smallest to the highest-order ones, act as corporate groups and are in balanced opposition in relation to other segments at the same level. Descent groups are jural entities that transcend residence, impose incest prohibitions, and define marriageability.

The Chestnut within the Model

Theoretical models are increasingly vulnerable to critiques as data on a wider range of facts accumulate. The African model of unilineal descent groups was no exception to that rule. As more information was gathered on African societies and new data were collected on non-African societies, the generalizability of the African model, and eventually its internal validity, were questioned by many. One important critique hinged on a fundamental distinction made by Fortes himself, that between filiation and descent, a distinction, wrote Scheffler, "that, once clearly stated and grasped, may seem virtually self-evident, commonsensical and perhaps even platitudinous," but which is nonetheless "profoundly significant" (1985, 1). In Fortes's terms, filiation refers to the simple fact of being the child of one's father (patrifiliation) or one's mother (matrifiliation). Descent, on the other hand, is defined "as a genealogical connection recognized between a person and any of his ancestors or ancestresses," that relation being mediated by the person's parent (Fortes 1959, 207; cited by Scheffler 1985, 2). Put differently, filiation refers to the recognition of a dyadic parental link, whereas descent refers to the recognition of a chain of dyadic parental links. Patrilineal descent thus implies that individuals define themselves in relation not only to their father's group but to their father's father's group, their great-grandfather's group, and so on.

The distinction between filiation and descent lay at the basis of an early critique concerning the applicability of the model outside of Africa, a critique that, paradoxically, will help us circumscribe the evolutionary foundation of descent. Barnes (1962) argued that various groups in the New Guinea Highlands, which had previously been ascribed to the category of patrilineal descent groups, lacked descent altogether and exhibited mere patrifiliation instead. In these societies each person is at birth identified with and belongs to his father's group. But individuals need not belong to their father's group for life. They may become members of other groups because criteria other than descent, namely affinal kinship, matrilateral kinship, and even mere coresidence, also confer membership in a local group. In Scheffler's terms, patrilineal descent is a *sufficient* condition for membership in these groups but not a *necessary* one, in contravention of the African model of unilineal descent. In these societies it is also the case that agnates who leave their local group may be lost to their father's group, implying that individuals do not necessarily belong to their grandfather's and great-grandfather's group. Thus there is patrifiliation, and patrifiliation across generations—what Barnes called "cumulative patrifiliation"—but there is no real patrilineal descent. Or, as stated by Scheffler, these societies have "patrifilial kin groups" but no patrilineal descent groups (1985, 16).

Another important critique concerned the extent to which lineage theory represented adequately the structure of the very societies whose description had served to elaborate the model, notably the Nuer. Central to that critique was the poor fit between actual social structures and the model of corporate segmentary structure, with its principle of balanced opposition. For example, referring to the evidence presented by Evans-Pritchard himself about the dynamics of conflicts and alliances among the Nuer, Holy concluded that "lineages of varying order of segmentation formed alliances and jointly fought lineages to which they were not related by the principle of balanced opposition" (1996, 82). Summarizing the evidence pertaining to a number of African societies, Holy emphasized that the groups that typically act collectively against other groups are the "lineages of the lowest degree of segmentation"; that higher-order groups of agnates crystallize in action extremely infrequently; and that when they do, they often assemble only a fraction of all the segments that, in theory, should get involved (88). Thus, although the corporate dimension of lineages may apply to lineages of the lower order of segmentation, it is doubtful that it applies to higher-order segments. In sum, some major elements of the African model of unilineal descent groups appear to lack empirical support even in African societies.

This brief account can hardly do justice to the long-standing debate surrounding lineage theory, but it gives a flavor of its content and serves to illustrate the kind of arguments that led a number of anthropologists to reject lineage theory altogether. Adam Kuper's conclusion, for example, was uncompromising: "My view is that the lineage model, its predecessors and its analogs, have no value for anthropological analysis . . . The efforts of generations of theorists have served only to buy time for the model in the face of its long-evident bankruptcy" (1982, 92–93). It may well be the case that segmentary lineage theory is inadequate as far as it is construed as an *analytical tool* with a high degree of generalizability and explanatory value across societies. But the temptation here is to confuse the model with the phenomenon it aims to account for, unilineal descent. However empirically founded the model's critiques may be, rejecting the concept of descent would amount to throwing away the baby with the bath water. Bloch and Sperber summed up the problem well in reference to another classic of anthropology, avuncular relations in patrilineal societies. The radical relativism of much of contemporary anthropology, they wrote, "negates what all those with a reasonable acquaintance with the ethnographic record know—which is that the regularities which have fascinated the discipline since its inception are suprisingly evident. Thus," they continued, "it is common for younger anthropologists reared on the diet of relativism . . . to be shocked by discovering the old chestnuts of traditional anthropology in their fieldwork just when they had been convinced that these were merely antique illusions" (2002, 726). This reasoning applies equally well to the concept of unilineal descent: it is such a chestnut and a most basic one. Two different claims may be made in relation to this assertion, a weaker one and a stronger one.

The weaker claim is that, even assuming that patrilineal descent groups as they were defined by lineage theory do not exist, that they are theoretical illusions, "patrifilial kin groups" do in fact exist (the same reasoning, of course, applies to matrilineal descent and matrifilial kin groups). In Scheffler's terms, patrifilial kin groups are groups whose members are related patrilineally across generations (through "cumulative patrifiliation"). In these groups paternal filiation does not necessarily, that is, obligatorily, confer group membership, as in the African model of unilineal descent, but it is nonetheless a sufficient condition for membership. Moreover, mere patrifiliation is a principle whose explanatory value is wide-ranging. Depending on the society, patrifiliation affects an individual's access to land, his marriageability to others, his rights regarding succession, his political alliances, and his religious practices. Thus few anthropologists would dispute the idea that patrifiliation is one significant organizational principle among others—coresidence, marriage bonds, affinal kinship, matrilateral kinship, and so forth—that pattern social relations and structure society as a whole. Minimally, therefore, unilineal descent explains a significant part of social variation in the corresponding societies, and as such it can hardly be discarded as a contributing variable. As we shall see, from an evolutionary viewpoint the existence of patrifilial kin groups is as significant as that of true unilineal descent groups.

But there is apparently more than "mere" cumulative patrifiliation in unilineal descent, and this is the stronger claim. It has often been reported that although the members of segmentary societies often act in ways that contradict the segmentary lineage model, they often *talk about* political relations in their society as if these were in agreement with the model. Thus the segmentary lineage model appears to be held by the people themselves, whether or not they behave in accordance with it. As summed up by Holy, "There seems to be almost a general consensus among anthropologists that the concept of the segmentary lineage structure is the actors' folk model" (1996, 80) and that it is "thus not a model which the actors operate in their actual political processes" but "merely a representation of the enduring form of their society, or, as it has often been expressed, a kind of ideology" (85). This helps explain the discrepancies between lineage theory and the actual working of segmentary societies. The African model of unilineal descent groups is a theoretical construct arrived at by anthropologists who apparently combined, and somewhat mixed, two distinct sources of data: (1) their own empirical observations of political reality in these societies and (2) their informants' ideological representations of it. With the benefit of hindsight, such a mixture was bound to produce inconsistencies between theory and real life.

But far from discrediting the unilineal descent model, this consideration, I believe, attests to its substance and points to its relevance for understanding human affairs. Again, substance and relevance not in the sense that the model constitutes a valuable analytical tool for understanding cultural variation: ideological representations alone can hardly have more than a poor explanatory value with regard to actual social patterns. But substance in the sense that the very existence of an ideology of unilineal descent organization begs the question why this particular kind of ideology is so common. Indeed, why do humans in so many different societies have a conception of their society as one that defines membership on the basis of lineal kinship links to common ancestors (real or mythic), one in which the transmission of rights and obligations is strongly sex-biased and in which political alliances, economic relations, marriage, and cult practices are organized by the principles of segmentation and corporateness? My response to this question, developed in the next chapter, is that such a conception is easily conjured up by the human mind in some social contexts because some of its underlying principles are part of our primate legacy.

To a majority of anthropologists unilineal descent groups are cultural constructs. To Radcliffe-Brown, for example, they were answers to some basic social necessities, such as preventing conflicts through the formulation of rights and obligations and preserving the continuity of social structure through entities that transcend the individual. Like Lévi-Strauss's interpretation of reciprocal exogamy, Radcliffe-Brown's conception of the unilineal descent organization saw it as emerging from an evolutionary vacuum and as a cultural response to some problems that challenged human society. According to this view, unilineal descent groups originated as symbolic constructions; hence they have no phylogenetic origins and should have no reality whatsoever in nonhuman primate societies. This is not the case, however. Some primate societies display several properties of the unilineal descent organization, and it is therefore likely that this type of organization thrived in various primitive forms well before it was institutionalized. To many anthropologists the fact that human descent groups are institutionalized entities existing in the *minds* of individuals as collective ideologies proved that the very existence of descent groups depends on their being internalized as symbolic representations. But the primate data belie this view, as I argue next. Descent predated its symbolic representation.

18

The Primate Origins of Unilineal Descent Groups

> The gens [clan] . . . does not include all the descendants of a common ancestor. It was for the reason that when the gens came in, marriage between single pairs was unknown, and descent through males could not be traced with certainty. Kindred were linked together chiefly through the bond of their maternity. In the ancient gens descent was limited to the female line.
>
> *Lewis Henry Morgan ([1877] 1974, 66)*

Nonhuman primates have no institutions, and hence no descent groups as such. Nevertheless, some species have social structures that exhibit several properties of human descent groups described in the previous chapter. These properties are manifest in, and emanate entirely from, recurrences in the social interactions of individuals. They are not normative or symbolically transmitted across generations, and therein lies much of their interest. They tell us that human and nonhuman forms of descent originate from a common set of principles grounded in the biology of primates, and that if it were not for this common evolutionary origin anthropologists would probably have had no descent groups to ponder. I focus here on a primate version of the human matrilineal descent group, or human matriclan. I do so by spelling out the properties of the primate model in reference to the properties of the African model of unilineal descent groups. The reason I focus on matrilineal rather than patrilineal descent is that there is no adequate primate model for a human patrilineal descent group. Most multimale-multifemale primate groups are female kin groups, and in male kin groups the inconsistency of paternity recognition prevents patrifiliation. As we shall see, however, the principles derived from the study of female kin groups readily apply to male kin groups and are crucial to understanding the evolutionary history of human descent groups.

By far the best evolutionary prototype of unilineal descent groups is found in female-philopatric primate species such as macaques and baboons. Because females reproduce in the group in which they are born, every female is potentially the founder of a multigenerational lineage, whose female members are resident and whose male members emigrate. Six properties ensue from this simple fact and a few others.

Group Membership through Birth

In these species matrifiliation (the kinship link between a mother and her offspring) is one of two factors that determine membership in a group. Any individual belongs to his mother's natal group, not his father's natal group. Relatedness through the mother confers membership in the mother's group, whereas relatedness through the father does not confer membership in the father's group. Accordingly, all individuals born in a given group are related to each other through a common *female* ancestor, not a male ancestor. For example, all natal group members might belong to any one of four different maternal lineages, each composed of three generations of individuals. The members of any one of these lineages would be related to one another through a common living female ancestor, for example their grandmother, or a deceased one, possibly their great-grandmother. And all four lineages would also be related to one another through a more remote common ancestress. The fact that matrifiliation confers group membership simply reflects the type of residence pattern: female philopatry.

If matrifiliation confers de facto group membership, it is not the only factor that does. Male immigration into the group also confers membership. Nonnatal males, especially those who have belonged to the group for some years and occupy a central position in it, enjoy several of the privileges of group membership, including access to the territory and physical resources controlled by the group. They do so by virtue of coresiding with the female kin group. Thus coresidence per se confers membership in the group, provided it is associated with sufficient length of tenure (seniority). Otherwise, the males are peripheral and do not enjoy all the privileges associated with group membership. In Scheffler's terms, in female-philopatric groups matrifiliation is a *sufficient* condition for group membership but not a *necessary* one. However, matrifiliation translates into social bonds and associated benefits that are not shared by nonnatal individuals. Bonds between related females are more enduring and more intimate than bonds between unrelated males. Put simply, matrifiliation confers "more membership" than mere coresidence.

Kinship-Based Segmentation

Another property of female kin groups is that matrilineages undergo "segmentation" along genealogical lines. We saw that human lineages grow in size and eventually split into segments, or sublineages, which in turn grow and divide into smaller sublineages. Lineage fissions thus occur along natural kinship lines. A similar process takes place in female-philopatric primate species when group fissions occur. A group fission refers to the division of a social group into two autonomous groups that remain permanently separated thereafter (Koyama 1970; Chepko-Sade and Sade 1979; Dittus 1988; Oi 1988; Prud'homme 1991; Ménard and Vallet 1993; Henzi, Lycett, and Piper 1997; Kuester and Paul 1997; Okamoto and Matsumura 2001; reviewed in Hill 2004; Okamoto 2004). One major factor observed to prompt group fission is an increase in group size. The fission probably relaxes levels of food competition within the group. Prior to fissioning, the two subgroups are often observed to range and forage independently of each other over several hours daily, if not over a few consecutive days, before merging back together. They do so progressively more often over a variable number of months until they eventually cease to come into contact and become independent groups.

A key feature of primate group fissions is of particular interest here. Fissions typically produce genealogically cohesive subgroups. In theory, a group split could result in close kin distributing themselves more or less randomly between the two subgroups. On the contrary, members of the same matriline leave together; that is, larger matrilines segment into smaller ones. Diane Chepko-Sade and Donald Sade (1979) analyzed a number of group fissions that took place in groups of rhesus monkeys whose uterine kinship structures were known. Each group was composed of a number of matrilines, whose matriarch could be dead or alive. In most cases, the matrilines comprised three generations; not infrequently they included only two generations, rarely four. Within a group all the matrilines were related to each other but through dead ancestors. In terms of genealogical structure, group fissions were of two related types. In the first type, a whole matriline left the parent group. Each matriline then included the matriarch, her immature offspring of both sexes, and her three or four adult daughters with their respective offspring of both sexes. Often the matriarch was already dead, so the offshoot included all individuals except her. In the second type of fission, a matriline would split, one part remaining in the parent group, the other part leaving it. Typically, the eldest daughter and her offspring separated from her younger sisters and the rest of the parent group. In this situation the death of the sisters' mother (the connecting female) was a factor increasing the likelihood that the matriline would split. In any case, all group splits had in common that they took place *between* matrilines, either between whole matrilines or between submatrilines.

The reason that female kin groups split along genealogical lines is relatively simple: fundamentally, it is because the strength of social bonds correlates with the degree of uterine kinship in these species. Bonds between mothers and daughters and bonds between sisters are stronger than those between more distant relatives. This is manifest in the well-documented fact that close uterine kin have higher rates of affiliative and cooperative interactions—proximity, huddling, grooming, aiding in conflicts, and cofeeding—than more distant kin or unrelated females (reviewed in Walters 1987; Bernstein 1991; Silk 2001; Berman 2004; Kapsalis 2004; see also Silk, Altmann, and Alberts 2006 a and b). On this basis alone, one would expect females to follow their mothers or sisters if the latter leave the group. Implied here is that if some other force of attraction engendered bonds stronger than those based on consanguinity, group fissions would most likely proceed along that line. For example, an attraction based primarily on age similarity would produce two age classes rather than two kin groups. Therefore, group fissions are particularly salient manifestations of the importance of kinship in patterning social structure in these societies.

Another factor adds its effect to consanguinity here, a factor that is itself a product of kinship. In several female-philopatric species, close kin are not only genetically related, but they belong to the same "class" in the matrilineal dominance structure (chapter 3, Figure 3.3). A matrilineal hierarchy is entirely predictable from the group's genealogical structure. Every female inherits her mother's rank and abides by it over her whole lifetime or a large portion of it. Dominance is known to confer various types of advantages. Above all, higher-ranking females enjoy priority of access to food in general, or to high-quality food in particular (Whitten 1983; Janson 1985; van Noordwijk and van Schaik 1987; Soumah and Yokota 1991; Barton and Whiten 1993; Saito 1996; Koenig 2000, 2002). This means that matrilines are more than consanguineal subgroups; they are, simultaneously, ranked "social classes" in terms of social status and access to resources. This factor is important in explaining why group fissions take place along genealogical lines. The observed group splits are indeed consistent with the hypothesis that they result from high levels of food competition within groups. First, group fissions are known to separate low-ranking from high-ranking matrilines. Second, the highest-ranking matrilines are the least likely to split. Third, when matrilines do split, they do so in a way that is consistent with the resource competition hypothesis: in rhesus macaques the eldest daughter—the female most likely to leave—is the lowest-ranking of the daughters. As such she is the matriarch of the lowest-ranking submatriline.

To summarize, female kin groups split along genealogical lines because close kin share two things that nonkin do not: (1) intimate and enduring social bonds involving cooperation, reciprocity, and altruism, and (2) a similar social status at the group level, a status affecting their competitive ability and access to resources. As a result, matrilines are naturally cohesive entities whose members stick together. In his criticism of the evidence for primate lineages, Fortes asserted that when "closely examined, such 'lineages' of three successive generations of living females and immature males split up not by fission between sibling lines . . . but by dispersion in the course of mating in every generation" (1983, 18). The available evidence rather indicates that matrilineages *do* split "by fission between sibling lines."

One important difference from segmentation in human groups is that after the fission is completed, members of the two female kin groups eventually cease to interact. They become two autonomous social groups that range independently and compete for resources. As is the case with primates in general, the two groups are not embedded in any sort of supragroup social structure, as human lineages are. For something like this to occur, pair-bonding and the pacification of intergroup relations—the primitive tribe—would be needed, as argued previously.

The Genealogical Boundaries of Exogamy

A third structural aspect of female kin groups is that kinship defines the boundaries of the "exogamous" unit, the largest social entity whose natal members do not breed with each other. In human descent groups it is the segmentary organization that delineates the exogamous unit, the frontier between marriageable and nonmarriageable individuals. That frontier is thus genealogical: the exogamous unit is a subgroup whose boundaries are defined by membership in a lineage. The size of the exogamous group may change with alterations in its segmentary structure, but its boundaries are still defined by membership in a lineage. Similarly, in female-philopatric primate groups, the "exogamous" unit, better called the outbreeding unit, is the local female kin group. Most individuals born in a given female kin group breed outside that group if they are males or within it if they are females, but with males born in distinct female kin groups. When a group splits to form two daughter groups, it produces two new exogamous units whose boundaries are defined by the parent group's genealogical structure. It is still the uterine kinship structure that defines the frontiers of the exogamous unit.

The Unisexual Transmission of Status

The sex-biased character of unilineal descent groups is among the most conspicuous manifestations of human descent. In patrilineal descent groups, succession to office and inheritance of property are transmitted from fathers to sons exclusively. Although daughters are members of their father's descent group, they are excluded from the chain of descent, as illustrated in Figure 18.1a. One might think that the reverse pattern would be the rule in matrilineal descent groups, that succession would be from mothers to daughters, with the sons excluded from the chain of descent, as illustrated in Figure 18.1b. But as alluded to earlier, matrilineal descent is not the exact mirror image of patrilineal descent. In matrilineal descent systems, descent is indeed matrilineal in that it is on the mother's side, but it runs through males. It is typically avuncular, from a woman's brother to her son or, stated differently, from the mater-

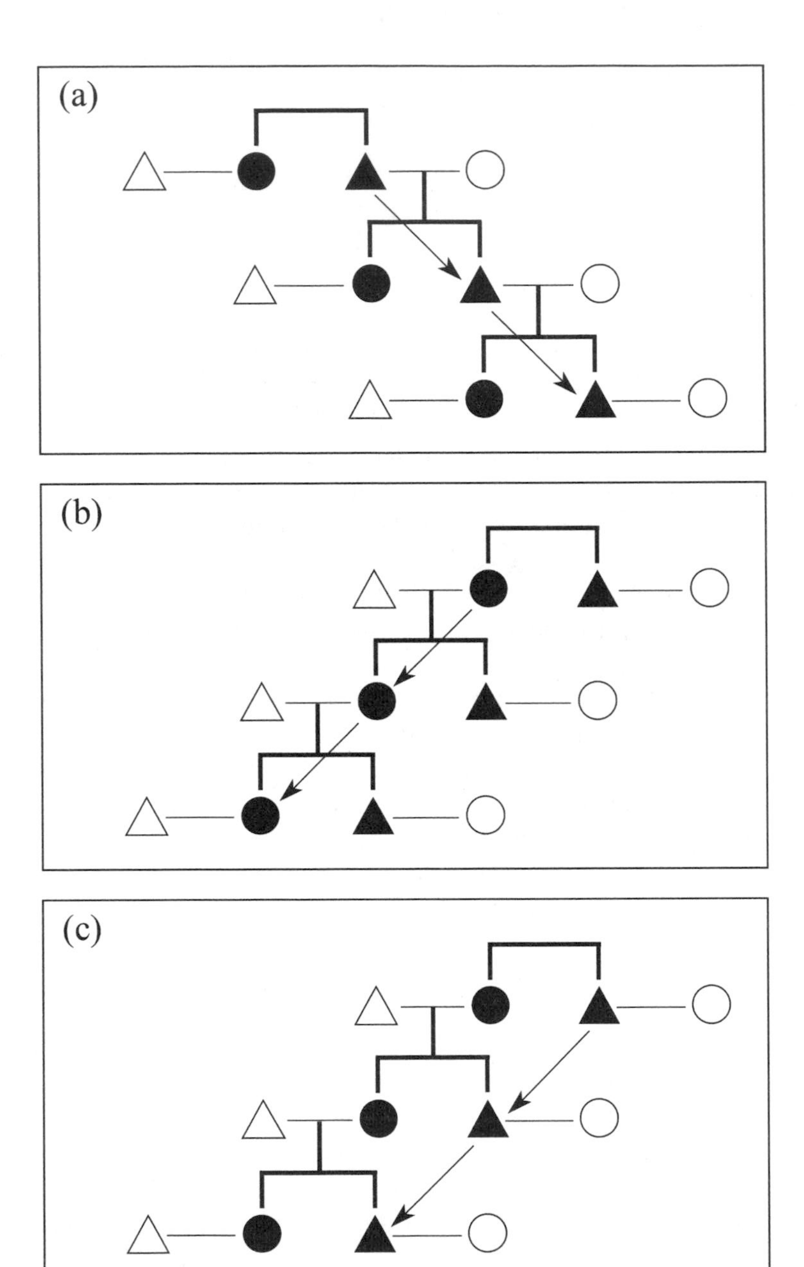

Figure 18.1. Three forms of unisexual transmission of status in primates. (a) Patrilineal transmission, from fathers to sons, in human societies. (b) Matrilineal transmission, from mothers to daughters, in female-philopatric primate societies. This is the mirror image of human patrilineality. (c) Matrilineal transmission in human societies, from mothers' brothers to sisters' sons (maternal uncles to sororal nephews), unknown in nonhuman primates. Triangles: males; circles: females. Black symbols represent members of the same patriline (a) or matriline (b and c). Arrows indicate the direction of transmission.

nal uncle to his sororal nephew, as shown in Figure 18.1c. Even though it is on the mother's side, it is under male control. Nonhuman primates also display unilineal patterns of status transmission, namely matrilineal ones. But, interestingly, transmission is from mothers to daughters; it is not avuncular. It is thus the mirror image of patrilineal descent, as in figure 18.1b.

There is overwhelming evidence that the matrilineal transmission of rank through generations in female-philopatric primate species is a social, not a genetic, process. Females acquire their birth rank mainly through the active help they receive from their mother and other close female kin who intervene on their behalf against lower born females. Experiments have established that any female, even the most subordinate one, could outrank higher-ranking females and assume her new status in a competent manner provided she was given allies that were collectively stronger than those of the formerly dominant females (Chapais 1988; Chapais and Gauthier 1993; Chapais, Prud'homme, and Teijeiro 1994; Chapais et al. 1997; Chapais, Savard, and Gauthier 2001). Correlatively, newly outranked females behaved in a typically submissive manner with newly dominant females. A female's social status is thus a socially transmittable attribute in these species. It is also one among several other such attributes. Nonhuman primates are known to acquire information from and transmit information to members of other generations through various social learning processes. Besides their dominance status, mothers may transmit to daughters various aspects of their maternal style or social networks, as well as food-processing techniques (for a review, see Chauvin and Berman 2004). Thus the matrilineal transmission of status is just one manifestation of general cognitive abilities that enable nonhuman primates to learn from others, either directly through interacting with them or indirectly through various learning processes, including exposure learning, observational conditioning, emulation, and imitation (McGrew 1998; Tomasello and Call 1997; Chauvin and Berman 2004).

As far as the present discussion is concerned, the important point is that both sons and daughters inherit their mother's position in the dominance order, but only the daughters transmit it to their offspring; males do not. That is, although rank inheritance is bisexual, status transmission is strictly unisexual. Males initially inherit their mother's rank through the same developmental processes that govern rank inheritance from mothers to daughters. But males leave their natal group around puberty and eventually join a new local group to breed. Upon entering the new group, they have no recognized social status in it; they start their "social career" all over. The reason that a male's dominance rank is not transferrable across groups is simply that it is an attribute of the male's relations with the individuals he grew up with. Because males do not carry their birth rank with them, they are not in a position to transmit it to their offspring, even assuming that they would recognize them.

In sum, the unilineal succession of status along the female line results from the co-occurrence of four factors in multimale-multifemale groups: (1) matrifiliation—the recognition of the mother–daughter link; (2) the existence of an entity that is transmittable from mothers to offspring, in the present case, a position in the order of access to resources—the transmissibility itself implying that matrifilial links persist long enough for the mother's influence to be effective; (3) the ability of nonhuman primates to learn from others, including members of the preceding generation; and (4) a pattern of female localization and male dispersal such that males, even if they are an integral part of their mother's kin group, are removed from the chain of descent. Fundamentally, then, unilineal transmission is a prerogative of the philopatric sex.

Primitive Corporateness

A fifth property of primate matrilineages is that they display some basic properties of corporateness. To Fortes, as we saw, the corporate character of lineages was the most important feature of unilineal descent groups. Corporateness is manifest in the fact that all members of a lineage are equal and undifferentiated from the viewpoint of individuals outside the lineage. Accordingly, the members of a lineage hold "collective responsibility in blood vengeance." Corporateness is certainly a cognitively sophisticated concept, but nonetheless it has a phylogenetic foundation. The most basic aspect of what I may call the "capacity for corporateness" is perhaps the very recognition of the lineage as a distinct social entity. Thanks to the symbolic capacity, humans label their mental representations. As a result, our concept of lineage is hardly dissociable from its corresponding symbolic labels. The question thus arises as to whether nonhuman primates might have a mental representation of the matriline as a distinct social entity, even though they are unable to symbolically label such a representation.

We saw previously that nonhuman primates recognize not only their own social bonds but those of others, notably associations between uterine kin. This is manifest, for example, in the fact that following a fight, the combatants are more likely to reconcile with their opponent's uterine kin than with other individuals (Cheney and Seyfarth 1989, 2004), indicating that they recognize the existence of *dyads* of uterine kin. This is an important primary step, but the question here is whether they recognize the whole matriline as an entity. Recent experimental evidence on nonhuman primates suggest that they may do so. Thore Bergman and his collaborators (2003) conducted experiments among wild baboons in which female subjects could hear prerecorded sequences of vocalizations that mimicked a fight between two females. The temporal sequence of screams was arranged to indicate that a formerly dominant female had been outranked by a formerly subordinate individual, in other words, that a rank reversal had taken place between two females. Interestingly, the subjects responded more strongly to rank reversals between members of two *different* matrilines than to rank reversals between members of the *same* matriline. This testifies to the monkeys' ability to classify others as belonging either to the same matriline or to different matrilines, which in turn suggests that they recognize matrilines as social entities.

There is also some evidence that from ego's viewpoint the members of another matriline are *interchangeable* in some contexts. This is manifest, for example, in patterns of redirected aggression among vervet monkeys. Following a fight between two individuals, one of the two combatants (ego) may redirect aggression onto a third, uninvolved individual. When this happens, the target is more likely than by chance to be a uterine kin of its opponent (Cheney and Seyfarth 1986; Aureli et al. 1992). Thus, from ego's point of view, it is as if any kin of the opponent would do. Behaviorally (rather than cognitively) speaking, ego is acting as if its opponent's relatives held a collective responsibility. Observations like these suggest that mere patterns of social interchangeability—members of a matriline sharing certain attributes that make them interchangeable from ego's viewpoint—may be the evolutionary precursors of the linguistically mediated concept of collective responsibility. Again we see that interactional recurrences may have paved the way for normative rules.

Lastly, to Radcliffe-Brown a corporate group was a group that "possesses or controls property which is collective," such as land (1950, 41).

Collective control of the territory is also a property of female kin groups in nonhuman primates. Females being the resident sex, they form the stable core of the group. In this sense the territory is passed on by adult females to their female descendants through generations. Females are born in the home range that their mothers inhabit and will later give birth to their own offspring in that same home range. They defend their home range collectively with the help of males against other local groups. The control of land among female uterine kin is thus de facto collective, even though it is not normatively so.

A Multilevel Structure of Solidarity

The last property of female kin groups discussed here is that matrifiliation gives rise to a multilevel structure of solidarity. We saw that the hierarchical nature of human segmentary organizations translates into a multilevel structure of solidarity and loyalty. For example, in a conflict between two individuals belonging to different lineages (1 and 2) of the same higher-order segment (A), the members of lineage 1 ally against the members of lineage 2. But in a conflict opposing individuals from two distinct higher-order segments (A and B), lineages 1 and 2 now ally with each other against the other higher-order segment. The same principle applies to higher-order segments. This is a *nested structure of loyalty.* In such a structure ego's allegiance toward a given individual is almost never absolute; it is relative, depending on the identity of the opponent. Ego would ally with his brother against his cousin but would side with the same cousin against a more distant kin. Individuals may thus be opponents or allies, depending on the context, and the whole structure of allegiance is genealogically patterned.

Structurally speaking, a similar pattern of nested allegiance prevails in female-philopatric primate species, especially those that exhibit a matrilineal dominance structure. In these societies all individuals are competitors for resources, and the dominance order regulates the dynamics of competition. Even close kin compete for resources. For example, sisters compete fiercely with each other for dominance by attempting to obtain their mother's support (Datta 1988; Chapais, Prud'homme, and Teijeiro 1994). But the same sisters commonly join forces against females belonging to other matrilines. As a rule, members of a matriline are loyal to each other against members of another matriline. This brings us to a more general principle. Nepotism in nonhuman primates is not synonymous with lack of competition and aggression between kin, nor is it synonymous with absolute and unfailing loyalty toward a given relative. Nepotism is more a matter of consistent relative loyalty, that is, consistent allegiance to the closest kin.

Another, more inclusive level of solidarity is found at the intermatriline level of alliances. A female's allegiance in conflicts is not limited to the members of her own matriline. Females may intervene in fights opposing individuals they are distantly related to. For example, members of different high-ranking matrilines (A and B) may ally with each other against members of lower-ranking matrilines (C or D). In experiments carried out on Japanese macaques, any member of the A or B matriline that was placed alone with the lowest-ranking matriline, C, was outranked by it. But any member of the A matriline could maintain its rank above the C matriline in the presence of the B matriline. And, reciprocally, any member of the B matriline could maintain its rank above the C matriline in the presence of the A matriline. This attested to the existence of alliances between the members of the A and B matrilines against the C matriline (Chapais, Girard, and Primi 1991).

Finally, a still more inclusive level of allegiance is that of the whole female kin group. In encounters between two local groups, all group members, from the highest-ranking to the lowest-ranking, may act as a group-wide coalition against the other group (reviewed by Cheney 1987; Cooper 2004). In sum, the dynamics of solidarity in female kin groups obey a nested structure of alliances that maps onto the group's genealogical structure.

Female primate kin groups display in embryonic form six basic properties of human matrilineal descent groups (or human matriclans): the birth criterion of membership in the group, segmentation along genealogical lines, a kinship-based delineation of the exogamous unit, the unisexual succession of status, some elements of corporateness, and a nested structure of solidarity. Obviously, each of these properties differs in some important respects from its human counterpart. But this is precisely what one would expect from a precultural, prenormative prototype of human descent groups, from what one might call in the present case a protomatriclan. The similarities between the structural properties of female kin groups in nonhuman primates and those of matrilineal descent groups in humans are not coincidental. But for all that, the evolutionary connection between the two realms is not direct: female kin groups are not the evolutionary precursors of matrilineal descent groups in humans. The relation between the two entities is indirect, as we shall see. Moreover, despite Morgan's assertion that "when the gens came in . . . descent was limited to the female line," there is every reason to believe that the protomatriclan was not the primeval form of human descent, as I argue in the next chapter.

19

The Evolutionary History of Human Descent

It is my hunch that, in an evolutionary sense, the formation of [residential] groups came before more abstract principles of kinship ideology.

Robin Fox (1967, 78)

Among the questions that have long fascinated anthropologists but basically eluded them are the priority in time of matrilineal descent groups (matriclans) versus patrilineal descent groups (patriclans), and that of unilineal versus bilineal descent. Bilineal descent, the combination of patrilineal and matrilineal descent in the same system, is thus double unisexual descent. In this section I provide evolutionarily informed answers to these questions, answers that follow directly from integrating the foregoing primate model of unilineal descent into hominid evolution.

Female Kin Groups as Precultural Matriclans

Members of the same primate matriline or the same female kin group obviously do not claim descent from a common ancestress (alive or dead). But one may infer that they would do something like that if they were endowed with the symbolic capacity and could communicate about their group's genealogical structure. Indeed, were the symbolic capacity to seize upon the six properties of female kin groups described in the previous chapter and produce a culturally elaborated version of them, the resulting social arrangement would have much in common with a humanlike matrilineal descent group. Genealogical segments—

matrilines, submatrilines, and so on—would have symbolic labels, as would the other subgroups derived from prior fissions. Ancestors, dead or alive, would have names. The kinship boundaries of the exogamous unit would be conceptually recognized. The unisexual transmission of status would be normative and, in all likelihood, would extend to other transmittable entities: physical objects (for example, tools or land) and other bases of social status besides dominance rank. Corporateness would reach higher levels of sophistication, and as a result collective responsibility would be better acted upon. The combination of all these features still does not amount, properly speaking, to a "matrilineal descent group." One important difference concerns the presence and active role played by kinsmen in human matrilineages. The inclusion of adult males in matrilineages is a phylogenetically more recent event, which I will discuss shortly.

The point at this stage in the discussion is that the cultural content and institutionalized dimension of human matrilineages, far from negating the phylogenetic connection between human matrilineages and primate matrilineages, help bridge the gap between the two realms. Some thirty years ago Fox (1975) noticed some of the similarities between primate matrilines and human descent groups. Referring to males circulating between two subgroups of macaques derived from the fission of their parent group, he formulated a bold analogy. "Had the males . . . ended up as permanent mates in a matriline of the other group," he wrote, "then there would have been little difference, except at the symbolic level, between this and an 'Iroquoian' system of two moieties each composed of ranked matrilineages, and a rule of moiety exogamy plus matrilocal residence" (28). This prompted Fortes to qualify Fox's position—namely, that "'the cumulative evidence shows clearly that [primate] matrilineages exist'" as "an astonishing statement for an anthropologist so expert in kinship theory to make, unless it is meant metaphorically." Fortes then spelled out the reasons why he thought that the similarities between human and nonhuman lineages were no more than metaphorical. His main points are embodied in the following quotation:

> Human lineages are groups of people conceptually rather than physically distinguished and united by recognized common descent from recollected or recorded deceased ancestors five or more generations

> back. They are not primarily breeding groups but political, economic and religious sub-divisions of a larger society, often exogamous by virtue of moral or juridical rules, not by fear of inbreeding. They exist not by virtue of the transitory dyadic relationships of successive generations of females but by reason of elaborately conceptualized and institutionalized models, rules and norms of social and personal relationships. Above all they represent a dimension of human social existence that is surely lacking in other primate species. The structure of the lineage represents the incorporation into the life of a human society of the sense of its continuity in time, more exactly, of its having a past and a future. (1983, 19)

Fortes is providing us here with a succinct summary of the uniquely human features of matrilineages. The differences are real and profound. True, humans conceptually define descent groups, identify themselves with dead ancestors, have a sense of their continuity in time, are exogamous by virtue of juridical rules, and so on. But these distinctive aspects of human descent groups are cultural extensions of the properties of primate kin groups, and therefore they vindicate the phylogenetic foundation of human descent groups. If such groups were cultural constructs obeying their own organizing principles and built from scratch, why would their structural properties so closely match those of primate kin groups? The likely answer is that nonhuman primates provided the initial blueprint of the lineage organization.

The key principle involved here is the *precedence of residence* over descent. What primate studies tell us about unilineal descent, essentially, is that the properties of descent groups have a residential basis, and more specifically, that these properties are structural correlates of sex-biased dispersal patterns. Here lies humankind's primate legacy in relation to descent.

The Residential Basis of Proto–Descent Groups

The direction of causality between residence patterns and descent patterns is another of those abiding topics of interest and dissent in the history of social anthropology. One major view holds that residence groups give rise to descent groups. For example, in his now classic *Kinship and Marriage*, Robin Fox discussed how matriclans might have developed out of subsistence activities that favored the localization of women (the cultivation of certain plants) and the dispersal of men and, similarly, how patriclans might have arisen from subsistence activities that favored the localization of men (hunting or herding). Matrilineality would be a correlate of matrilocality, and patrilineality a correlate of patrilocality. As emphasized by Fox, ideally one would test this sort of hypothesis using hard data on the history of specific residence groups and descent groups. But the required diachronic evidence is scanty, and therefore anthropologists must satisfy themselves with studying societies in which the patterns of residence and descent are already well established. As a result, anthropologists must "take the principle of descent as a given," Fox wrote, "and ask about its implications. But often this leads them to put the cart before the horse and impute causality to the [descent] principle . . . Thus," he continued, "it is often thought that in patrilineal societies the rule of patrilocal residence at marriage is derived from the principle of descent. In fact the reverse is probably true" (1967, 95).

But although Fox believed in the precedence of residence over descent, he was of the opinion that causality could go both ways, that changes in descent could bring about changes in residence. "Once a form of descent is adopted," he wrote, "the fact that there is a *rule* of unilineal descent is of enormous importance and will 'feedback' onto behavior and action" (96). In short, causality might be bidirectional, but the residence → descent causal sequence would be prevalent. The primate evidence concurs with the view that it is residence that sets constraints on descent, not the other way around. It suggests that unilineal descent groups are derived from sex-biased residence patterns. Indeed, all six properties of female kin groups are structural correlates of female philopatry and enduring matrifilial bonds. Applying this principle to hominid evolution, it follows that female localization in association with male dispersal generated the first protomatriclans, and male localization in association with female dispersal produced the first protopatriclans.

The next logical question concerns the very origin of sex-biased residence patterns. If descent does not determine residence, what then does? In the view espoused by Fox and many others, including Murdock, sex-biased localization patterns reflect the advantages for members of the residing sex of cooperating in the context of subsistence activities (hunting, herding, cultivation, and so on). Sex-biased patterns of cooperation are seen as imposing strong constraints on residence patterns. Combining all three aspects, one obtains the following causal sequence: subsistence-related patterns of cooperation determine residence patterns, which determine descent patterns, or *cooperation* → *residence* → *descent*.

The primate data concur with this hypothesis: a similar causal sequence was described in relation to the origin of *male* kin groups, specifically in chimpanzees and bonobos (chapter 10). The current explanation of male philopatry in these species is that males cooperatively defend a feeding territory for their own sake and for the females of their community and that because males benefit by being philopatric, the females disperse to avoid inbreeding. It is thus the males' cooperative drive that creates the sex-biased residence pattern. Significantly, a similar argument has been made regarding the origin of *female* kin groups in nonhuman primates. Current explanations of female localization have in common the idea that a female's foraging efficiency increases with her familiarity with the environment in which she lives. All else being equal, females should stay and forage in their natal range. But this does not necessarily mean that they should form groups. At least three hypotheses have been proposed to account for the existence of female kin groups and, interestingly, the three explanations boil down to some common principles. The *intergroup competition hypothesis,* formulated by Richard Wrangham (1980) states that females form groups to better compete for patches of food (for example, fruit trees) that can be defended against other groups. It further assumes that females benefit more from forming groups with their kin than with nonrelatives because by doing so they increase their inclusive fitness. This hypothesis thus conceives of the whole social group as a permanent coalition in the context of feeding competition between groups. Another explanation, the *intragroup competition hypothesis,* was proposed a few years later by Carel van Schaik (1989; Sterck, Watts, and van Schaik 1997). It states that females form groups as a means to better resist predators, and that they form kin groups because by staying with their relatives they can help them in contests over food within the group. Both hypotheses are founded on the principle that females form kin groups as a means to cooperatively defend resources. But while the first hypothesis places the emphasis on between-group competition, the second places it on within-group competition.

Mothers who tolerate daughters in their home range may incur potential *costs,* because their daughters are themselves food competitors who may reduce their mothers' future reproduction. In connection with this idea, Lynne Isbell (2004) recently proposed the *foraging efficiency hypothesis*, which states that mothers should tolerate their daughters' presence only (1) if they can accommodate them by expanding their home range, hence without sacrificing their own foraging efficiency, and (2) if the costs of dispersal incurred by their daughters are so high as to thwart the daughter's chance of reproducing. Kin groups would form in situations where the mother's tolerance is both possible and profitable. The model thus explains female philopatry without requiring or excluding the benefits of cooperating with coresident kin. These benefits are seen merely as secondary.

The three models are compatible. The bottom line is that females do better by remaining in their natal area; that they may do so provided this does not hinder their mother's reproduction; and that by being philopatric they are in a position to cooperate with their kin, both within their group and against other groups. Given that females are philopatric and reluctant to mate with their kin, the males would increase their reproductive success by dispersing and reproducing in other groups (Pusey and Packer 1987; Isbell 2004). One thereby obtains a pattern of female philopatry. Primate studies thus concur that the prime mover of female localization lies in its impact on the acquisition of resources by females and that one major benefit of female philopatry in this context is that it allows females to cooperate with their kin in subsistence activities.

Taken together, the primate data suggest the following causal chain underlying the evolutionary origin of proto-unilineal descent groups: ranging patterns and subsistence-related patterns of cooperation determine philopatry patterns, which in turn determine the structural properties of kin groups. This is the same causal sequence as that advocated by a number of social anthropologists, but with an important nuance: social anthropologists are concerned with the history of residence and descent, primatologists with its evolutionary origin.

The Latent Patriclan

We can now apply the foregoing causal chain to the evolutionary history of human descent. The ancestral male kin group hypothesis posits that immediately after the *Pan–Homo* split early hominids formed chimpanzee-like male kin groups. Genealogically speaking, male philopatric groups are the mirror image of female philopatric groups. But while female kin groups amount to proto-matriclans, chimpanzee-like male kin groups do not approximate proto-patriclans, because paternity is not recognized in any consistent manner in this species. In the absence of patrifiliation, there can be no kinship-based social structure matching the existing genealogical structure. Hence there can be no segmentation between patrilines, no patrilineal transmission of status, no kinship-based corporateness, and no patrilineally nested structure of alliances. The remaining properties of descent groups in a chimpanzee community are that birth confers membership in the father's group, the genealogical (agnatic) structure defines the boundaries of the exogamous unit, and males exercise collective control over the territory.

The evolution of pair-bonds altered that picture substantially. In the newly evolved multifamily group, pair-bonding entailed paternity recognition, hence patrifiliation and the potential for agnatic kinship structures. In this context all properties of our primate model of matrilineal descent groups are expected to unfold. The key element is the evolution of paternal nepotism and long-term bonds between fathers and sons, for the same reason mother–daughter bonds are the key feature of uterine kinship structures. The prevalence of various forms of patrilineal inheritance, patrifilial groups, and patrilineal descent groups in the ethnographic record amply testifies to the human propensity for paternal nepotism. I have already discussed in some detail the factors underlying the evolution of cooperative partnerships between fathers and sons among hominids, as well as what the chimpanzee data suggest about the developmental processes that might have been involved. I also mentioned that in primate societies in which paternity is recognized, some forms of paternal nepotism are observed (Watts and Pusey 1993; Stewart 2001; Buchan et al. 2003). Another particularly significant point here is that a form of patrilineal transmission of status is documented in our third closest relative, the gorilla. Not surprisingly, this species combines the two basic conditions for the occurrence of patrilineality: male philopatry and enduring breeding bonds.

Most groups of mountain gorillas include a single adult male (silverback), but some groups have two or more adult males, and many groups include sexually mature adolescent males. Maturing males may either stay in their natal group and reach adulthood there—they are called followers—or they may leave their natal group to become either solitary or members of an all-male group, in which case they are called bachelors. Silverback males tolerate followers. They try to prevent them from mating, but they do not attempt to evict them through persistent and intense aggression. David Watts calculated that followers have an 82–90 percent chance of replacing the silverback either by outranking him or by retaining his females after his death, and that followers have a much higher estimated reproductive success than bachelors. The reason that not all males elect to be followers is probably that in certain circumstances males obtain a higher reproductive success by leaving their natal group. Indeed, bachelors leave groups that contain relatively few females and in which their waiting time before becoming dominant would be relatively long because they occupy a low position in the queue of followers, and/or because the dominant male is young. In any case, the end result is that the silverback's sons are most likely to inherit their father's breeding unit. Because older sons may tolerate younger brothers as followers, the breeding unit may also be transferred between paternal brothers (Watts 1996). The gorilla example shows that mere selective tolerance toward one's close paternal relatives, accompanied by intolerance toward outside males, translates into a form of patrilineal transmission of status. If such paternal favoritism were to take place in a multimale primate group in which paternity is recognized, one would obtain a form of patrilineal status transmission mapping onto the whole genealogical structure.

Let us now come back to the evolutionary history of human descent. Prior to the evolution of pair-bonds, there could be no form of descent in hominid societies: patrilineal descent was prevented by the absence of recognized patrifilial links, and matrilineal descent was precluded by female dispersal and the termination of matrifilial links. The first descent groups could emerge only after the evolution of pair-bonding. And the primeval form of descent would have been patrilineal. But even if some sort of protopatriclan *had* been able to evolve, this does not mean that it did. This is best illustrated by the fact that patrilocal groups of hunter-gatherers have no real descent groups. However, their kinship structure lends itself readily to the formation of patrilineal descent groups. As stated by Steward, "patrilocal residence . . . produces the fact or fiction that all members of the band are patrilineally related" (1955, 125). In all likelihood, therefore, the patriclan lay dormant within male kin groups long before it developed into the full-fledged patrilineal descent group.

A majority of nineteenth-century historical evolutionists believed that matrilineal descent predated patrilineal descent, contrary to the present view. It is interesting to note that in spite of this contradiction, the evolutionists' basic argument in support of the priority of matrilineality is consistent with the above line of reasoning. The evolutionists held that patrilineal descent could not evolve prior to the advent of pair-bonding between particular males and females, as argued here. But they inferred from this that matrilineal descent must have been primeval. They, of course, had no reason to assume that before humans formed pair-bonds, males were localized while females dispersed, a situation that precluded matrilineal descent. Had they known this, they would have had to conclude that patrilineality came first.

Evolutionists like Henry Morgan and John McLennan were quite explicit about why pair-bonds had to be a prerequisite to patrilineality. Morgan argued that patrilineality had to await the advent of "the pairing of a male with a female under the form of marriage . . . the germ of the Monogamian Family" ([1877] 1974, 28). Prior to that period, the prevalent form of marriage, according to Morgan's evolutionary scheme, was "founded upon the intermarriage of several brothers [this term including cousins of various degrees] to each other's wives in a group; and of several sisters [and cousins] to each other's husbands in a group" (27), so group marriage rendered paternity recognition impossible. Thus, Morgan reasoned, when the first clan evolved, "marriage between single pairs was unknown, and descent through males could not be traced with certainty. Kindred were linked together chiefly through the bond of their maternity" (66). At about the same time, McLennan was making a very similar argument: "There could be no system of kinship through males," he wrote, "if paternity was usually, or in a great proportion of cases, uncertain. The requisite degree of certainty can be had only when the mother is appropriated to a particular man as his wife, or to men of one blood as wife, and when women thus appropriated are usually found faithful to their lords" ([1865] 1970, 65).

Clearly, then, Morgan's and McLennan's point was simply that patrilineal descent requires long-term breeding bonds as opposed to some sort of sexual promiscuity. Nevertheless, they have been widely misunderstood on this point by later generations of scholars, who thought that the evolutionists meant that patrilineal descent required an understanding of *physical* paternity. George Murdock stated that chief among the evolutionists' arguments for the precedence of the matriclan over the patriclan was "the presumed ignorance of physical paternity in primitive times," which contrasted with "the biological inevitability of the association of mother and child." Murdock rightly pointed out that ignorance of paternity was wholly irrelevant to the issue. "The fact that certain Australian tribes who are quite ignorant of physical paternity nevertheless recognize patrilineal descent is conclusive," he wrote (1949, 186). The same argument was put forward by Marvin Harris: "The idea that matrilineality is the result of confusion concerning paternity," he reasoned, "is wholly confounded by the numerous cases of primitive peoples who deny that the male is necessary for conception but who regard themselves as descended from a line of males" (1968, 187).

But Morgan's and McLennan's argument was not that patrilineal descent is precluded when paternity is uncertain or when the father's role in procreation is negated. It was that patrilineality requires a form of marriage based on long-term bonds between particular men and women because such bonds are a necessary condition for the recognition of patrifilial links. In all the examples given by Murdock and Harris, as is the case in human societies in general, pair-bonding, not promiscuous group marriage, is the rule. Morgan and McLennan would certainly have concurred with Murdock that although Australian tribes are ignorant of physical paternity they are nonetheless in a position to recognize patrifilial links. Ironically, then, many scholars have rejected Morgan's and McLennan's arguments about the priority of the matriclan partly for the wrong reasons.

Matrilineality as a Male Affair

If Morgan, McLennan, and others were right in arguing that descent could not be patrilineal prior to the advent of pair-bonding, they were wrong about the anteriority of matrilineality. Matrilineal descent as we know it in human societies was a latecomer in human evolution. Somewhat paradoxically, therefore, even though female kin groups and several structural correlates of matrilineal descent groups are found in nonhuman primates, humanlike matrilineal descent is not primitive; evolutionarily speaking, it is relatively recent.

Primate matrilines are composed of females of all ages and preadult males. Older males emigrate, losing membership in their matrilineage in the process, and as a result brothers do not recognize their sisters' offspring. Humanlike matrilineal descent groups, in which brothers maintain intimate bonds with their sisters, and maternal uncles with their sororal nephews, are a structural impossibility in primate groups. Matrilineality in nonhuman primates is necessarily a female affair. For humanlike matrilineality to become a possibility, male and female relatives had to be in a position to maintain lifetime bonds. Human matrilineality had to await the evolution of the tribe. Henceforward the conditions were set for the evolution of the brother–sister kinship complex and its various components, including avuncular relations and cross-cousin marriage.

There is thus no direct phylogenetic continuity between primatelike female kin groups and humanlike matrilineal descent groups. The continuity, rather, goes from the ancestral male kin group (and associated latent patriclan) to the true unilineal descent group, patrilineal or matrilineal. In keeping with the principle of precedence of residence over descent, the passage from patrilineal to matrilineal descent in some hominid populations necessarily involved a change in residence pattern, namely, a change from patrilocality to some form of matrilocality. When such changes took place, the tribal level of organization ensured that female localization and male transfer between groups did not cut males off from their kinswomen. The involvement of males in the politics of their matrilineage was now a structural possibility. Male-controlled matrilineality, avuncular relations, and the like were practicable.

This brings us to a significant critique by some social anthropologists concerning the meaning of the expression "primate matrilineage." The following quotation by Maurice Godelier sums up the issue:

> Some authors [primatologists] use the terms "maternal lineage" and "matrilineage" to describe the descendants stemming from an aged female, still present in the group, and surrounded by her daughters and their offspring. These terms are inadequate. In anthropology, matrilineage refers to a human group composed both of men and women who descend from a common *ancestress* through the women only. In such groups, political and economic authority is generally in the hands of men, that is of the women's brothers, who exert their authority upon their sisters and their offspring. . . . No such structure is found among macaques. (2004b, 322)

Some twenty years earlier Fortes had made the same point, remarking that "the males in particular [in macaque societies], unlike their human counterparts, sever all connections with their mothers and sisters when they reach maturity" (1983, 18). To Fortes—but possibly not to Godelier—this important difference was one element among others attesting to the mere metaphorical resemblance of primate matrilines to human matrilineages. The active role of adult males in human matrilineages and their absence in primate matrilineages supposedly proved that the two entities had different origins, cultural and biological, respectively. But as just argued, the role of adult males in matrilineal descent groups is a relatively recent event in the evolution of human descent. It is a refinement made possible by the pacification of relationships between local groups and the emergence of nepotism between brothers and sisters.

The same reasoning applies to bilineal, or double, descent, a rare form of unilineal descent that combines both the matrilineal and the patrilineal lines. In a double-descent system, every individual belongs to both a matriclan and a patriclan, but as noted by Fox, "any double-descent system can work provided it gives each principle [the patrilineal and the matrilineal) different work to do" (1967, 137). For example, land and houses (immovables) may be under the control of the patriclan, while movable goods such as livestock are under the control of the matriclan. Like matrilineality, bilineal descent could not appear prior to the evolution of the tribe, and one may conclude that unilineal descent predated bilineal descent.

20

Conclusion: Human Society as Contingent

"Knowledge about the great apes' present situation," asserted Lévi-Strauss (2000, 494), "teaches us little about man's past, close or distant." Ironically, not only does knowledge about nonhuman primates teach us a lot about man's past, but it vindicates much of Lévi-Strauss's insight about what makes human society unique in the animal world. Lévi-Strauss uncovered what I called the "deep structure" of human societies by abstracting their lowest common denominator, the "atom of kinship," a structure he described as "the most elementary form of kinship that can exist," a "primitive and irreducible unit" upon which "all kinship systems are constructed." The atom of kinship rests upon four terms—brother, sister, sister's husband, and sister's child—which together embody sister exchange, and thus reciprocal exogamy. To explain the universal character of female exchange, Lévi-Strauss posited the necessity for men to build alliances with other men. To achieve that primary goal, men could make use of mental structures predisposing them to engage in reciprocity-based relations, and they could call upon their most precious possession, their daughters and sisters. The three points taken together—the need for alliances, a propensity for reciprocity, and the control by men of their kinswomen—readily produce female exchange—unless, of course, fathers are sexually attracted to their daughters, and brothers to their sisters. Because Lévi-Strauss believed in the natural character of incest, he had to supplement his argument with an incest impediment: in order to exchange their sisters, men first had to renounce marrying them. The incest taboo is thus inherent in female exchange in Lévi-Strauss's scheme; it is an integral part of the atom of kinship itself, and it stands as the cornerstone of human society.

Having identified the smallest unit of all kinship systems, Lévi-Strauss attempted to explain its origin. But the comparative analysis of human societies—social anthropology as it is conceived of by some—can hardly go beyond what it has identified as its lowest common denominator, whether that is the atom of kinship or any other structure. Beyond that denominator, or underneath it, lies the realm of mental processes and psychological explanations. Lévi-Strauss did allude to some of the universal mental structures underlying reciprocal exogamy, namely the constituent properties of reciprocity and the synthetic nature of the gift. But he did not consider the possibility that several other aspects of reciprocal exogamy might have their own biological underpinnings and evolutionary history. Reducing the universal mental structures of reciprocal exogamy to its reciprocity dimension led him to assert that the human brain had generated most of the phenomenon from scratch. Lévi-Strauss set aside the possibility that the evolved human mind might impose a rich array of constraints affecting how humans generate their social environment and react to it. In the absence of such constraints, human society could only be the creation of culture.

The comparative analysis of human and nonhuman primate societies reveals that a substantial portion of the exogamy configuration existed in the form of behavioral regularities well before humans could make use of the symbolic capacity to communicate about their social relations and institute rules of conduct. Table 20.1 summarizes the evolutionary history of reciprocal exogamy. Reading it columnwise, Phase I gives the main features of hominid society right after the *Pan–Homo* split. Compared to other primate societies, especially human society, the ancestral male kin group was a society with minimal kinship or, more appropriately, a society with a dormant agnatic kinship structure. The evolution of pair-bonding, which marked the onset of Phase II, revealed that structure and deeply transformed the ancestral male kin group. Henceforth hominid groups were composed of several biparental families embedded in a group-wide genealogical web, and they exhibited primitive patrilocality, strong brotherhoods, and behavioral exogamy, as well as the basic structural prerequisites of patrilineal descent groups. Yet at this stage the local group was still the largest organized social entity. Supragroup levels of social organization were absent. But thanks to paternity recognition and its correlate, agnatic kinship, pair-bonded females could now act as peacemakers between their natal kinsman and their "husbands." Intergroup pacification coincided with the development of affinal brotherhoods and marked the onset of Phase III. It set the stage for the brother–sister kinship complex, the first rules of exogamy—or kinship constraints on marital unions—and the diversification of postmarital residence patterns. That is, Phase III saw the emergence of the primitive, prelinguistic tribe whose existence at that time was manifest merely as a state of mutual tolerance between "intermarrying" groups.

Table 20.1 thus substantiates the claim that the deep structure of human society resulted, basically, from the integration of pair-bonding with the ancestral male kin group, a claim embodied in the following "equation": *male kin group + pair-bonding → exogamy configuration.* The key to the deep structure of human society lies in the extensive repercussions of pair-bonding on the ancestral male kin group. At that point hominid society entered a runaway evolutionary process, in the course of which some of the most important aspects of the exogamy configuration unfolded or even snowballed. This reasoning fits rather well with what we know about the workings of biological evolution in general. Evolution is, fundamentally, a process of cumulative integration: novel combinations of older material or otherwise ordinary features generate systems with new properties—emergent properties—which in turn promote further evolutionary change. Earlier I gave the example of bipedal locomotion, which upon merging with the hand's older adaptations opened up a whole new range of situations in which the hand could be used. Similarly, I have argued here (1) that the sexual division of labor was the fortuitous byproduct of bipedalism (gathering) combining with pair-bonding (a mate-guarding tactic) and a chimpanzee-like male hunting bias; (2) that generalized monogamy was the outcome of generalized polygyny combining with the rise of technology, which equalized the competitive power of males; (3) that group-wide uterine kinship structures were a correlate of female localization and lifetime mother–daughter bonds; (4) that agnatic kinship structures were bound to emerge from the conjunction of kin-recognition abilities, male localization, pair-bonding (paternity recognition), and lengthened father–son bonds; (5) that kin-group exogamy and postmarital residence patterns were the outcome of pair-bonding merging with kin-group dispersal; (6) that bilateral kinship networks appeared as soon as uterine kin recognition processes combined with consistent paternity recognition (pair-bonding) and the tribal level of organization; (7) that the bilateral recognition of affines emanated from the co-occurrence of consanguineal kin-

Table 20.1. Three phases in the construction of the exogamy configuration. Phase I (the ancestral male kin group) extended from the *Pan–Homo* split to the evolution of pair-bonding. Phase II followed the evolution of pair-bonding and ended just before intergroup pacification and the emergence of the tribe, which marked the onset of Phase III.

	Phase I	Phase II	Phase III
Type of local group	Multimale-multifemale	Multifamily	Multifamily
Supragroup structure	No	No	Primitive tribe
Relations between local groups	Avoidance/hostility	Avoidance/hostility	Within-tribe tolerance and between-tribe hostility
Residence pattern	Male philopatry	Primitive patrilocality	Diversification within tribe
Mating system	Sexual promiscuity	Pair-bonding (monogamous and/or polygynous)	Pair-bonding (monogamous and/or polygynous)
Kinship boundaries of exogamy	Kin-group outbreeding	Kin-group "exogamy"	Within-tribe kin-group exogamy
Domain of uterine kindred	Primary kin	Primary kin	Primary kin and matrilateral kin
Paternity recognition	Inconsistent at best	Consistent	Consistent
Domain of agnatic kindred	Nil	Primary kin and patrilateral kin	Primary kin and patrilateral kin
Kinship (laterality)	Unilateral and limited	Bilateral but asymmetric	Fully bilateral
Minimal domain of incest avoidance	Primary uterine kin	Primary kin including father–daughter dyads	Primary kin including father–daughter dyads

Table 20.1 (continued)

	Phase I	Phase II	Phase III
Patrilineal descent	Not possible	Latent or actual patriclan	Latent or actual patriclan
Matrilineal descent	Not possible	Not possible	Possible, depending on residence
Affinal kinship	Absent	Unilateral	Bilateral
Family	Monoparental	Biparental	Biparental
Relations between primary agnates	Limited to brothers and inconsistent	Strong brotherhoods including father	Strong brotherhoods, including father
Relations between brothers and sisters	Weak, inconsistent, and temporary	Strong, consistent, but temporary	Strong, consistent, and lifetime
Marriage between affines	Nil	No bias	Positive bias within affinal brotherhoods
Cross-cousin marriage	Nil	Asymmetric availability of cross- and parallel cousins	Positive bias within affinal brotherhoods

ship, pair-bonding, and intergroup pacification; and (8) that the most basic structural aspects of unilineal descent were emergent properties of groups combining sex-biased dispersal, matrifiliation (or patrifiliation), and socially transmittable entities (for example, social status). All these instances may serve to illustrate the parsimony principle in relation to how one accounts for evolutionary change. The idea that novel traits need not be the product of specific selective pressures has been long recognized. But a recurrent and striking characteristic of evolutionary scenarios about hominid evolution is how readily they resort to hypothetical selective pressures to explain any single trait, as if most traits necessarily originated as biological adaptations. The examples listed above furnish several counterexamples.

They also serve to exemplify another evolutionary principle. Any complex trait, anatomical or behavioral, is the outcome of a phylogenetic sequence whose constituent steps obviously had nothing to do, initially, with the final outcome. Evolution is fundamentally contingent and opportunistic. To illustrate this point further, Table 20.2 describes the "cumulative" construction of female exchange, the core process of reciprocal exogamy, over the three phases described in Table 20.1. Each phase contributed a fraction of the constitutive elements of affinal brotherhoods, but every single step evolved for reasons wholly unrelated to the end result: a social context conducive to female exchange. In Phase I the brother–sister bond, one of the two constituent bonds of the elementary unit of exogamy, was already present. Moreover, brothers and sisters ended up in distinct local groups. Phase II contributed the other constituent bond of the elementary unit of exogamy: the pair-bond. It further added stronger brother–sister bonds and brotherhoods and the unilateral recognition of affines. It is only with Phase III that the two types of bonds were connected in the form of affinal brotherhoods. Had any one of these steps not materialized, some of the evolutionary prerequisites of female exchange would not have been met and reciprocal exogamy would likely not have evolved. By Phase III hominids had reached a stage where female exchange between male kin groups was "in the air." Language, along with other cognitive abilities, would eventually do the rest.

In Lévi-Strauss's scheme the atom of kinship and its correlate, the incest taboo, epitomized the "transition from nature to culture." But the atom

Table 20.2. The cumulative construction of female exchange in the course of hominid evolution. The three phases are defined as in Table 1.

	Phase I	Phase II	Phase III
Males are localized.	X		
Females circulate between local groups.	X		
Brothers and sisters recognize each other.	X		
Brothers and sisters breed in distinct local groups.	X		
Males and females breed in the context of lasting pair-bonds.		X	
Primary agnates form strong, cohesive brotherhoods.		X	
Brothers and sisters have strong bonds consistently.		X	
"Wives" recognize their affines.		X	
Males of distinct intermarrying groups tolerate each other.			X
Brothers and sisters maintain lifetime bonds.			X
"Husbands" recognize their affines (bilateral affinity).			X
Affines experience pair-bonding biases.			X
Cross-cousins experience pair-bonding biases.			X

of kinship is not a cultural construct. It is composed of three basic bonds that have deep evolutionary roots: a kinship bond, a sexual bond, and a parental bond. This basic kinship structure is an integral part of *human nature,* along with such features as bipedal locomotion, language, and morality. But the atom of kinship does not exist as an inborn mental structure; it is a facultative response. Nonetheless, it is the natural outcome of a combined set of features that themselves have biological underpinnings. The atom of kinship emerged when the evolution of the tribe brought about a structural connection between the brother–sister bond and the pair-bond through the mediation of the sister/wife. Each constituent element of the atom of kinship—kin-group outbreeding, incest avoidance, kin recognition, kin favoritism, and pair-bonding—does have a biological foundation. That is why human beings are naturally led to form relationships with their affines and to forge social ties beyond their local group, engendering supragroup social entities in the process. Chimpanzees, gorillas, and macaques are not led to do so. Even if they have some of the necessary prerequisites, they lack several others. Accordingly, a social life confined to the boundaries of the local group is part of their biological nature. In humans the tribal level of organization is natural. Tribal life and exogamy are unavoidable outcomes of our evolutionary heritage.

The atom of kinship—or its close kin, which I have called the atom of exogamy—is perhaps mankind's greatest innovation in the social sphere. It marks the distinctiveness of human society's deep social structure. It embodies the very substance of the "genealogical unity of mankind," whose existence was negated by many. It stands alongside the thousands of other kinship structures of other species, and it stands in filiation with them—like Schultz's human foot among other primate feet. If the promiscuous male kin group is the paradigmatic structure of chimpanzee society, the atom of kinship is the paradigmatic structure of human society. This archetypal kinship structure was to beget thousands of cultural variants, whose diversity has challenged social anthropologists until today, almost obliterating its very unity.

References

Index

References

Aiello, L. C., and Key, C. 2002. "Energetic consequences of being a *Homo erectus* female." *American Journal of Biology* 14: 551–565.

Alberts, S. C. 1999. "Paternal kin discrimination in wild baboons." *Proceedings of the Royal Society of London,* series B: *Biological Sciences* 266: 1501–1506.

Alexander, R. D. 1979. *Darwinism and Human Affairs.* Seattle: University of Washington Press.

Alexander, R. D., Hoogland, J. L., Howard, R. D., Noonan, K. M., and Sherman, P. W. 1979. "Sexual dimorphism and breeding systems in pinnipeds, ungulates, primates, and humans." In *Evolutionary Biology and Human Social Behavior: An Anthropological Perspective,* ed. N. A. Chagnon and W. Irons, 432–435. North Scituate, Mass.: Duxbury Press.

Alexander, R. D., and Noonan, K. M. 1979. "Concealment of ovulation, parental care, and human social evolution." In *Evolutionary Biology and Human Social Behavior: An Anthropological Perspective,* ed. N. A. Chagnon and W. Irons, 436–453. North Scituate, Mass.: Duxbury Press.

Altmann, S. A., and Altmann, J. 1979. "Demographic constraints on behavior and social organization." In *Primate Ecology and Human Origins: Ecological Influences on Social Organization,* ed. I. S. Bernstein and E. O. Smith, 47–63. New York: Garland STPM Press.

Alvarez, H. P. 2004. "Residence groups among hunter-gatherers: A view of the claims and evidence for patrilocal bands." In *Kinship and Behavior in Primates,* ed. B. Chapais and C. M. Berman, 420–442. New York: Oxford University Press.

Arnhart, L. 2004. "The incest taboo as Darwinian natural right." In *Inbreeding, Incest, and the Incest Taboo: The State of Knowledge at the Turn of the Century,* ed. A. P. Wolf and W. H. Durham, 190–218. Stanford: Stanford University Press.

Asfaw, B., White, T., Lovejoy, O., Latimer, B., Simpson, S., and Suwa, G. 1999. "*Australopithecus garhi:* A new species of early hominid from Ethiopia." *Science* 284: 629–635.

Aureli, P., Cozzolino, R., Cordishi, C., and Scucchi, S. 1992. "Kin-oriented redirection among Japanese macaques: An expression of a revenge system?" *Animal Behaviour* 44: 283–291.

Barnard, A. 1983. "Contemporary hunter-gatherers: Current issues in ecology and social organization." *Annual Review of Anthropology* 12: 193–214.

Barnard, C. J., and Aldhous, P. 1991. "Kinship, kin discrimination and mate choice." In *Kin Recognition*, ed. P. G. Hepper, 125–147. Cambridge: Cambridge University Press.

Barnes, J. A. 1962. "African models in the New Guinea Highlands." *Man* 62: 5–9.

Barrett, L., Dunbar, R., and Lycett, J. 2002. *Human Evolutionary Psychology.* Princeton: Princeton University Press.

Barton, R. A. 1999. "Socioecology of baboons: The interaction of male and female strategies." In *Primate Males: Causes and Consequences of Variation and Group Composition*, ed. P. M. Kappeler, 97–107. Cambridge: Cambridge University Press.

Barton, R. A., and Whiten, A. 1993. "Female competition among female olive baboons, *Papio anubis*." *Animal Behaviour* 46: 777–789.

Bateson, P. 1983. "Optimal outbreeding." In *Mate Choice*, ed. P. Bateson, 257–277. Cambridge: Cambridge University Press.

———. 2004. "Inbreeding avoidance and incest taboos." In *Inbreeding, Incest, and the Incest Taboo: The State of Knowledge at the Turn of the Century*, ed. A. P. Wolf and W. H. Durham, 24–38. Stanford: Stanford University Press.

Baxter, M. J., and Fedigan, L. 1979. "Grooming and consort partner selection in a troop of Japanese monkeys *(Macaca fuscata)*." *Archives of Sexual Behavior* 8: 445–458.

Bélisle, P., and Chapais, B. 2001. "Tolerated co-feeding in relation to degree of kinship in Japanese macaques." *Behaviour* 138: 487–509.

Bergman, T. J., Beehner, J. C., Cheney, D. L., and Seyfarth, R. M. 2003. "Hierarchical classification by rank and kinship in baboons." *Science* 302: 1234–1236.

Berman, C. M. 2004. "Developmental aspects of kin bias in behavior." In *Kinship and Behavior in Primates*, ed. B. Chapais and C. M. Berman, 317–346. New York: Oxford University Press.

Berman, C. M., and Kapsalis, E. 1999. "Development of kin bias among rhesus monkeys: Maternal transmission or individual learning?" *Animal Behaviour* 58: 883–894.

Berman, C. M., Rasmussen, K. L. R., and Suomi, S. J. 1997. "Group size, infant development and social networks in free-ranging rhesus monkeys." *Animal Behaviour* 53: 405–421.

Bernstein, I. S. 1991. "The correlation between kinship and behaviour in nonhuman primates." In *Kin Recognition*, ed. P. G. Hepper, 6–29. Cambridge: Cambridge University Press.

Betzig, L. 1991. Comment about "Violence and sociality in human evolution" by B. M. Knauft. *Current Anthropology* 32: 410–411.

Bevc, I., and Silverman, I. 1993. "Early proximity and intimacy between siblings and incestuous behavior: A test of the Westermarck hypothesis." *Ethology and Sociobiology* 14: 171–181.

———. 2000. "Early separation and sibling incest: A test of the revised Westermarck theory." *Evolution and Human Behavior* 21: 151–161.

Bird, R. 1999. "Cooperation and conflict: The behavioral ecology of the sexual division of labor." *Evolutionary Anthropology* 8: 65–75.

Bischof, N. 1971. "The biological foundations of the incest taboo." *Social Sciences Information* 11: 7–36.

———. 1975. "Comparative ethology of incest avoidance." In *Biosocial Anthropology,* ed. R. Fox, 37–67. New York: Halsted Press.

Bittles, A. H. 2004. "Genetic aspects of inbreeding and incest." In *Inbreeding, Incest, and the Incest Taboo: The State of Knowledge at the Turn of the Century,* ed. A. P. Wolf and W. H. Durham, 38–60. Stanford: Stanford University Press.

Bloch, M., and Sperber, D. 2002. "Kinship and evolved psychological dispositions: The mother's brother controversy reconsidered." *Current Anthropology* 43: 723–748.

Boesch, C. 1994. "Cooperative hunting in wild chimpanzees." *Animal Behaviour* 48: 653–667.

Boesch, C., and Boesch, H. 1984. "Possible causes of sex differences in the use of natural hammers by wild chimpanzees." *Journal of Human Evolution* 13: 415–440.

———. 1989. "Hunting behavior of wild chimpanzees in the Taï National Park." *American Journal of Physical Anthropology* 78: 547–573.

———. 1990. "Tool use and tool making in wild chimpanzees." *Folia Primatologica* 54, 86–99.

Boesch, C., and Boesch-Achermann, H. 2000. *The Chimpanzees of the Taï Forest: Behavioural Ecology and Evolution.* New York: Oxford University Press.

Boesch, C., Boesch-Achermann, H., and Vigilant, L. 2006. "Cooperative hunting in chimpanzees: Kinship or mutualism?" In *Cooperation in Primates and Humans: Mechanisms and Evolution,* ed. P. M. Kappeler and C. P. van Schaik, 139–150. Heidelberg: Springer.

Boesch, C., and Tomasello, M. 1998. "Chimpanzee and human cultures." *Current Anthropology* 39: 591–614.

Boyd, R. 1992. "The evolution of reciprocity when conditions vary." in *Coalitions and Alliances in Humans and Other Animals,* ed. A. H. Harcourt and F. B. M. de Waal, 473–489. New York: Oxford University Press.

Bradley, B. J., Doran-Sheehy, D. M., Lukas, D., Boesch, C., and Vigilant, L. 2004. "Dispersed male networks in western gorillas." *Current Biology* 14: 510–513.

Brotherton, P. M. N., and Komers, P. E. 2003. "Mate guarding and the evolution of social monogamy in mammals." In *Monogamy: Mating Strategies and Partnerships in Birds, Humans and Other Mammals,* ed. U. H. Reichard and C. Boesch, 42–58. Cambridge: Cambridge University Press.

Brunet, M., and thirty-seven other authors. 2002. "A new hominid from the Upper Miocene of Chad, Central Africa." *Nature* 418: 145–152.

Buchan, J. C., Alberts, S. C., Silk, J. B., and Altmann, J. 2003. "True paternal care in a multi-male primate society." *Nature* 425: 179–181.

Burnstein, E. 2005. "Altruism and genetic relatedness." In *The Handbook of Evolutionary Psychology,* ed. D. M. Buss, 528–551. Hoboken, N.J.: John Wiley and Sons.

Byrne, R. W. 1995. *The Thinking Ape: Evolutionary Origins of Intelligence.* Oxford: Oxford University Press.

———. 2001. "Social and technical forms of primate intelligence." In *Tree of Origin: What Primate Behavior Can Tell Us about Human Social Evolution,* ed. F. B. M. de Waal, 145–172. Cambridge, Mass.: Harvard University Press.

Byrne, R. W., and Whiten, A., eds. 1988. *Machiavellian Intelligence: Social Expertise and the Evolution of Intellect in Monkeys, Apes, and Humans.* Oxford: Oxford University Press.

Campbell, B. G. 1966. *Human Evolution: An Introduction to Man's Adaptation.* Chicago: Aldine.

Carpenter, C. R. 1934. "A field study of the behavior and social relations of the howling monkeys *(Alouatta palliata).*" *Comparative Psychological Monographs* 10: 1–168.

———. 1935. "Behavior of the red spider monkeys *(Ateles geoffroyi)* in Panama." *Journal of Mammalogy* 16: 171–180.

———. 1940. "A field study in Siam of the behavior and social relations of the gibbon *(Hylobates lar).*" *Comparative Psychological Monographs* 16: 1–212.

———. 1942. "Societies of monkeys and apes." *Biological Symposia* 8: 177–204. Reprinted in *Primate Social Behavior,* ed. C. Southwick, 24–51. Princeton, N.J.: van Nostrand.

Chagnon, N. A., and Irons, W., eds. 1979. *Evolutionary Biology and Human Social Behavior: An Anthropological Perspective.* North Scituate, Mass.: Duxbury Press.

Chapais, B. 1983. "Matriline membership and male rhesus reaching high ranks in the natal troops." In *Primate Social Relationships: An Integrated Approach,* ed. R. A. Hinde, 171–175. Oxford: Blackwell.

———. 1988. "Experimental matrilineal inheritance of rank in female Japanese macaques." *Animal Behaviour* 36: 1025–1037.

———. 1991. "Primates and the origins of aggression, power, and politics among humans." In *Understanding Behavior: What Primate Studies Tell Us about Human Behavior,* ed. J. D. Loy and C. B. Peters, 190–228. New York: Oxford University Press.

———. 1992. "The role of alliances in social inheritance of rank among female primates." In *Coalitions and Alliances in Humans and Other Animals,* ed. A. H. Harcourt and F. B. M. de Waal, 29–59. Oxford: Oxford University Press.

———. 2001. "Primate nepotism: What is the explanatory value of kin selection?" *International Journal of Primatology* 22: 203–229.

———. 2004. "How kinship generates dominance structures: A comparative perspective." In *Macaque Societies: A Model for the Study of Social Organization,* ed. B. Thierry, M. Singh, and W. Kaumanns, 186–204. Cambridge: Cambridge University Press.

———. 2006. "Kinship, competence and cooperation in primates." In *Cooperation in Primates and Humans: Mechanisms and Evolution,* ed. P. M. Kappeler and C. P. van Schaik, 47–64. Berlin: Springer.

Chapais, B., and Bélisle, P. 2004. "Constraints on kin selection in primate groups." In *Kinship and Behavior in Primates,* ed. B. Chapais and C. M. Berman, 365–386. New York: Oxford University Press.

Chapais, B., and Berman, C. M. 2004a. "Introduction: The kinship black box." In *Kinship and Behavior in Primates,* ed. B. Chapais and C. M. Berman, 3–11. New York: Oxford University Press.

———, eds. 2004b. *Kinship and Behavior in Primates.* New York: Oxford University Press.

Chapais, B., and Gauthier, C. 1993. "Early agonistic experience and the onset of matrilineal rank acquisition." In *Juvenile Primates: Life History, Development, and Behavior,* ed. M. E. Pereira and L. A. Fairbanks, 245–258. New York: Oxford University Press.

Chapais, B., Gauthier, C., Prud'homme, J., and Vasey, P. 1997. "Relatedness threshold for nepotism in Japanese macaques." *Animal Behaviour* 53: 1089–1101.

Chapais, B., Girard, M., and Primi, G. 1991. "Non-kin alliances, and the stability of matrilineal dominance relations in Japanese macaques." *Animal Behaviour* 41: 481–491.

Chapais, B., and Mignault, C. 1991. "Homosexual incest avoidance among females in captive Japanese macaques." *American Journal of Primatology* 23: 171–183.

Chapais, B., Prud'homme, J., and Teijeiro, S. 1994. "Dominance competition among siblings in Japanese macaques: Constraints on nepotism." *Animal Behaviour* 48: 1335–1347.

Chapais, B., Savard, L., and Gauthier, C. 2001. "Kin selection and the distribution of altruism in relation to degree of kinship in Japanese macaques (*Macaca fuscata*)." *Behavioral Ecology and Sociobiology* 49: 493–502.

Charlesworth, D. 1987. "Inbreeding depression and its evolutionary consequences." *Annual Review of Ecology and Systematics* 18: 237–268.

Chauvin, C., and Berman, C. M. 2004. "Intergenerational transmission of behavior." In *Macaque Societies: A Model for the Study of Social Organization,* ed. B. Thierry, M. Singh, and W. Kaumanns, 209–230. Cambridge: Cambridge University Press.

Chen, F. C., and Li, W. H. 2001. "Genomic divergences between humans and other hominoids and the effective population size of the common ancestor of humans and chimpanzees." *American Journal of Human Genetics* 68: 444–456.

Cheney, D. L. 1987. "Interactions and relationships between groups." In *Primate Societies,* ed. B. B. Smuts, D. L. Cheney, R. M. Seyfrath, R. W. Wrangham, and T. T. Struhsaker, 267–281. Chicago: University of Chicago Press.

Cheney, D. L., and Seyfarth, R. M. 1980. "Vocal recognition in free-ranging vervet monkeys." *Animal Behaviour* 28: 362–367.

———. 1986. "The recognition of social alliances by vervet monkeys." *Animal Behaviour* 34: 1722–1731.

———. 1989. "Redirected aggression and reconciliation among vervet monkeys, *Cercopithecus aethiops.*" *Behaviour* 110: 258–275.

———. 1990. *How Monkeys See the World: Inside the Mind of Another Species.* Chicago: University of Chicago Press.

———. 1999. "Recognition of other individuals' social relationships by female baboons." *Animal Behaviour* 58: 67–75.

———. 2004. "The recognition of other individuals' kinship relationships." In *Kinship and Behavior in Primates,* ed. B. Chapais and C. M. Berman, 347–364. New York: Oxford University Press.

———. 2005. "Constraints and preadaptations in the earliest stages of language evolution." *Linguistic Review* 22: 35–159.

Chepko-Sade, B. D., and Sade, D. S. 1979. "Patterns of group splitting within matrilineal kinship groups: A study of social group structure in *Macaca mulatta* (Cercopithecidae: Primates)." *Behavioral Ecology and Sociobiology* 5: 67–86.

Chism, J. 2000. "Allocare patterns among cercopithecines." *Folia Primatologica* 71, 55–66.

Clutton-Brock, T. H. 1989a. "Female transfer and inbreeding avoidance in social mammals." *Nature* 337: 70–71.

———. 1989b. "Mammalian mating systems." *Proceedings of the Royal Society of London* B, 236: 339–372.

———. 1991. *The evolution of parental care.* Princeton, N.J.: Princeton University Press.

———. 2004. "What is sexual selection?" In *Sexual Selection in Primates: New and Comparative Perspectives,* ed. P. M. Kappeler and C. van Schaik, 24–36. New York: Cambridge University Press.

Clutton-Brock, T. H., and Harvey, P. H. 1976. "Evolutionary rules and primate societies." In *Growing Points in Ethology,* ed. P. P. G. Bateson and R. A. Hinde, 195–237. Cambridge: Cambridge University Press

Clutton-Brock, T. H., Harvey, P., and Rudder, D. 1977. "Sexual dimorphism, socioeconomic sex-ratio, and body weight in primates." *Nature* 269: 797–800.

Clutton-Brock, T., and Parker, G. A. 1992. "Potential reproductive rates and the operation of sexual selection." *Quarterly Review of Biology* 67: 437–456.

Colmenares, F. 2004. "Kinship structure and its impact on behavior in multilevel societies." In *Kinship and Behavior in Primates,* ed. B. Chapais and C. M. Berman, 242–270. New York: Oxford University Press.

Constable, J. L., Ashley, M. V., Goodall, J., and Pusey, A. E. 2001. "Noninvasive paternity assignment in Gombe chimpanzees." *Molecular Ecology* 10: 1279–1300.

Cooper, M. A. 2004. "Inter-group relationships." In *Macaque Societies: A Model for the Study of Social Organization,* ed. B. Thierry, M. Singh, and W. Kaumanns, 204–208. Cambridge: Cambridge University Press.

Crnokrak, P., and Roff, D. A. 1999. "Inbreeding depression in the wild." *Heredity* 83: 260–270.

Darwin, C. 1859. *On the Origin of Species by Means of Natural Selection.* London: John Murray.

———. 1981. *The Descent of Man and Selection in Relation to Sex.* Princeton, N.J.: Princeton University Press. Photoreproduction of the 1871 edition. London: J. Murray.

Dasser, V. 1988a. "A social concept in Java monkeys." *Animal Behaviour* 36: 225–230.

———. 1988b. "Mapping social concepts in monkeys." In *Machiavellian Intelligence: Social Expertise and the Evolution of Intellect in Monkeys, Apes, and Humans,* ed. R. W. Byrne and A. Whiten, 85–93. Oxford: Oxford University Press.

Datta, S. B. 1983. "Relative power and the acquisition of rank." In *Primate Social Relationships: An Integrated Approach,* ed. R. A. Hinde, 93–103. Oxford: Blackwell.

———. 1988. "The acquisition of dominance among free-ranging rhesus monkey siblings." *Animal Behaviour* 36: 754–772.

de Heinzelin, J., Clark, J. D., White, T., Hart, W., Renne, P., WoldeGabriel, G., Beyene, Y., and Vrba, E. 1999. "Environment and behavior of 2.5-million-year-old Bouri hominids." *Science* 284: 625–629.

Delson, E. 1980. "Fossil macaques, phyletic relationships and a scenario of deployment." In *The Macaques: Studies in Ecology, Behavior and Evolution*, ed. D. G. Lindburg, 10–30. New York: Van Nostrand Rheinhold.

de Waal, F. B. M. 1996. *Good Natured: The Origin of Right and Wrong in Humans and other Animals*. Cambridge, Mass.: Harvard University Press.

———. 2001. *The Ape and the Sushi Master: Cultural Reflections by a Primatologist*. London: Allen Lane, Penguin Press.

Dittus, W. P. J. 1988. "Group fission among wild toque macaques as a consequence of female resource competition and environmental stress." *Animal Behaviour* 36: 1626–1645.

Divale, W. 1984. *Matrilocal Residence in Pre-Literate Society*. Ann Arbor, Mich.: UMI Research Press.

Dixson, A. F. 1998. *Primate Sexuality: Comparative Studies of the Prosimians, Monkeys, Apes and Human Beings*. Oxford: Oxford University Press.

Doran, D. M., and McNeilage, A. 1998. "Gorilla ecology and behavior." *Evolutionary Anthropology* 6: 120–131.

Doran, D. M., Jungers, W. L., Sugiyama, Y., Fleagle, J. G., and Heesy, C. P. 2002. "Multivariate and phylogenetic approaches to understanding chimpanzee and bonobo behavioral diversity." In *Behavioural Diversity in Chimpanzees and Bonobos*, ed. C. Boesch, G. Hohmann, and L. F. Marchant, 14–34. New York: Cambridge University Press.

Dufour, D. L., and Sauther, M. L. 2002. "Comparative and evolutionary dimensions of the energetics of human pregnancy and lactation." *American Journal of Human Biology* 14: 584–602.

Dumont, L. 1997. *Groupes de Filiation et Alliance de Mariage*. Paris: Gallimard.

Dunbar, R. I. M. 1993. "The coevolution of neocortical size, group size and language in humans." *Behavioral and Brain Sciences* 16: 681–735.

———. 1995. "The mating system of Callitrichid primates: I. Conditions for the coevolution of pair bonding and twinning." *Animal Behaviour* 50: 1057–1070

———. 2001. "Brains on two legs: Group size and the evolution of intelligence." In *Tree of Origin: What Primate Behavior Can Tell Us about Human Social Evolution*, ed. F. B. M. de Waal, 173–191. Cambridge, Mass.: Harvard University Press.

Eaton, G. G. 1978. "Longitudinal studies of sexual behavior in the Oregon troop of Japanese macaques." In *Sex and Behavior*, ed. T. E. Gill, D. A. Dewsbury, and B. D. Sachs, 35–59. New York: Plenum.

Ember, C. 1975. "Residential variation among hunter-gatherers." *Behavior Science Research* 3: 199–227.

———. 1978. "Myths about hunter-gatherers." *Ethnology* 17: 439–448.

Enard, W., and Pääbo, S. 2004. "Comparative primate genomics." *Annual Review of Genomics and Human Genetics* 5: 351–378.

Erhart, E. M., Coelho, A. M. Jr., and Bramblett, C. A. 1997. "Kin recognition by paternal half-siblings in captive *Papio cynocephalus.*" *American Journal of Primatology* 43: 147–157.

Erickson, M. T. 2004. "Evolutionary thought about the current clinical understanding of incest." In *Inbreeding, Incest, and the Incest Taboo: The State of Knowledge at the Turn of the Century,* ed. A. P. Wolf and W. H. Durham, 161–189. Stanford: Stanford University Press.

Evans-Pritchard, E. E. 1940. *The Nuer: A Description of the Modes of Livelihood and Political Institutions of a Nilotic People.* London: Oxford University Press.

Fa, J. E. 1989. "The genus *Macaca:* A review of taxonomy and evolution." *Mammal Reviews* 19: 45–81.

Fa, J. E., and Lindburg, D. G., eds. 1996. *Evolution and Ecology of Macaque Societies.* New York: Cambridge University Press.

Fairbanks, L. A. 1990. "Reciprocal benefits of allomothering for female vervet monkeys." *Animal Behaviour* 40: 553–562.

———. 2000. "Maternal investment throughout the life span in Old World monkeys." In *Old World Monkeys,* ed. P. F. Whitehead and C. J. Jolly, 341–367. Cambridge: Cambridge University Press.

Fisher, H. 1992. *Anatomy of Love: The Natural History of Monogamy, Adultery, and Divorce.* New York: W. W. Norton.

———. 2006. *Why We Love: The Nature and Chemistry of Romantic Love.* New York: Henry Holt.

Flinn, M. V., Ward, C. V., and Noone, R. J. 2005. "Hormones and the human family." In *The Handbook of Evolutionary Psychology,* ed. D. M. Buss, 552–583. Hoboken, N.J.: John Wiley and Sons.

Foley, R. A. 1989. "The evolution of hominid social behaviour." In *Comparative Socioecology: The Behavioural Ecology of Humans and Other Mammals,* ed. V. Standen and R. A. Foley, 473–494. Oxford: Blackwell Scientific Publications.

———. 1996. "An evolutionary and chronological framework for human social behavior." *Proceedings of the British Academy* 88: 95–117.

———. 1999. "Hominid behavioural evolution: Missing links in comparative primate socioecology." In *Comparative Primate Socioecology,* ed. P. C. Lee, 363–386. Cambridge: Cambridge University Press.

Foley, R. A., and Lee, P. C. 1989. "Finite social space, evolutionary pathways, and reconstructing hominid behaviour." *Science* 243: 901–906.

Fooden, J. 1980. "Classification and distribution of living macaques (*Macaca* Lacépède, 1799)." In *The Macaques: Studies in Ecology, Behavior and Evolution,* ed. D. G. Lindburg, 1–9. New York: van Nostrand Reinhold.

Fortes, M. 1953. "The structure of unilineal descent groups." *American Anthropologist* 55: 17–41.

———. 1983. *Rules and the Emergence of Society.* London: Royal Anthropological Institute of Great Britain and Ireland.

Fox, R. 1962. "Sibling incest." *British Journal of Sociology* 13: 128–150.

———. 1967. *Kinship and Marriage: An Anthropological Perspective.* Baltimore: Penguin Books.

———. 1972. "Alliance and constraint: Sexual selection and the evolution of human kinship systems." In *Sexual Selection and the Descent of Man,* ed. B. Campbell, 282–331. Chicago: Aldine.

———. 1975. "Primate kin and human kinship." In *Biosocial Anthropology,* ed. R. Fox, 9–35. New York: Halsted Press.

———. 1979. "Kinship categories as natural categories." In *Evolutionary Biology and Human Behavior,* ed. N. A. Chagnon and W. Irons, 132–144. North Scituate, Mass.: Duxbury Press.

———. 1980. *The Red Lamp of Incest.* New York: E. P. Dutton.

———. 1993. *Reproduction and Succession: Studies in Anthropology, Law and Society.* New Brunswick, N.J.: Transaction Publishers.

Fredrickson, W. T., and Sackett, G. P. 1984. "Kin preferences in primates *(Macaca nemestrina):* Relatedness or familiarity?" *Journal of Comparative Psychology* 98: 29–34.

Fruth, B., and Hohmann, G. 2002. "How bonobos handle hunts and harvests: Why share food?" In *Behavioural Diversity in Chimpanzees and Bonobos,* ed. C. Boesch, G. Hohmann, and L. F. Marchant, 231–243. Cambridge: Cambridge University Press.

Furuichi, T. 1989. "Social interactions and the life history of female *Pan paniscus* in Wamba, Zaire." *International Journal of Primatology* 10: 173–197.

———. 1997. "Agonistic interactions and matrifocal dominance rank of wild bonobos *(Pan paniscus)* at Wamba." *International Journal of Primatology* 18: 855–875.

Futuyma, D. G. 1998. *Evolutionary Biology.* Sunderland, Mass.: Sinauer Associates.

Galdikas, B. M. F., and Teleki, G. 1981. "Variations in subsistence activities of female and male pongids: New perspectives on the origins of hominid labor divisions." *Current Anthropology* 22: 241–256.

Geary, D. C. 2005. "Evolution of paternal investment." In *The Handbook of Evolutionary Psychology.* ed. D. M. Buss, 483–505. Hoboken, N.J.: John Wiley and Sons.

Gerloff, U., Hartung, B., Fruth, B., Hohmann, G., and Tautz, D. 1999. "Intracommunity relationships, dispersal pattern and paternity success in a wild living community of bonobos *(Pan paniscus)* determined from DNA analysis of faecal samples." *Proceedings of the Royal Society of London,* series B: *Biological Sciences* 266: 1189–1195.

Ghiglieri, M. P. 1987. "Sociobiology of the great apes and the hominid ancestor." *Journal of Human Evolution* 16: 319–357.

Godelier, M. 1986. *The Making of Great Men: Male Domination and Power among the New Guinea Baruya.* Cambridge: Cambridge University Press.

———. 2004a. *Métamorphose de la parenté.* Paris: Fayard.

———. 2004b. "An anthropologist among macaques." In *Macaque Societies: A Model for the Study of Social Organization,* ed. B. Thierry, M. Singh, and W. Kaumanns, 321–328. Cambridge: Cambridge University Press.

Goldberg, T. L., and Wrangham, R. W. 1997. "Genetic correlates of social behaviour in wild chimpanzees: Evidence from mitochondrial DNA." *Animal Behaviour* 54: 559–579.

Goodall, J. 1986. *The Chimpanzees of Gombe: Patterns of Behavior.* Cambridge, Mass.: Harvard University Press.

———. 1990. *Through a Window: My Thirty Years with the Chimpanzees of Gombe.* Boston: Houghton Mifflin.

Goodman, M., Grossman, L. I., and Wildman, D. E. 2005. "Moving primate genomics beyond the chimpanzee genome." *Trends in Genetics* 21: 511–517.

Goodman, M., Porter, C. A., Czelusniak, J., Page, S. L., Schneider, H., Shoshani, J., Gunnnell, G., and Groves, C. P. 1998. "Toward a phylogenetic classification of primates based on DNA evidence complemented by fossil evidence." *Molecular Phylogenetics and Evolution* 9: 585–598.

Goodman, M., Tagle, D. A., Fitch, D. H., Bailey, W., Czelusniak, J., Koop, B. F., Benson, P., and Slighton, J. L. 1990. "Primate evolution at the DNA level and a classification of hominoids." *Journal of Molecular Evolution* 30: 260–266.

Gough, K. 1971. "The origin of the family." *Journal of Marriage and the Family* 33: 760–771.

Gouzoules, S. 1984. "Primate mating systems, kin associations, and cooperative behavior: Evidence for kin recognition?" *Yearbook of Physical Anthropology* 27: 99–134.

Greenwood, P. J. 1980. "Mating systems, philopatry and optimal dispersal in birds and mammals." *Animal Behaviour* 28: 1140–1162.

Gurven, M. 2004. "To give and to give not: The behavioral ecology of human transfers." *Behavioral and Brain Sciences* 27: 543–583.

Gurven, M., and Hill, K. n.d. "Hunting as subsistence and mating effort? Re-evaluation of 'Man the Hunter,' the sexual division of labor and the evolution of the nuclear family." Forthcoming.

Gust, D. A., McCaster, T., Gordon, T. P., Gergits, W. F., Casna, N. J., and McClure, H. M. 1998. "Paternity in sooty mangabeys." *International Journal of Primatology* 19: 83–94.

Hamilton, W. D. 1964. "The genetical theory of social behavior." *Journal of Theoretical Biology* 7: 1–52.

———. 1993. "Inbreeding in Egypt and in this book: A childish perspective." In *The Natural History of Inbreeding and Outbreeding: Theoretical and Empirical Perspectives,* ed. N. W. Thornhill, 429–450. Chicago: University of Chicago Press.

Hammerschmidt, K., and Fisher, J. 1998. "Maternal discrimination of offspring vocalizations in Barbary macaques *(Macaca sylvanus)*." *Primates* 39: 231–236.

Harris, M. 1968. *The Rise of Anthropological Theory.* New York: Thomas Y. Crowell.

Hashimoto, C., Furuichi, T., and Takenaka, O. 1996. "Matrilineal kin relationship and social behavior of wild bonobos *(Pan paniscus)*: Sequencing the D-loop region of mitochondrial DNA." *Primates* 37: 305–318.

Hauser, M. D., and Fitch, W. T. 2003. "What are the uniquely human components of the language faculty?" In *Language Evolution,* ed. M. H. Christiansen and S. Kirby, 158–181. New York: Oxford University Press.

Hawkes, K. 1991." Showing off: tests of an hypothesis about men's foraging goals." *Ethology and Sociobiology* 12: 29–54.

———. 1993. "Why hunter-gatherers work: An ancient version of the problem of public goods." *Current Anthropology* 34: 341–361.

———. 2003. "Grandmothers and the evolution of human longevity." *American Journal of Human Biology* 15: 380–400.

———. 2004. "Mating, parenting, and the evolution of human pair bonds." In *Kinship and Behavior in Primates,* ed. B. Chapais and C. M. Berman, 443–473. New York: Oxford University Press.

Hawkes, K., O'Connell, J. F., and Blurton Jones, N. G. 2001. "Hunting and nuclear families: Some lessons from the Hadza about men's work." *Current Anthropology* 42(5): 681–709.

———. 2003. "Human life histories: Primate trade-offs, grandmothering, socioecology, and the fossil record." In *Primate Life Histories and Socioecology,* ed. P. M. Kappeler and M. E. Pereira, 204–231. Chicago: University of Chicago Press.

Hawkes, K., O'Connell, J. F., Blurton-Jones, N. G., Alvarez, H., and Charnov, E. L. 1998. "Grandmothering, menopause, and the evolution of human life histories." *Proceedings of the National Academy of Sciences* 95: 1336–1339.

Hawkes, K., Rogers, A. R., and Charnov, E. L. 1995. "The male's dilemma: Increased offspring production is more paternity to steal." *Evolutionary Ecology* 9: 662–677.

Henzi, S. P., Lycett, J. E., and Piper, S. E. 1997. "Fission and troop size in a mountain baboon population." *Animal Behaviour* 53: 525–535.

Hill, D. A. 2004. "The effects of demographic variation on kinship structure and behavior in cercopithecines." In *Kinship and Behavior in Primates,* ed. B. Chapais and C. M. Berman, 132–150. New York: Oxford University Press.

Hill, K. 1982. "Hunting and human evolution." *Journal of Human Evolution* 11: 521–544.
Hinde, R. A. 1979. *Toward Understanding Relationships.* London: Academic Press.
———. 1987. "Can nonhuman primates help us understand human behavior?" In *Primate Societies,* ed. B. B. Smuts, D. L. Cheney, R. M. Seyfarth, R. W. Wrangham, and T. T. Struhsaker, 413–420. Chicago: University of Chicago Press.
Holmes, W. G., and Sherman, P. W. 1982. "The ontogeny of kin recognition in two species of ground squirrel." *American Zoologist* 22: 491–517.
Holy, L. 1996. *Anthropological Perspectives of Kinship.* London: Pluto Press.
Hopkins, K. 1980. "Brother-sister marriage in Roman Egypt." *Comparative Studies in Society and History* 22: 303–354.
Hrdy, S. B. 1979. "Infanticide among animals: A review, classification, and examination of the implications for reproductive strategies of females." *Ethology and Sociobiology* 1: 13–40.
———. 1997. "Raising Darwin's consciousness: Female sexuality and the prehominid origin of patriarchy." *Human Nature* 1: 1–49.
———. 1999. *Mother Nature: A History of Mothers, Infants and Natural Selection.* New York: Random House.
———. 2005. "Comes the child before man: How cooperative breeding and prolonged postweaning dependence shaped human potential." In *Hunter-Gatherer Childhoods: Evolutionary, Developmental and Cultural Perspectives,* ed. B. S. Hewlett and M. E. Lamb, 65–91. New Brunswick, N.J.: Aldine Transaction.
Hunt, K. D., and McGrew, W. C. 2002. "Chimpanzees in the dry habitats of Assirik, Senegal and Semliki Wildlife Reserve, Uganda." In *Behavioural Diversity in Chimpanzees and Bonobos,* ed. C. Boesch, G. Hohmann, and L. F. Marchant, 35–51. New York: Cambridge University Press.
Hurford, J. R. 2003. "The language mosaic and its evolution." *Language Evolution,* ed. M. H. Christiansen and S. Kirby, 38–57. New York: Oxford University Press.
Idani, G. 1990. "Relations between unit-groups of bonobos at Wamba, Zaire: Encounters and temporary fusions." *African Study Monographs* 11: 153–186.
Imanishi, K. 1965. "The origin of the human family: A primatological approach." In *Japanese Monkeys: A Collection of Translations,* ed. K. Imanishi and S. A. Altmann, 113–140. Edmonton, Alberta: S. A. Altmann. Originally published in Japanese in the *Japanese Journal of Ethnology* 25: 119–138, 1961.
Isaac, G. L. 1978. "The food-sharing behaviour of proto-human hominids." *Scientific American* 238: 90–108.
Isbell, L. A. 2004. "Is there no place like home? Ecological bases of female dispersal and philopatry and their consequences for the formation of kin

groups." In *Kinship and Behavior in Primates,* ed. B. Chapais and C. M. Berman, 71–108. New York: Oxford University Press.

Isbell, L. A., and Young, T. P. 2002. "Ecological models of female social relationships in primates: Similarities, disparities, and some directions for future clarity." *Behaviour* 139: 177–202.

Janson, C. H. 1985. "Aggressive competition and individual food consumption in wild brown capuchin monkeys *(Cebus apella).*" *Behavioral Ecology and Sociobiology* 18: 125–138.

Kalin, N. H., and Shelton, S. E. 2003. "Nonhuman primate models to study anxiety, emotion regulation, and psychopathology." *Annals of the New York Academy of Sciences* 1008: 189–200.

Kano, T. 1992. *The Last Ape: Pygmy Chimpanzee Behavior and Ecology.* Stanford, Calif.: Stanford University Press.

Kaplan, H. S., Hill, K. R., Lancaster, J. B., and Hurtado, A. M. 2000. "A theory of human life history evolution: Diet, intelligence, and longevity." *Evolutionary Anthropology* 9:156–185.

Kappeler, P. M., and van Schaik, C., eds. 2004. *Sexual Selection in Primates: New and Comparative Perspectives.* New York: Cambridge University Press.

———, eds. 2006. *Cooperation in Primates and Humans: Mechanisms and Evolution.* Berlin: Springer-Verlag.

Kapsalis, E. 2004. "Matrilineal kinship and primate behavior." In *Kinship and Behavior in Primates,* ed. B. Chapais and C. M. Berman, 153–176. New York: Oxford University Press.

Kapsalis, E., and Berman, C. M. 1996a. "Models of affiliative relationships among free-ranging rhesus monkeys *(Macaca mulatta).* I. Criteria for kinship." *Behaviour* 133: 1209–1234.

———. 1996b. "Models of affiliative relationships among free-ranging rhesus monkeys *(Macaca mulatta).* II. Testing predictions for three hypothesized organizing principles." *Behaviour* 133: 1235–1263.

Kawanaka, K., and Nishida, T. 1974. "Recent advances in the study of inter-unit-group relationships and social structure of wild chimpanzees of the Mahali mountains." In *Proceedings from the Symposia of the 5th Congress of the International Primatological Society,* ed. S. Kondo, M. Kawai, A. Ehara, and S. Kawamura, 173–186. Tokyo: Japan Science Press.

Keenan, J. P., Gallup, G. G., and Falk, D., eds. 2003. *The Face in the Mirror: The Search for the Origins of Consciousness.* New York: Harper Collins.

Kelly, R. C. 2000. *Warless Societies and the Origin of War.* Ann Arbor, Mich.: University of Michigan Press.

Key, C. A. 2000. "The evolution of human life history." *World Archaeology* 31: 329–350.

Killen, M., and de Waal, F. B. M. 2000. "The evolution and development of morality." In *Natural Conflict Resolution,* ed. F. Aureli and F. B. M. de Waal, 352–372. Berkeley: University of California Press.

King, B. J., ed. 1999. *The Origins of Language: What Nonhuman Primates Can Tell Us*. Santa Fe, N.M.: SAR Press.

Knauft, B. M. 1991. "Violence and sociality in human evolution." *Current Anthropology* 32: 391–428.

Koenig, A. 2000. "Competitive regimes in forest-dwelling Hanuman langur females *(Semnopithecus entellus)*." *Behavioral Ecology and Sociobiology* 48: 93–109.

———. 2002. "Competition for resources and its behavioral consequences among female primates." *International Journal of Primatology* 23: 759–783.

Koyama, N. 1970. "Changes in dominance rank and division of a wild Japanese monkey troop in Arashiyama." *Primates* 11: 335–390.

Kuester, J., and Paul, A. 1997. "Group fission in Barbary macaques *(Macaca sylvanus)* at Affenberg Salem." *International Journal of Primatology* 18: 941–966.

Kuester, J., Paul, A., and Arnemann, J. 1994. "Kinship, familiarity and mating avoidance in Barbary macaques, *Macaca sylvanus*." *Animal Behaviour* 48: 1183–1194.

Kummer, H. 1979. "On the value of social relationships to nonhuman primates: A heuristic scheme." In *Human Ethology: Claims and Limits of a New Discipline,* ed. M. von Cranach, K. Foppa, W. Lepenies, and D. Ploog, 381–395. Cambridge: Cambridge University Press.

Kuper, L. 1982. "Lineage theory: A critical retrospective." *Annual Review in Anthropology* 11: 71–95.

Kurland, J. A., and Gaulin, S. J. C. 2005. "Cooperation and conflict among kin." In *The Handbook of Evolutionary Psychology,* ed. D. M. Buss, 447–482. Hoboken, N.J.: John Wiley and Sons.

Langergraber, K. E. Mitani, J. C., and Vigilant, L. 2007. The limited impact of kinship on cooperation in wild chimpanzees. *Proceedings of the National Academy of Sciences* 104: 7786–7790.

Leach, E. 1991. "The social anthropology of marriage and mating." In *Mating and Marriage,* ed. V. Reynolds and J. Kellett, 91–110. Oxford: Oxford University Press.

Lee, P. C., and Johnson, J. A. 1992. "Sex differences in alliances and the acquisition and maintenance of dominance status among immature primates." In *Coalitions and Alliances in Humans and Other Animals,* ed. A. H. Harcourt and F. B. M. de Waal, 391–414. Oxford: Oxford University Press.

Lee, R. B., and DeVore, I. 1968. "Problems in the study of hunters and gatherers." In *Man the Hunter,* ed. R. B. Lee and I. DeVore, 4–12. Hawthorne, N.Y.: Aldine.

Lévi-Strauss, C. 1949. *Les structures élémentaires de la parenté.* Paris: Presses Universitaires de France.

———. 1958. "Intervention de M. Lévi-Strauss après la communication de M. Zuckerman." In *Les Processus de l'Hominisation, Colloques Internationaux du Centre National de la Recherche Scientifique,* 160–161. Paris: Centre National de la Recherche Scientifique.

———. 1963. *Structural Anthropology.* New York: Basic Books. Translation of *Anthropologie structurale,* 1958.

———. 1967. *Les Structures Élémentaires de la Parenté,* 2nd ed. Paris: Mouton.

———. 1969. *The Elementary Structures of Kinship.* Boston: Beacon Press. Translation of *Les structures élémentaires de la parenté,* 1967.

———. 1985. *The View from Afar.* Oxford: Basil Blackwell. Translation of *Le regard éloigné,* 1983, Librairie Plon, Paris.

———. 2000. "Apologie des amibes." In *En substance: Textes pour Françoise Héritier,* ed. J.-L. Jamard, E. Terray, and M. Xanthakou, 493–496. Paris: Fayard.

Lieberman, D., Tooby, J., and Cosmides, L. 2003. "Does morality have a biological basis? An empirical test of the factors governing moral sentiments relating to incest." *Proceedings of the Royal Society of London,* series B: 270: 819–826.

Lockwood, C. A., Richmond, B. G., Jungers, W. L., and Kimbel, W. H. 1996. "Randomization procedures and sexual dimorphism in *Australopithecus afarensis.*" *Journal of Human Evolution* 31: 537–548.

Lovejoy, O. 1981. "The origin of man." *Science* 211: 341–350.

Maestripieri, D., and Roney, J. R. 2006. "Evolutionary developmental psychology: Contributions from comparative research with nonhuman primates." *Developmental Review* 26: 120–137.

Maestripieri, D., and Wallen, K. 2003. "Nonhuman primate models of psychopathology: Problems and prospects." In *Neurodevelopmental Mechanisms in Psychopathology,* ed. D. Cicchetti and E. F. Walker, 187–214. New York: Cambridge University Press.

Malinowski, B. 1929. *The Sexual Life of Savages in North-Western Melanesia.* London: Routledge and Kegan Paul.

Manson, J. H., and Perry, S. E. 1993. "Inbreeding avoidance in rhesus macaques: Whose choice?" *American Journal of Physical Anthropology* 90: 335–344.

Manson, J. H., and Wrangham, R. W. 1991. "Intergroup aggression in chimpanzees and humans." *Current Anthropology* 32: 369–390.

Marlowe, F. W. 2004. "Marital residence among foragers." *Current Anthropology* 45: 277–284.

———. 2005. "Reply to Otterbein's comment on my analysis of forager marital residence." *Current Anthropology* 46: 126–127.

Masset, C. 1986. "Préhistoire de la famille." In *Histoire de la famille 1: Mondes Lointains,* ed. A. Burguière, C. Klapisch-Zuber, M. Segalen, and F. Zonabend, 101–125. Paris: Arman Colin.

Mauss, M. 1923. "Essai sur le don." *L'année sociologique*, 2ième série.
McGrew, W. C. 1979. "Evolutionary implications of sex differences in chimpanzee predation and tool use." In *The Great Apes*, ed. D. A. Hamburg and E. R. McCown, 440–463. Menlo Park, Calif.: Benjamin/Staples.
———. 1992. *Chimpanzee Material Culture: Implications for Human Evolution.* Cambridge: Cambridge University Press.
———. 1998. "Culture in non-human primates?" *Annual Review of Anthropology* 27: 301–328.
———. 2001. "The nature of culture: Prospects and pitfalls of cultural primatology." In *Tree of Origin: What Primate Behavior Can Tell Us about Human Social Evolution*, ed. F. B. M. de Waal, 229–254. Cambridge, Mass.: Harvard University Press.
McHenry, H. M. 1992. "Body size and proportions in early hominids." *American Journal of Physical Anthropology* 87: 407–431.
———. 1994. "Behavioral ecological implications of early hominid body size." *Journal of Human Evolution* 27: 77–87.
———. 1996. "Sexual dimorphism in fossil hominids and its socioecological implications." In *The Archeology of Human Ancestry: Power, Sex and Tradition*, ed. J. Steele and S. Shennan, 91–109. New York: Routledge.
McLennan, J. F. 1970. *Primitive Marriage: An Inquiry into the Origin of the Form of Capture in Marriage Ceremonies.* Chicago: University of Chicago Press. Reprint of 1865 edition published by A. and C. Black, Edinburgh.
Ménard, N., and Vallet, D. 1993. "Dynamics of fission in a wild Barbary macaque group *(Macaca sylvanus)*." *International Journal of Primatology* 14: 479–500.
Ménard, N., von Segesser, F., Scheffrahn, W., Pastorini, J., Vallet, D., Gagi, B., Martin, R. D., and Gautier-Hion, A. 2001. "Is male-infant caretaking related to paternity and/or mating activities in wild Barbary macaques *(Macaca sylvanus)*?" *Comptes rendus de l'Académie des Sciences, Paris*, série III: *Sciences de la vie/Life Sciences* 324: 601–610.
Mesnick, S. L. 1997. "Sexual alliances: Evidence and evolutionary implications." In *Feminism and Evolutionary Biology: Boundaries, Intersections and Frontiers*, ed. P. A. Gowaty, 207–260. New York: Chapman and Hall.
Mitani, J. C., and Watts, D. P. 1999. "Demographic influences on the hunting behaviour of chimpanzees." *American Journal of Physical Anthropology* 109: 439–454.
———. 2001. "Why do chimpanzees hunt and share meat?" *Animal Behaviour* 61: 915–924.
Mitani, J. C., Merriwether, D. A., and Zhang, C. B. 2000. "Male affiliation, cooperation and kinship in wild chimpanzees." *Animal Behaviour* 59: 885–893.
Mitani, J. C., Watts, D. P., Pepper, J. W., and Merriwether, D. A. 2002. "Demo-

graphic and social constraints on male chimpanzee behaviour." *Animal Behaviour* 64: 727–737.

Moore, J. 1993. "Inbreeding and outbreeding in primates: What's wrong with 'the dispersing sex'?" In *The Natural History of Inbreeding and Outbreeding: Theoretical and Empirical Perspectives,* ed. N. W. Thornhill, 392–426. Chicago: University of Chicago Press.

———. 1996. "Savanna chimpanzees, referential models and the last common ancestor." In *Great Ape Societies,* ed. W. C. McGrew, L. F. Marchant, and T. Nishida, 275–292. Cambridge: Cambridge University Press.

Morgan, L. H. 1871. *Systems of Consanguinity and Affinity of the Human Family.* Washington: Smithsonian Institution.

———. 1974. *Ancient Society.* Gloucester, Mass.: Peter Smith. Reprint of 1877 edition published by C. H. Kerr, Chicago.

Morin, P. A., and Goldberg, T. L. 2004. "Determination of genealogical relationships from genetic data: A review of methods and applications." In *Kinship and Behavior in Primates,* ed. B. Chapais and C. M. Berman, 15–45. New York: Oxford University Press.

Murdock, G. P. 1949. *Social Structure.* New York: Macmillan.

———. 1967. *Ethnographic Atlas.* Pittsburgh: University of Pittsburgh Press.

Murray, M. A. 1934. "Marriage in ancient Egypt." In *Congrès International des Sciences Anthropologiques et Ethnologiques, Compte-rendu de la Première Session.* Royal Institute of Anthropology, London.

Nadler, R. D., and Phoenix, C. H. 1991. "Male sexual behavior: Monkeys, men, and apes." In *Understanding Behavior: What Primate Studies Tell Us about Human Behavior,* ed. J. D. Loy and C. B. Peters, 152–189. New York: Oxford University Press.

Nakamichi, M., and Yoshida, A. 1986. "Discrimination of mother by infant among Japanese macaques *(Macaca fuscata).*" *International Journal of Primatology* 7: 481–489.

Needham, R. 1971. "Remarks on the analysis of kinship and marriage." In *Rethinking Kinship and Marriage,* ed. R. Needham, 1–34. London: Tavistock Publications.

Newton-Fisher, N. E. 1999. "Infant killers of Budongo." *Folia Primatologica* 70: 167–169.

Nicolson, N. A. 1987. "Infants, mothers, and other females." In *Primate Societies,* ed. B. B. Smuts, D. L. Cheney, R. M. Seyfarth, R. W. Wrangham, and T. T. Struhsaker, 330–342. Chicago: University of Chicago Press.

———. 1991. "Maternal behavior in human and nonhuman primates." In *Understanding Behavior: What Primate Studies Tell Us about Human Behavior,* ed. J. D. Loy and C. B. Peters, 17–50. New York: Oxford University Press.

Nishida, T. 1979. "The social structure of chimpanzees of the Mahale Mountains." In *The Great Apes: Perspectives on Human Evolution,* vol. 5, ed. D. A. Hamburg and E. R. McCown, 73–121. Menlo Park, Calif.: Benjamin/Cummings.

———. 1990. *The Chimpanzees of the Mahale Mountains: Sexual and Life History Strategies.* Tokyo: Tokyo University Press.

Nishida, T., Corp, N., Hamai, M., Hasegawa, T., Hiraiwa-Hasegawa, M., Hosaka, K., Hunt, K. D., Itoh, N., Kawanaka, K., Matsumoto-Oda, A., Mitani, J. C., Nakamura, M. 2003. "Demography, female life-history, and reproductive profiles among the chimpanzees of Mahale." *American Journal of Primatology* 59: 99–121.

Nissen, H. W. 1931. "A field study of the chimpanzee." *Comparative Psychology Monographs* 8: 122.

Norikoshi, K., Sakamaki, T., Turner, L., Uehara, S., and Zamma, K. 2003. "Demography, female life history, and reproductive profiles among the chimpanzees of Mahale." *American Journal of Primatology* 59: 99–121.

O'Connell, J. F., Hawkes, K., and Blurton-Jones, N. G. 1999. "Grandmothering and the evolution of *Homo erectus.*" *Journal of Human Evolution* 36: 461–485.

Oi, T. 1988. "Sociological study on the troop fission of wild Japanese monkeys *(Macaca fuscata yakui)* on Yakushima Island." *Primates* 29: 1–19.

Okamoto, K. 2004. "Patterns of group fission." In *Macaque Societies: A Model for the Study of Social Organization,* ed. B. Thierry, M. Singh, and W. Kaumanns, 112–116. New York: Cambridge University Press.

Okamoto, K., and Matsumara, S. 2001. "Group fission in moor macaques *(Macaca maurus).*" *International Journal of Primatology* 22: 481–493.

Otterbein, K. F. 2005. "On hunting and virilocality." *Current Anthropology* 46: 124–127.

Packer, C. 1979. "Inter-troop transfer and inbreeding avoidance in *Papio anubis.*" *Animal Behaviour* 27: 1–36.

Palombit, R. A. 1999. "Infanticide and the evolution of pair bonds in nonhuman primates." *Evolutionary Anthropology* 7: 117–129.

Parker, S. T., Mitchell, R. W., and Boccia, M. L., eds. 1994. *Self-awareness in Animals and Humans.* Cambridge: Cambridge University Press.

Parr, L. A., and de Waal, F. B. M. 1999. "Visual kin recognition in chimpanzees." *Nature* 399: 647–648.

Paul, A., and Kuester, J. 2004. "The impact of kinship on mating and reproduction." In *Kinship and Behavior in Primates,* ed. B. Chapais and C. M. Berman, 271–291. New York: Oxford University Press.

Paul, A., Kuester, J., and Arnemann, J. 1992. "DNA fingerprinting reveals that infant care by male Barbary macaques *(Macaca sylvanus)* is not paternal investment." *Folia Primatologica* 58: 93–98.

———. 1996. "The sociobiology of male-infant interactions in Barbary macaques, *Macaca sylvanus.*" *Animal Behaviour* 51: 155–170.

Pereira, M. E. 1986. "Maternal recognition of juvenile offspring coo vocalizations in Japanese macaques." *Animal Behaviour* 34: 935–937.

———. 1989. "Agonistic interactions of juvenile savanna baboons. II. Agonistic support and rank acquisition." *Ethology* 80, 152–71.

———. 1992. "The development of dominance relations before puberty in cercopithecine societies." In *Aggression and Peacefulness in Humans and Other Primates,* ed. J. Silverberg and J. P. Gray, 117–149. New York: Oxford University Press.

Plavcan, J. M. 2001. "Sexual dimorphism and primate evolution." *Yearbook of Physical Anthropology* 44: 25–53.

Plavcan, J. M., Lockwood, C. A., Kimbel, W. H., Lague, M. R., and Harmon, E. H. 2005. "Sexual dimorphism in *Australopithecus afarensis* revisited: How strong is the case for a human-like pattern of dimorphism?" *Journal of Human Evolution* 48: 313–320.

Power, M. 1991. *The Egalitarians—Human and Chimpanzee: An Anthropological View of Social Organization.* Cambridge: Cambridge University Press.

Preuschoft, S. 1997. "The social function of 'smile' and 'laughter': Variations across primate species and societies." In *Nonverbal Communication: Where Nature Meets Culture,* ed. U. Segerstråle and P. Molnár, 171–189. Mahwah, N.J.: Lawrence Erlbaum.

Preuschoft, S., and Paul, A. 1999. "Dominance, egalitarianism, and stalemate: An experimental approach to male-male competition in Barbary macaques." in *Primate Males: Causes and Consequences of Variation in Group Composition,* ed. P. N. Kappeler, 205–216. Cambridge: Cambridge University Press.

Prud'homme, J. 1991. "Group fission in a semifree-ranging population of Barbary macaques *(Macaca sylvanus).*" *Primates* 32: 9–22.

Pruetz, J. D., and Bertolani, P. 2007. "Savanna chimpanzees, *Pan troglodytes verus,* hunt with tools." *Current Biology* 17: 1–6.

Prum, R. O. 1990. "Phylogenetic analysis of the evolution of display behavior in the neotropical manakins (Aves: Pipridae)." *Ethology* 84: 202–231.

Pryce, C. R., Martin, R. D., and Skuse, D., eds. 1995. *Motherhood in Human and Nonhuman Primates: Biosocial Determinants.* Basel: Karger.

Pusey, A. E. 1980. "Inbreeding avoidance in chimpanzees." *Animal Behaviour* 28: 543–552.

———. 1983. "Mother-offspring relationships in chimpanzees after weaning." *Animal Behaviour* 31: 363–377.

———. 1987. "Sex-biased dispersal and inbreeding avoidance in birds and mammals." *Trends in Ecology and Evolution* 2: 295–299.

———. 1990a. "Mechanisms of inbreeding avoidance in nonhuman primates."

In *Pedophilia: Biosocial Dimensions*, ed. J. R. Feierman, 201–220. New York: Springer-Verlag.

———. 1990b. "Behavioral changes at adolescence in chimpanzees." *Behaviour* 115: 203–246.

———. 2001. "Of genes and apes: Chimpanzee social organization and reproduction." In *Tree of Origin: What Primate Behavior Can Tell Us about Human Social Evolution*, ed. F. B. M. de Waal, 9–37. Cambridge, Mass.: Harvard University Press.

———. 2004. "Inbreeding avoidance in primates." In *Inbreeding, Incest, and the Incest Taboo: The State of Knowledge at the Turn of the Century*, ed. A. P. Wolf and W. H. Durham, 61–75. Stanford, Calif.: Stanford University Press.

Pusey, A. E., and Packer, C. 1987. "Dispersal and philopatry." In *Primate Societies*, ed. B. B. Smuts, D. L. Cheney, R. M. Seyfarth, R. W. Wrangham, and T. T. Struhsaker, 250–266. Chicago: University of Chicago Press.

———. 1997. "The ecology of relationships." In *Behavioural Ecology: An Evolutionary Approach*, ed. J. R. Krebs and N. B. Davies, 254–283. Oxford: Blackwell Scientific Publications.

Pusey, A. E., and Wolf, M. 1996. "Inbreeding avoidance in animals." *Trends in Ecology and Evolution* 11: 201–206.

Radcliffe-Brown, A. R. 1931. "Social organization of Australian tribes." *Oceania Monographs* 1: 1–124.

———. 1950. *African Systems of Kinship and Marriage*. London: Oxford University Press.

Ralls, K., Ballou, J. D., and Templeton, A. 1988. "Estimates of lethal equivalents and the cost of inbreeding in mammals." *Conservation Biology* 2: 185–193.

Rendall, D. 2004. "'Recognizing' kin: Mechanisms, media, minds, modules, and muddles." In *Kinship and Behavior in Primates*, ed. B. Chapais and C. M. Berman, 295–316. New York: Oxford University Press.

Reno, P. L., Mindl, R. S., McCollum, M. A., and Lovejoy, C. O. 2003. "Sexual dimorphsim in *Australopithecus afarensis* was similar to that of modern humans." *Proceedings of the National Academy of Sciences* 100: 9404–9409.

Reynolds, V. 1994. "Kinship in nonhuman and human primates." In *Hominid Culture in Primate Perspective*, ed. D. Quiatt and J. Itani, 137–165. Niwot: University Press of Colorado.

Robbins, M. M. 1995. "A demographic analysis of male life-history and social structure of mountain gorillas." *Behaviour* 132: 21–47.

———. 2001. "Variation in the social system of mountain gorillas: The male perspective." In *Mountain Gorillas: Three Decades of Research at Karisoke*, ed. M. M. Robbins, P. Sicotte, and K. J. Stewart, 29–58. New York: Cambridge University Press.

Robson, S. L. 2004. "Breast milk, diet, and large human brains." *Current Anthropology* 45: 419–425
Rodseth, L., and Novak, S. A. 2006. "The impact of primatology on the study of human society." In *Missing the Revolution: Darwinism for Social Scientists,* ed. J. H. Barkow, 187–220. New York: Oxford University Press.
Rodseth, L., and Wrangham, R. 2004. "Human kinship: A continuation of politics by other means?" In *Kinship and Behavior in Primates,* ed. B. Chapais and C. M. Berman, 389–419. New York: Oxford University Press.
Rodseth, L., Wrangham, R. W., Harrigan, A. M., and Smuts, B. B. 1991. "The human community as a primate society." *Current Anthropology* 32: 221–254.
Ross, C., and MacLarnon, A. 2000. "Evolution of non-maternal care in anthropoid primates: A test of the hypotheses." *Folia primatologica* 71: 93–113.
Ruvolo, M. 1997. "Molecular phylogeny of the hominoids: Inferences from multiple independent DNA sequence data sets." *Molecular Biology and Evolution* 14: 248–265.
Sackett, G. P., and Fredrikson, W. T. 1987. "Social preferences by pigtailed macaques: Familiarity versus degree and type of kinship." *Animal Behaviour* 35: 603–606.
Sade, D. S. 1968. "Inhibition of son-mother mating among free-ranging rhesus." *Science and Psychoanalysis* 12: 18–38.
Sahlins, M. D. 1960. "The origin of society." *Scientific American* 203: 76–87.
———. 1977. *The Use and Abuse of Biology: An Anthropological Critique of Sociobiology.* London: Tavistock.
Saito, C. 1996. "Dominance and feeding success in female Japanese macaques, *Macaca fuscata:* Effects of food patch size and inter-patch distance." *Animal Behaviour* 51: 967–980.
Scheffler, H. 1985. "Filiation and affiliation." *Man* 20: 1–21.
Scheidel, W. 2004. "Ancient Egyptian sibling marriage and the Westermarck effect." In *Inbreeding, Incest, and the Incest Taboo: The State of Knowledge at the Turn of the Century,* ed. A. P. Wolf and W. H. Durham, 93–108. Stanford, Calif.: Stanford University Press.
Schmitt, S. 2006. *Aux Origines de la Biologie Moderne: L'Anatomie Comparée d'Aristote à la Théorie de l'Évolution.* Paris: Belin.
Schneider, D. M. 1961. "Introduction: The distinctive features of matrilineal descent groups." In *Matrilineal Kinship,* ed. D. M. Schneider and K. Gough, 1–29. Berkeley: University of California Press.
———. 1971. "What is kinship all about?" Paper read to the Anthropological Society of Washington.
———. 1984. *A Critique of the Study of Kinship.* Ann Arbor: University of Michigan Press.

Schneider, D. M., and Gough, K. 1961. *Matrilineal Kinship.* Berkeley: University of California Press.

Schultz, A. H. 1969. *The Life of Primates.* London: Weidenfield and Nicolson.

Service, E. R. 1962. *Primitive Social Organization: An Evolutionary Perspective.* New York: Random House.

Shaw, B. D. 1992. "Explaining incest: brother-sister marriage in Greco-Roman Egypt." *Man* 27: 267–299.

Shepher, J. 1971. "Mate selection among second generation kibbutz adolescents and adults: Incest avoidance and negative imprinting." *Archives of Sexual Behavior* 1: 293–307.

———. 1983. *Incest: A Biosocial View.* New York: Academic Press.

Sibley, C. G., and Ahlquist, J. E. 1987. "DNA hybridization evidence of hominoid phylogeny: Results from an expanded data set." *Journal of Molecular Evolution* 26: 99–121.

Sicotte, P. 1993. "Inter-group interactions and female transfer in mountain gorillas: Influence of group composition and male behavior." *American Journal of Primatology* 30: 21–36.

Sigg, H., and Falett, J. 1985. "Experiments on possession and property in hamadryas baboons *(Papio hamadryas).*" *Animal Behaviour* 33: 978–984.

Sillén-Tullberg, B., and Møller, A. P. 1993. "The relationship between concealed ovulation and mating systems in anthropoid primates: A phylogenetic analysis." *American Naturalist* 141: 1–25.

Silk, J. 2001. Ties that bond: The role of kinship in primate societies. In *New Directions in Anthropological Kinship,* ed. L. Stone, 71–92. Lanham, Md.: Rowman and Littlefield.

———. 2002. "Kin selection in primate groups." *International Journal of Primatology* 23: 849–875.

———. 2006. "Practicing Hamilton's rule: Kin selection in primate groups." In *Cooperation in Primates and Humans: Mechanisms and Evolution,* ed. P. M. Kappeler and C. P. van Schaik, 25–46. Berlin: Springer.

Silk, J. B., Altmann, J., and Alberts, S. C. 2006a. "Social relationships among adult female baboons *(Papio cynocephalus).* I. Variation in the strength of social bonds." *Behavioral Ecology and Sociobiology* 61: 183–195.

———. 2006b. "Social relationships among adult female baboons *(Papio cynocephalus).* II. Variation in the quality and stability of social bonds." *Behavioral Ecology and Sociobiology* 61: 197–204.

Smith, D. G. 1995. "Avoidance of close consanguineous inbreeding in captive groups of rhesus macaques." *American Journal of Primatology* 35: 31–40.

Smuts, B. B. 1985. *Sex and Friendship in Baboons.* Chicago, University of Chicago Press.

———. 1992. "Male aggression against women: An evolutionary perspective." *Human Nature* 3: 1–44.

———. 1995. "The Evolutionary Origins of Patriarchy." *Human Nature* 6: 1–32.

Smuts, B. B., and Gubernick, D. 1992. "Male-infant relationships in non-human primates: Paternal investment or mating effort?" In *Father-Child Relations: Cultural and Biosocial Contexts,* ed. B. S. Hewlett, 1–30. New York: Aldine de Gruyter.

Snowdon, C. T. 2001. "From primate communication to human language." In *Tree of Origin: What Primate Behavior Can Tell Us about Human Social Evolution,* ed. F. B. M. de Waal, 193–227. Cambridge, Mass.: Harvard University Press.

Soumah, A. G., and Yokota, N. 1991. "Female rank and feeding strategies in a free-ranging provisioned troop of Japanese macaques." *Folia Primatologica* 57: 191–200.

Stanford, C. B. 1996. "The hunting ecology of wild chimpanzees: Implications for the evolutionary ecology of Pliocene hominids." *American Anthropologist* 98: 96–113.

———. 1999. *The Hunting Apes: Meat Eating and the Origins of Human Behavior.* Princeton: Princeton University Press.

———. 2001. "The ape's gift: Meat-eating, meat-sharing, and human evolution." In *Tree of Origin: What Primate Behavior Can Tell Us about Human Social Evolution,* ed. F. B. M., de Waal, 95–117. Cambridge, Mass.: Harvard University Press.

———. 2003. *Upright: The Evolutionary Key to Becoming Human.* Boston: Houghton Mifflin.

Stanford, C. B., Wallis, J., Matama, H., and Goodall, J. 1994. "Patterns of predation by chimpanzees on red colobus monkeys in Gombe National Park, 1982–1991." *American Journal of Physical Anthropology* 94: 213–228.

Sterck, E. H. M., Watts, D. P., and van Schaik, C. P. 1997. "The evolution of female social relationships in nonhuman primates." *Behavioral Ecology and Sociobiology* 41: 291–309.

Steward, J. H. 1955. *Theory of Culture Change: The Methodology of Multilinear Evolution.* Urbana: University of Illinois Press.

Stewart, K. J. 2001. "Social relationships of immature gorillas and silverbacks." *In Mountain Gorillas: Three Decades of Research at Karisoke,* ed. M. M. Robbins, P. Sicotte, and K. J. Stewart, 183–213. Cambridge: Cambridge University Press.

Stewart, K. J., and Harcourt, A. H. 1987. "Gorillas: Variation in female relationships." In *Primate Societies,* ed. B. B. Smuts, D. L. Cheney, R. M. Seyfarth, R. W. Wrangham, and T. T. Struhsaker, 155–164. Chicago: University of Chicago Press.

Stone, L., ed. 2001a. *New Directions in Anthropological Kinship*. Oxford: Rowman and Littlefield.

———. 2001b. "Introduction: Theoretical implications of new directions in anthropological kinship." In *New Directions in Anthropological Kinship*, ed. L. Stone, 1–20. Oxford: Rowman and Littlefield.

Strier, K. B. 2004. "Patrilineal kinship and primate behavior." In *Kinship and Behavior in Primates*, ed. B. Chapais and C. M. Berman, 177–199. New York: Oxford University Press.

Sussman, R. W. 1999. "The myth of man the hunter, man the killer, and the evolution of human morality." *Zygon, Journal of Religion and Science* 34: 453–471.

Symmes, D., and Biben, M. 1985. "Maternal recognition of individual infant squirrel monkeys from isolation call playbacks." *American Journal of Primatology* 9: 39–46.

Takahata, Y. 1982. "The socio-sexual behavior of Japanese monkeys." *Zeitschrift für Tierpsychologie* 59: 89–108.

Takahata, Y., Huffman, M. A., and Bardi, M. 2002. "Long-term trends on matrilineal inbreeding among the Japanese macaques of Arashiyama B Troop." *International Journal of Primatology* 23: 399–410.

Talmon, G. Y. 1964. "Mate selection in collective settlements." *American Sociological Review* 29: 408–491.

Tanner, N. M. 1987. "Gathering by females: The chimpanzee model revisited and the gathering hypothesis." In *The Evolution of Human Behavior: Primate Models*, ed. W. G. Kinzey, 3–27. Albany: State University of New York Press.

Teleki, G. 1973. *The Predatory Behavior of Chimpanzees*. Lewisburg, Pa.: Bucknell University Press.

Thierry, B. 2000. "Covariation of conflict management patterns across macaque species." In *Natural Conflict Resolution*, ed. F. Aureli and F. B. M. de Waal, 106–128. Berkeley: University of California Press.

Thierry, B., Iwaniuk, A. N., and Pellis, S. M. 2000. "The influence of phylogeny on the social behaviour of macaques (Primates: Cercopithecidae, genus *Macaca*)." *Ethology* 106: 713–728.

Thornhill, N. W. 1993. *The Natural History of Inbreeding and Outbreeding: Theoretical and Empirical Perspectives*. Chicago: University of Chicago Press.

Tomasello, M., and Call, J. 1997. *Primate Cognition*. Oxford: Oxford University Press.

Tooby, J., and DeVore, I. 1987. "The reconstruction of hominid behavioral evolution through strategic modeling." In *The Evolution of Human Behavior: Primate Models*, ed. W. G. Kinzey, 183–237. Albany: State University of New York Press.

Trivers, R. L. 1972. "Parental investment and sexual selection." In *Sexual Selec-*

tion and the Descent of Man, 1871–1971, ed. B. Campbell, 136–179. Chicago: Aldine-Atherton.

Turke, P. 1984. "Effects of ovulatory concealment and synchrony on proto-hominid mating system and parental roles." *Ethology and Sociobiology* 5,: 33–44.

Tylor, E. B. 1889a. "On a method of investigating the development of institutions; applied to laws of marriage and descent." *Journal of the Anthropological Institute of Great Britain and Ireland* 18.

———. 1889b. *Primitive Culture: Researches into the Development of Mythology, Philosophy, Religion, Language, Art and Custom.* 3rd American ed., from 1873 English ed. New York: Henry Holt.

van den Berghe, P. L. 1979. *Human Family Systems: An Evolutionary View.* New York: Elsevier.

van Hooff, J. A. R. A. M. 1972. "A comparative approach to the phylogeny of laughter and smiling." In *Non-verbal Communication,* ed. R. A. Hinde, 209–241. Cambridge: Cambridge University Press.

———. 1999. "Relationships among non-human primate males: a deductive framework." in *Primate Males: Causes and Consequences of Variation in Group Composition,* ed. P. N. Kappeler, 183–191. Cambridge: Cambridge University Press.

———. 2001. "Conflict, reconciliation and negotiation in non-human primates: The value of long-term relationships." In *Economics in Nature: Social Dilemmas, Mate Choice and Biological Markets,* ed. R. Noe, J. A. R. A. M. van Hooff, and P. Hammerstein, 67–92. Cambridge: Cambridge University Press.

van Hooff, J. A. R. A. M., and van Schaik, C. P. 1994. "Male bonds: Affiliative relationships among nonhuman primate males." *Behaviour* 130: 309–337.

van Noordwijk, M. A., and van Schaik, C. P. 1987. "Competition among female long-tailed macaques, *Macaca fascicularis.*" *Animal Behaviour* 35: 577–589.

van Schaik, C. P. 1989. "The ecology of social relationships amongst female primates." In *Comparative Socioecology: The Behavioural Ecology of Humans and other Mammals,* ed. V. Standen and R. A. Foley, 195–218. Oxford: Blackwell Scientific Publications.

———. 1996. "Social evolution in primates: the role of ecological factors and male behavior." *Proceedings of the British Academy* 88: 9–31.

———. 2004. *Among Orangutans: Red Apes and the Rise of Human Culture.* Cambridge, Mass.: Harvard University Press.

van Schaik, C., Ancrenaz, M., Borgen, G., Galdikas, B., Knott, C. D., Singleton, I., Suzuki, A., Utami, S. S., and Merrill, M. 2003. "Orangutan cultures and the evolution of material culture." *Science* 299: 102–105.

van Schaik, C., and Dunbar, R. I. M. 1990. "The evolution of monogamy in large primates: A new hypothesis and some crucial tests." *Behaviour* 115: 30–62.

van Schaik, C., and Kappeler, P. M. 1997. "Infanticide risks and the evolution of male-female association in primates." *Proceedings of the Royal Society of London,* series B, 264: 1687–1694.

———. 2003. "The evolution of social monogamy in primates." In *Monogamy: Mating Strategies and Partnerships in Birds, Humans and Other Mammals,* ed. U. H. Reichard and C. Boesch, 59–80. Cambridge: Cambridge University Press.

van Schaik, C., and Paul, A. 1996. "Male care in primates: Does it ever reflect paternity?" *Evolutionary Anthropology* 5: 152–156.

Varki, A., and Altheide, T. K. 2005. "Comparing the human and chimpanzee genomes: Searching for needles in a haystack." *Genome Research* 15: 1746–1758.

Vasey, P. L. 1995. "Homosexual behavior in primates: A review of evidence and theory." *International Journal of Primatology* 16: 173–204.

Vigilant, L., Hofreiter, M., Siedel, H., and Boesch, C. 2001. "Paternity and relatedness in wild chimpanzee communities." *Proceedings of the National Academy of Sciences of the USA* 98: 12890–12895.

Walters, J. R. 1987. "Kin recognition in nonhuman primates." In *Kin Recognition in Animals,* ed. D. J. C. Fletcher and C. D. Michener, 359–393. New York: John Wiley and Sons.

Washburn, S. L., and DeVore, I. 1961. "Social behavior of baboons and early man." In *Social Life of Early Man,* ed. S. L. Washburn, 91–105. Chicago: Aldine.

Washburn, S. L., and Lancaster, C. 1968. "The evolution of hunting." In *Man the Hunter,* ed. R. B. Lee and I. DeVore, 293–303. Chicago: Aldine.

Watts, D. P. 1989. "Infanticide in mountain gorillas: New cases and a reconsideration of the evidence." *Ethology* 81: 1–18.

———. 1994. "The influence of male mating tactics on habitat use in mountain gorillas *(Gorilla gorilla beringei)*." *Primates* 35: 35–47.

———. 1996. "Comparative socio-ecology of gorillas." In *Great Ape Societies,* ed. W. C. McGrew, L. F. Marchant, and T. Nishida, 16–28. Cambridge: Cambridge University Press.

———. 2001. "Social relationships of female mountain gorillas." In *Mountain Gorillas: Three Decades of Research at Karisobe,* ed. M. M. Robbins, P. Sicotte, and K. J. Stewart, 215–240. Cambridge: Cambridge University Press.

Watts, D. P., and Mitani, J. C. 2001. "Boundary patrols and intergroup encounters in wild chimpanzees." *Behaviour* 138: 299–327.

———. 2002. "Hunting behavior of chimpanzees at Ngogo, Kibale National Park, Uganda." *International Journal of Primatology* 23: 1–28.

Watts, D. P., Muller, M., Amsler, S. J., Mbabazi, G., and Mitani, J. C. 2006. "Lethal intergroup aggression by chimpanzees in Kibale National Park, Uganda." *American Journal of Primatology* 68: 161–180.

Watts, D. P., and Pusey, A. E. 1993. "Behavior of juvenile and adolescent great apes." In *Juvenile Primates: Life History, Development, and Behavior,* ed. M. E. Pereira, and L. A. Fairbanks, 148–171. New York: Oxford University Press.

Welker, C., Schwibbe, M. H., Schäefer-Witt, C., and Visalberghi, E. 1987. "Failure of kin recognition in *Macaca fascicularis.*" *Folia Primatologica* 49: 216–221.

Westermarck, E. 1891. *The History of Human Marriage.* London: Macmillan.

———. 1926. *A Short History of Marriage.* London: Macmillan.

White, F. J. 1996. "Comparative socio-ecology of *Pan paniscus.*" In *Great Ape Societies,* ed. W. C. McGrew, L. F. Marchant, and T. Nishida, 29–41. Cambridge: Cambridge University Press.

White, L. A. 1949. *The Science of Culture: A Study of Man and Civilization.* New York: Farrar, Strauss.

———. 1959. *The Evolution of Culture: The Development of Civilization to the Fall of Rome.* New York: McGraw-Hill.

Whiten, A., Goodall, J., McGrew, W. C., Nishida, T., Reynolds, V., Sugiyama, Y., Tutin, C. E. G., Wrangham, R. W., and Boesch, C. 1999. "Cultures in chimpanzees." *Nature* 399: 682–685.

Whitten, P. L. 1983. "Diet and dominance among female vervet monkeys *Cercopithecus aethiops.*" *American Journal of Primatology* 5: 139–159.

Widdig, A. 2002. "Paternal kinship among adult female rhesus macaques *(Macaca mulatta).*" Ph.D. thesis, Institut für Biologie der Humboldt-Universität zu Berlin.

———. 2007. "Paternal kin discrimination: The evidence and likely mechanisms." *Biological Review* 82: 319–334.

Widdig, A., Nürnberg, P., Krawczak, M., Streich, W. J., and Bercovitch, F. B. 2001. "Paternal relatedness and age proximity regulate social relationships among adult female rhesus macaques." *Proceedings of the National Academy of Sciences of the USA* 98: 13769–13773.

Wildman, D. E., Uddin, M., Grossman, L. I., and Goodman, M. 2003. "Implications of natural selection in shaping 99.4% nonsynonymous DNA identity between humans and chimpanzees: Enlarging genus *Homo.*" *Proceedings of the National Academy of Sciences of the USA* 100: 7181–7188.

Williams, B. 1983. "Evolution, ethics, and the representation problem." In *Evolution from Molecules to Men,* ed. D. S. Bendall, 555–566. Cambridge: Cambridge University Press.

Williams, J. M., Oehlert, G. W., Carlis, J. V., and Pusey, A. E. 2004. "Why do male chimpanzees defend a group range?" *Animal Behaviour* 68: 523–532.

Williams, J. M., Pusey, A. E., Carlis, J. V., Farm, B. P., and Goodall, J. 2002. "Female competition and male territorial behaviour influence female chimpanzees' ranging patterns." *Animal Behaviour* 63: 347–360.

Wilson, M. L., and Wrangham, R. W. 2003. "Intergroup relations in chimpanzees." *Annual Review of Anthropology* 32: 363–392.

Wolf, A. P. 1966. "Childhood association, sexual attraction, and the incest taboo: A Chinese case." *American Anthropologist* 68: 883–898.

———. 1970. "Childhood association and sexual attraction: A further test of the Westermarck hypothesis." *American Anthropologist* 72: 503–515.

———. 1995. *Sexual Attraction and Childhood Association: A Chinese Brief for Edward Westermarck*. Stanford, Calif.: Stanford University Press.

———. 2004a. "Introduction." In *Inbreeding, Incest, and the Incest Taboo: The State of Knowledge at the Turn of the Century*, ed. A. P. Wolf and W. H. Durham, 1–23. Stanford, Calif.: Stanford University Press.

———. 2004b. "Explaining the Westermarck effect, or, what did natural selection select for?" In *Inbreeding, Incest, and the Incest Taboo: The State of Knowledge at the Turn of the Century*, ed. A. P. Wolf and W. H. Durham, 76–92. Stanford, Calif.: Stanford University Press.

Wolfe, L. D. 1984. "Japanese macaque female sexual behavior: A comparison of Arashiyama East and West." In *Female Primates: Studies by Women Primatologists*, ed. M. F. Small, 141–157. New York: Alan R. Liss.

———. 1991. "Human evolution and the sexual behavior of female primates." In *Understanding Behavior: What Primate Studies Tell Us about Human Behavior*, ed. J. D. Loy and C. B. Peters, 121–151. New York: Oxford University Press.

Wood, B. M. 2006. "Prestige or provisioning: a test of foraging goals among the Hadza." *Current Anthropology* 47: 383–387.

Woodruff, D. S. 2004. "Noninvasive genotyping and field studies of free-ranging nonhuman primates." In *Kinship and Behavior in Primates*, ed. B. Chapais and C. M. Berman, 46–68. New York: Oxford University Press.

Wrangham, R. 1979. "On the evolution of ape social systems." *Social Sciences Information* 18: 334–368.

———. 1980. "An ecological model of female-bonded primate groups." *Behaviour* 75: 262–300.

———. 1987. "The significance of African apes for reconstructing human social evolution." In *The Evolution of Human Behavior: Primate Models*, ed. W. G. Kinzey, 51–71. Albany: State University of New York Press.

———. 1993. "The evolution of sexuality in chimpanzees and bonobos." *Human Nature* 4: 47–79.

———. 1999. "Evolution of coalitionary killing." *Yearbook of Physical Anthropology* 42: 1–30.

———. 2000. "Why are male chimpanzees more gregarious than mothers: A scramble competition hypothesis." In *Primate Males: Causes and Consequences of Variation and Group Composition*, ed. P. M. Kappeler, 248–258. Cambridge: Cambridge University Press.

———. 2001. "Out of the *Pan*, into the fire: How our ancestors' evolution depended on what they ate." In *Tree of Origin: What Primate Behavior Can Tell*

Us about Human Social Evolution, ed. F. B. M. de Waal, 119–143. Cambridge, Mass.: Harvard University Press.

———. 2002. "The cost of sexual attraction: Is there a trade-off in female *Pan* between sex appeal and received coercion?" In *Behavioural Diversity in Chimpanzees and Bonobos,* ed. C. Boesch, G. Hohmann, and L. F. Marchant, 204–215. Cambridge: Cambridge University Press.

Wrangham, R. W., Chapman, C. A., Clark-Arcadi, A. P., and Isabirye-Basuta, G. 1996. "Social ecology of Kanyawara chimpanzees: implications for understanding the costs of great ape groups." In *Great Ape Societies,* ed. W. C. McGrew, L. F. Marchant, and T. Nishida, 45–57. Cambridge: Cambridge University Press.

Wrangham, R. W., Jones, J. H., Laden, G., Pilbeam, D., and Conklin-Brittain, N. 1999. "The raw and the stolen: Cooking and the ecology of human origins." *Current Anthropology* 40: 567–594.

Wrangham, R. W., and Peterson, D. 1996. *Demonic Males: Apes and the Origins of Human Violence.* Boston: Houghton Mifflin.

Wrangham, R. W., and Pilbeam, D. 2001. "African apes as time machines." In *All Apes Great and Small,* vol. 1: *African Apes,* ed. B. M. F. Galdikas, N. E. Briggs, L. K. Sheeran, G. L. Shapiro, and J. Goodall, 5–17. New York: Kluwer Academic/Plenum.

Yanagisako, S. J. 1979. "Family and the household: The analysis of domestic groups." *Annual Review of Anthropology* 8: 161–205.

Zihlman, A. 1996. "Reconstructions reconsidered: Chimpanzee models and human evolution." In *Great Ape Societies,* ed. W. C. McGrew, L. F. Marchant, and T. Nishida, 293–304. Cambridge: Cambridge University Press.

Zonabend, F. 1986. "De la famille: Regard ethnologique sur la parenté et la famille." In *Histoire de la famille,* vol. 1: *Mondes Lointains,* ed. A. Burguière, C. Klapisch-Zuber, M. Segalen, and F. Zonabend, 19–97. Paris: Armand Colin.

Zuckerman, S. 1932. *The Social Life of Monkeys and Apes.* London: Routledge and Kegan Paul.

Index

Note: References are indexed based on first author's name only.